Nematodes

Structure, Development, Classification, and Phylogeny

V. V. MALAKHOV

Translated by George V. Bentz

Edited by W. Duane Hope

Nematodes

Structure, Development, Classification, and Phylogeny

Smithsonian Institution Press
Washington and London

Production Editor: Eileen D'Araujo
Designer: Alan Carter

The original volume was edited by O. I. Belogurov, Chief Editor, Doctor of Biological Sciences; reviewed by K. M. Ryzhikov, now deceased, and V. A. Sveshnikov; and published by the Institute of Marine Biology, Far East Scientific Center of the USSR Academy of Sciences. Copyright 1986 by Izdatel'stvo "Nauka."

Library of Congress Cataloging-in-Publication Data

Malakhov, V. V. (Vladimir Vasil'evich)
[Nematody. English]
Nematodes : structure, development, classification, and phylogeny
/ V. V. Malakhov ; translated by George V. Bentz ; edited by W. Duane Hope.
p. cm.
Includes bibliographical references (p.) and index.
ISBN 1-56098-255-1 (alk. paper)
1. Nematoda. 2. Nematoda—Classification. I. Hope, W. Duane, 1935– . II. Title.
QL391.N4M22513 1993
595.1'82—dc20 92-38962
CIP

British Library Cataloguing-in-Publication data available

Manufactured in the United States of America
97 96 95 94 5 4 3 2 1

∞ The paper used in this publication meets the minimum requirements of the American National Standard for Permanence of Paper for Printed Library Materials Z39.48–1984.

Contents

Foreword vii

Abstract ix

Preface xi

1 Structure 1

Cuticle 1

Hypodermis 24

Somatic Musculature 34

Nervous System 46

Sensory Organs and Glands 61

Body Cavity and Phagocytic Cells 109

Digestive System 115

Reproductive System 129

2 Development 134

Embryonic Development 134

Postembryonic Development 169

3 Classification of Nematodes and Phylogenetic Relationships within the Class 175

General Classification 176

Characteristics of Higher Taxa 178

Phylogenetic Relationships within the Class 187

4 Origin of Nematodes and Origins of the Nematode Biological Process 202

Contemporary Information on the Structure of Gastrotrichs 202

Comparison of Nematode and Gastrotrich Structure 213

Biological Advancement of Nematodes 219

5 Classification of the Pseudocoelomates 224

Rotifera 226

Acanthocephala 229

Nematoda and Gastrotricha 231

Priapulida, Kinorhyncha, Nematomorpha, and Loricifera 236

Pseudocoelomate Classification 249

Bibliography 254

Index 282

Foreword

Several books concerned with various aspects of the classification and biology of nematodes are currently available. Although there is some overlap in the topics covered by many of these books, each has its unique emphasis on certain taxa or certain aspects of nematode biology, each approach reflecting objectives set by the author.

The objective of this book is an attempt to better understand evolutionary relationships among all taxa of nematodes, free-living as well as parasitic, and between nematodes and other groups of "pseudocoelomate" animals. Malakhov approaches this task with a review of the morphology and development of free-living and parasitic nematodes. The review of literature is extensive, with many references to Soviet literature not otherwise readily available. With regard to morphological data, Malakhov includes the older information based solely on light microscopy, as well as more contemporary work in which scanning electron microscopy or transmission electron microscopy have been employed. He also includes his own extensive morphological observations, although it is sometimes difficult to distinguish between that which is presented as new and that which is already documented in the literature. For inclusion in this translation, Malakhov has provided some corrections and made available some new information that is not part of the original book.

The reviews of the morphology and development are followed by a summary of the characteristics of each taxon from subclass to order and a discussion of the phylogenetic relationships within the Nematoda.

Malakhov reviews the morphology of the gastrotrichs and compares it with that of the nematodes and ends with a discussion of the classification of the "pseudocoelomates."

Although the review of literature and Malakhov's own empirical data are valuable, the theories of evolutionary relationships are derived in a traditional manner, so they are not as acceptable as would have been the case had they been derived by the testable methods of phylogenetic systematics (cladistics). Nonetheless, these ideas, and the taxonomic changes upon which they are based, must be available for consideration and evaluation by nematologists and invertebrate zoologists alike.

This translation would not have been possible had it not been for the enormous generosity and language skills of George V. Bentz, who contributed many hours as a volunteer translator. Ellen Unumb and Anne Sustrick also volunteered many hours skillfully proofing the translation. The services of these and other volunteers have been made available through the dedicated and capable assistance of Nancy T. Hinton formerly of the Smithsonian Institution's Behind-the-Scenes Volunteer Program. Abbie Yorkoff of the Department of Invertebrate Zoology, National Museum of Natural History, devoted many hours to proofreading the manuscript and confirming literature references. To all of these people I owe my sincerest gratitude. Finally, I wish to gratefully acknowledge the Atherton Seidell Endowment Fund for a grant (Fund No. 14540206) that has subsidized costs of text and copyediting and printing.

W. Duane Hope

Abstract

This monograph, based upon original research and a review of the literature, provides an analysis of the structure of all nematode organ systems and constructs comparative morphological categories from the most primitive free-living marine nematodes to extremely specialized parasitic forms. It examines information on the embryonic development of representatives of all nematode orders and analyzes evolutionary paths of nematode embryogeny. It proposes a new classification for the class Nematoda, including three subclasses; provides diagnoses of the subclasses and orders; and examines the phylogenetic relationships within the class. It analyzes the morphology of all groups of pseudocoelomates and proposes a new classification for these animals, which includes four groups: Rotifera, Acanthocephala, Nematoda, and Rhyncocoela. This monograph is intended for zoologists, helminthologists, embryologists, and specialists in the comparative anatomy and phylogeny of animals.

Preface

Nematodes constitute one of the most numerous and widespread groups of animals. Inhabiting seas, fresh water, soil, and living as parasites of animals and plants, nematodes have exploited a tremendous range of habitats. As A. A. Paramonov (1962) correctly noted, nematodes represent a group that is living through a state of rapid biological progress during the present period of biosphere development. Despite the immense significance that nematodes have in nature and in the economic activities of man, the study of this group still proceeds at a very low level. At present, descriptions have been made of about 20,000 species of nematodes, 2,000 species being phytohelminths, 5,000 species being animal parasites, and 13,000 species being free-living nematodes in the ocean, fresh water, and soil. The total number of nematode species living on our planet is estimated by various authors to be from 80,000–100,000 to 1 million. Thus, the nematodes that have been described up to now are not more than 25% of the total number of the species of this class. There is no other group of animals where the number of undescribed species is so large.

For a long time the study of nematodes was conducted exclusively on parasitic forms. This is natural, because nematodes are among the most important parasites of man and agricultural animals and plants.

In our time the study of parasitic nematodes in animals, plants, and man has become an important branch of helminthology—the study of parasitic worms—that has very important applied significance. However, the study of parasitic nematodes, although extremely important for the understanding of various aspects of parasitology and for the development of methods for combating parasitic diseases, cannot fully reveal the chief characteristics of the structure, development, classification, and phylogeny of nematodes in general.

Despite the large number of parasitic species, nematodes fundamentally are a free-living group of animals. Parasitic forms constitute about 35% of the total nematode species described. Clearly, future additions to the number of described species will largely be the free-living forms. As a whole, parasitic nematodes probably constitute no more than a tenth of the total number of species of this group.

The study of free-living nematodes is important from several points of view. First, the theoretical basis of applied nematology, the most important branch of helminthology, was and remains the natural classification of nematodes. At present, the necessity for such a classification is very evident. Many variations in the classification of nematode parasites of animals, and even in nematode parasites of vertebrates, have appeared in the literature. Such systems do not appear to be natural, and they arise from insufficient knowledge concerning the structure and development of free-living nematodes from which came all of the parasitic representatives of this class. The absence of a natural nematode classification encompassing representatives of all ecological groupings of this class significantly hampers the development of nematology as a division of helminthology and affects the further development of all its applied aspects.

Second, the significant role of nematodes in the total matter and energy cycle in the biosphere is now becoming obvious. The quantity of nematodes in natural biocenoses is enormous: the total number of nematodes in 1 m^2 of surface soil or of marine or freshwater benthic sediment is in the millions of specimens. The high intensity of exchange and the short life cycles in combination with great abundance make nematodes the paramount group of consumers of organic material. The study of marine, freshwater, and soil communities in our time is inconceivable without consideration of the very important role of nematodes.

Third, at present, research in experimental biology and medicine increasingly focuses on free-living nematodes. Such characteristics of nematodes as the constancy of cell composition (and, in small forms, the total number of cells is 600–800), determinant development, short life cycles (which, in some forms, is about a day), and the simplicity of breeding in artificial environments make them ideal animals for experiments. Many laboratories in the world now conduct research using nematodes in such key areas as genetics, cell differentiation, and aging.

The free-living marine nematodes are the most ecologically varied group of the class and include the most primitive nematodes. In all previous contemporary compendia and textbooks, nematode organization has been examined on the basis of very specialized parasitic or saprobiotic species. In this work, free-living marine nematodes are used as the basis for examining nematode organization and for tracing morphological series from the most primitive marine forms to very specialized parasitic forms. In addition to examining questions of morphology and phylogeny of nematodes, we will also treat the broader problem of the classification and phylogeny of all worms that have a primary body cavity. This will permit the correct understanding of the position of nematodes—one of the largest classes of animals—in the overall classification of the animal kingdom.

The present work contains new data derived from the author's original research, especially in chapters on the structure of the cuticle, the hypodermis, the nervous system, and the sensory organs of nematodes, and on nematode embryological development. Original data have also been used in sections that analyze the structure and development of Priapulida, Kinorhyncha, and Nematomorpha.

1

Structure

Cuticle

The cuticle is one of the nematode's most important structural components. Its acquisition largely established the organization of nematodes and predetermined their basic evolutionary pathways.

The structure of the nematode cuticle has attracted the attention of many researchers. The classic investigations by Bömmel (1894), Goldschmidt (1904, 1905, 1906), Toldt (1899, 1904, 1905), and other authors have provided a detailed picture of the cuticle structure of *Ascaris*, which has subsequently been elaborated upon at the ultrastructural level (Hinz 1963; Watson 1965a; Bird 1971). Investigations of animal nematode parasites have demonstrated great variation in the structure of the cuticle of these animals (Chitwood and Chitwood 1950; Bird 1971; Bogoyavlenskii 1973). Light microscopy, which has done much to reveal the structure of the large zooparasitic nematodes with a thick cuticle, has not proved to be effective for studying the thin cuticle of phytonematodes and saprobiotic and free-living forms. But in the last decade, electron microscopy of nematode cuticle has developed rapidly. With the aid of ultrastructural methods, the structure of the cuticle of dozens of species of nematode parasites of animals and plants, saprobiotic and soil species has been studied, primarily from the subclass Rhabditia (Bird and Rogers 1965; Eckert and Schwarz 1965; Anya 1966; Lee 1966a, b; Roggen et al. 1967; Yuen 1967; Poinar and Leutenegger 1968; Wisse and Daems 1968; Raski et al. 1969; Johnson et al. 1970; Bird 1971, 1980; Epstein et al. 1971; Zuckerman et al. 1973; Croll 1976; and others).

The cuticle of free-living marine nematodes has been studied significantly less than the cuticle of representatives of other ecological groups of nematodes, but here also we have knowledge available of the detailed structure of the cuticle for a majority of orders.

In examining the structure of nematode cuticle, it is necessary to use nomenclature to designate its layers. Traditionally, the nematode cuticle has been regarded as a three-layered structure, composed of the outer (cortical), middle (homogeneous or matrix), and inner (basal) layers. However, ultrastructural research in recent years has shown that it is more appropriate to use a cuticle model of four layers, which are designated the epi-, exo-, meso-, and endocuticle, and each of which, in turn, consists of several sublayers (Maggenti 1979). But this represents the most complex type of nematode cuticle organization, many species being without one or two of these layers.

The outer layer of nematode cuticle, the epicuticle, has a thickness of 6–45 nm. An indispensable component of this layer is a thin osmiophilic surface membrane with a thickness of 6–10 nm, to the outer surface of which is joined a layer of glycocalyx. In many species, the osmiophilic membrane is the only component of the epicuticle. In other forms, the epicuticle includes, in addition to the outer layer, an inner osmiophilic layer. Between the two osmiophilic layers of epicuticle is a more or less pronounced clear layer. The outer osmiophilic layer, in turn, has a three-layered structure and, according to certain morphological and functional characteristics, is like an elementary cell membrane (Bird 1980). Some writers (Bonner et al. 1970) have proposed that the outer membrane comes from the cytoplasmatic membrane of the hypodermis with recurrent shedding. More recent work, however, indicates that new cuticle is formed in the shedding process outside of the cytoplasmatic membrane in a process of secretion (Bonner and Weinstein 1972; Dick and Wright 1973). Thus, the direct identification of the surface membrane of the nematode cuticle with the cytoplasmic membrane of the hypodermis is hardly justified. However, it is not ruled out that, phylogenetically, it was formed from cytoplasmic membrane of ectodermal epithelium and modified during the process of evolution.

There are pores or striations at points where the epicuticular layer is broken in nematodes of the enoplid branch. The cuticle among members of Enoplia is usually characterized as smooth, but its annulation is distinctly visible under higher magnification with a scanning electron microscope. This annulation arises from deep transverse grooves that cut across the epi- and exocuticle (Figure 1). Curiously, the transverse grooves in *Enoplus* reach the cavity lying in the mesocuticle and, at least in some places, connect it with the outside environment (Figure 2).

The striation of the cuticle of this type is characteristic of representatives of Enoplida and also of Dorylaimida (Roggen et al. 1967; Aboul-Eid 1969; Raski et al. 1969; Grotaert and Lippens 1974; Lippens et al. 1974; Siddiqui and Viglierchio 1977). In Mermithida, the cuticle has numerous pores from which canals extend into the epi- and exocuticle (Lee 1970).

A clearly annulated cuticle is characteristic of nematodes of the chromadorid branch (orders Monhysterida, Chromadorida, Desmodorida, Desmoscolecida, Araeolaimida, and Plectida). In most cases the annulation is provided by the structure of the cuticle's surface, which sometimes is accompanied by sclerites in the inner layers of the cuticle. The outer surface of the cuticle of members of the chromadorid branch may have the appearance of simple annulation, scales, or complicated longitudinal ribs (Figure 4).

Among nematodes of the Chromadorid-Monhysterid branch, the type of cuticle structure most similar to the enoplid type is that among members of Monhysterida (Figure 5). Here the annulation also is created by striations that cut across the surface layers of the cuticle (epi- and exocuticle). Among nematodes of the orders Chromadorida and Desmodorida, striation involves layers of the epicuticle, exocuticle, and sometimes the mesocuticle, which curve following the surface sculpture. The nature of the annulation in the cuticle of the Rhabditia is the same as among nematodes of the chromadorid-monhysterid branch. The epicuticle consists of the same three layers, but can be significantly thicker.

The exocuticle is a layer that varies greatly in thickness from 0.2 μm among small, free-living nematodes to several tens of micrometers among large parasitic forms. This layer includes sublayers that vary in structure, i.e., striated, homogeneous, and/or layers of longitudinal and transverse fibers. The striated layer is the most constant and important layer of the exocuticle. It consists of parallel rachises oriented perpendicularly to the surface of the cuticle (Figure 6). The periodicity of stri-

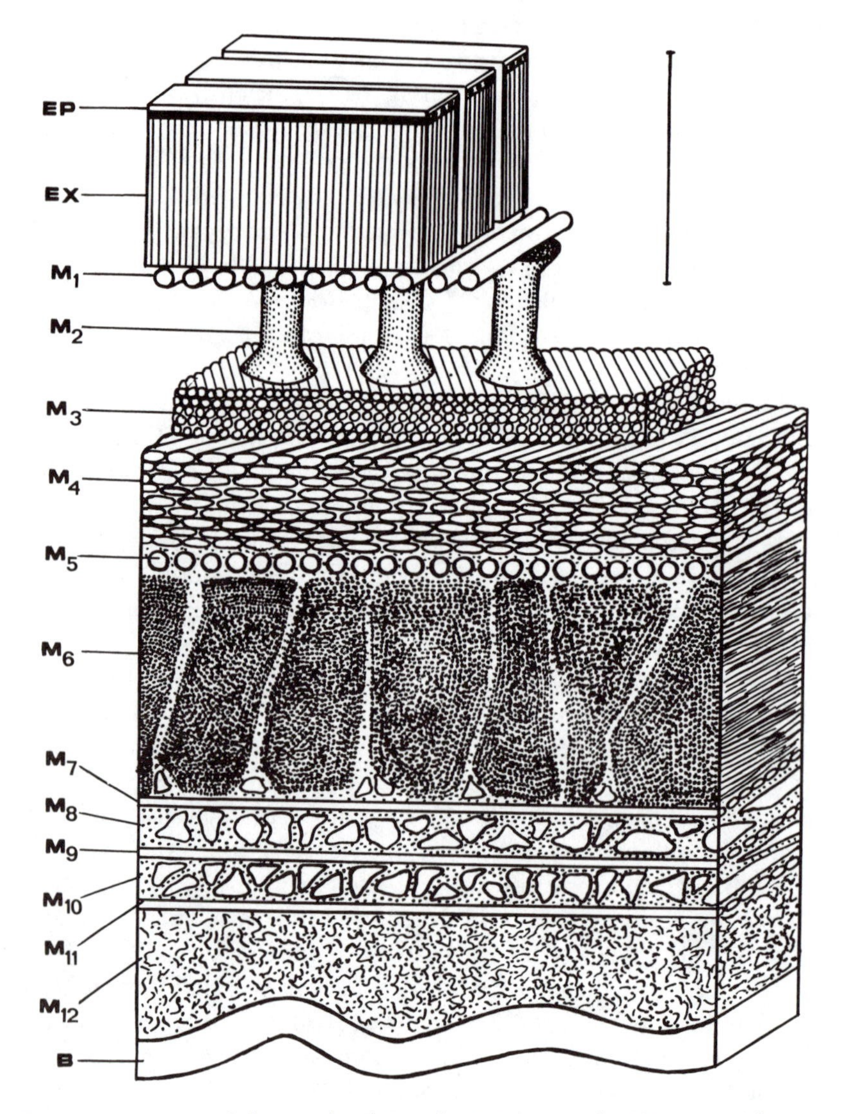

Figure 1. Diagram of the cuticle of *Enoplus*. EP, epicuticle; EX, exocuticle; M_1–M_{12}, layers of mesocuticle; B, endocuticle. Scale = 1 μm.

ation in longitudinal sections is 10–17 nm and, in transverse sections, 20–27 nm. The striated layer probably consists of uniformly oriented protein molecules forming a crystalline structure.

Among nematodes of the orders Enoplida, Dorylaimida, and Mermithida, the striated layer closely adjoins the epicuticle (Wright 1965; Roggen et al. 1967; Aboul-Eid 1969; Lee 1970; Taylor et al. 1970; Grotaert and Lippens 1974; Lippens et al. 1974; Siddiqui and Viglierchio 1977).

Among *Euchromadora* (Watson 1965b), *Acanthonchus* (Wright and Hope 1968), and *Paracanthonchus* (order Chromadorida), researched by the present author, the striated layer also closely adjoins the epicuticle. Among representatives of the order Monhysterida, i.e., *Theristus* and *Sphaerolaimus*, which were also investigated by the author, the

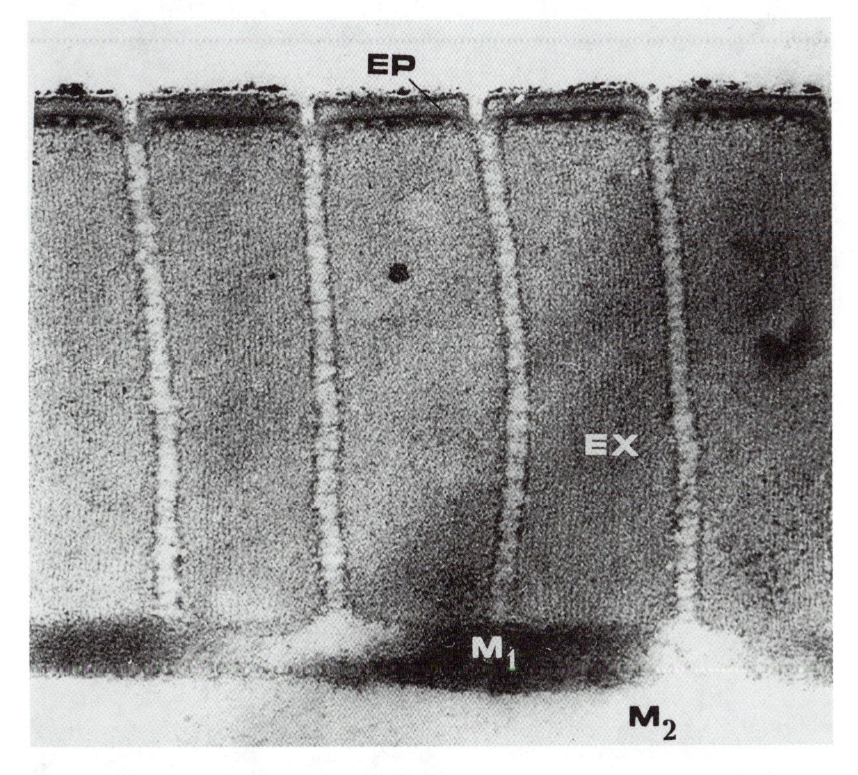

Figure 2. Longitudinal section of the epi- and exocuticle of *Enoplus*. Labels as in Figure 1. Magnification = 70,000.

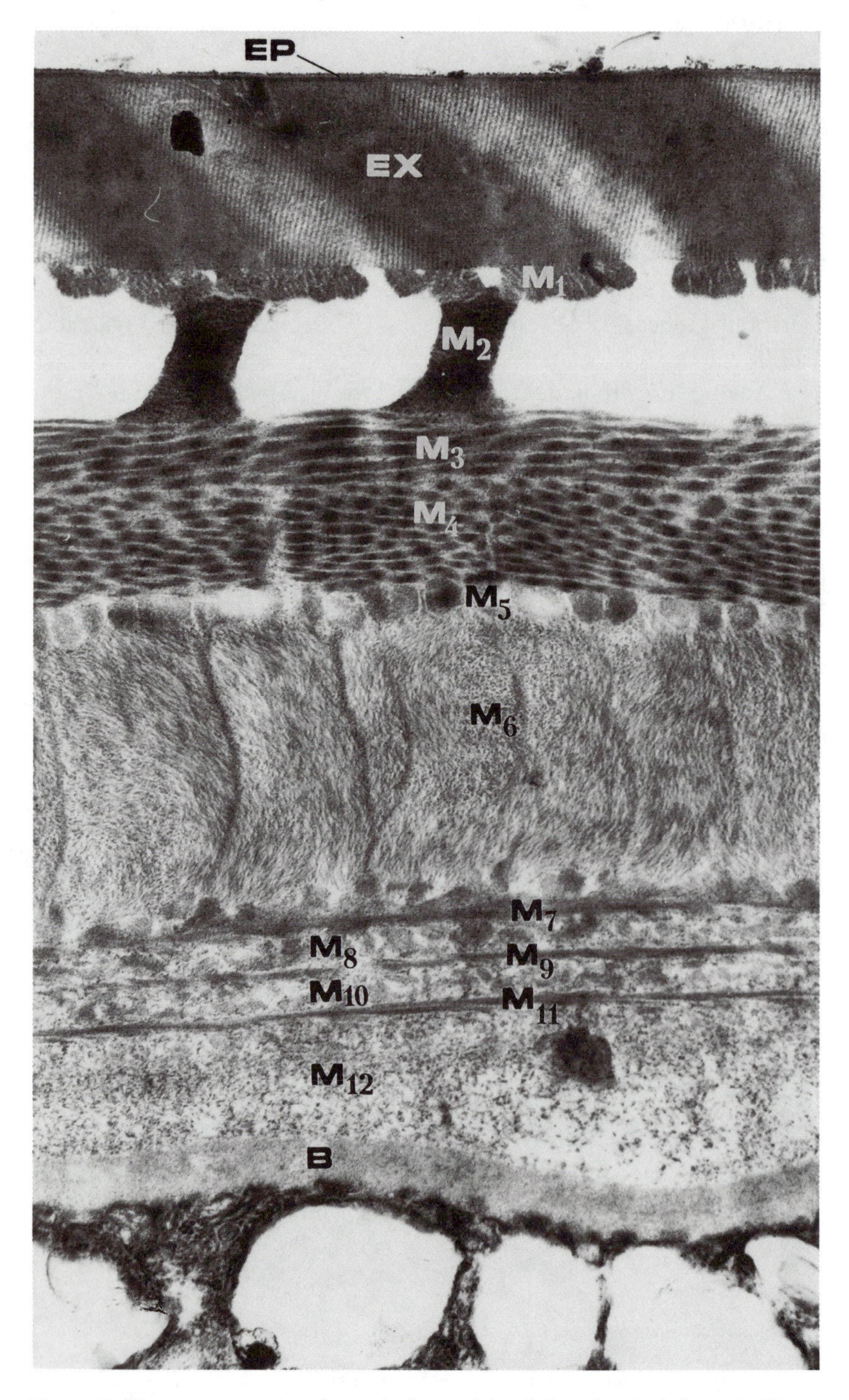

Figure 3. Transverse section through the cuticle of *Enoplus*. Labels as in Figure 1. Magnification = 24,500.

Figure 4. Surface of the cuticle of *Monoposthia costata*. Magnification = 10,000.

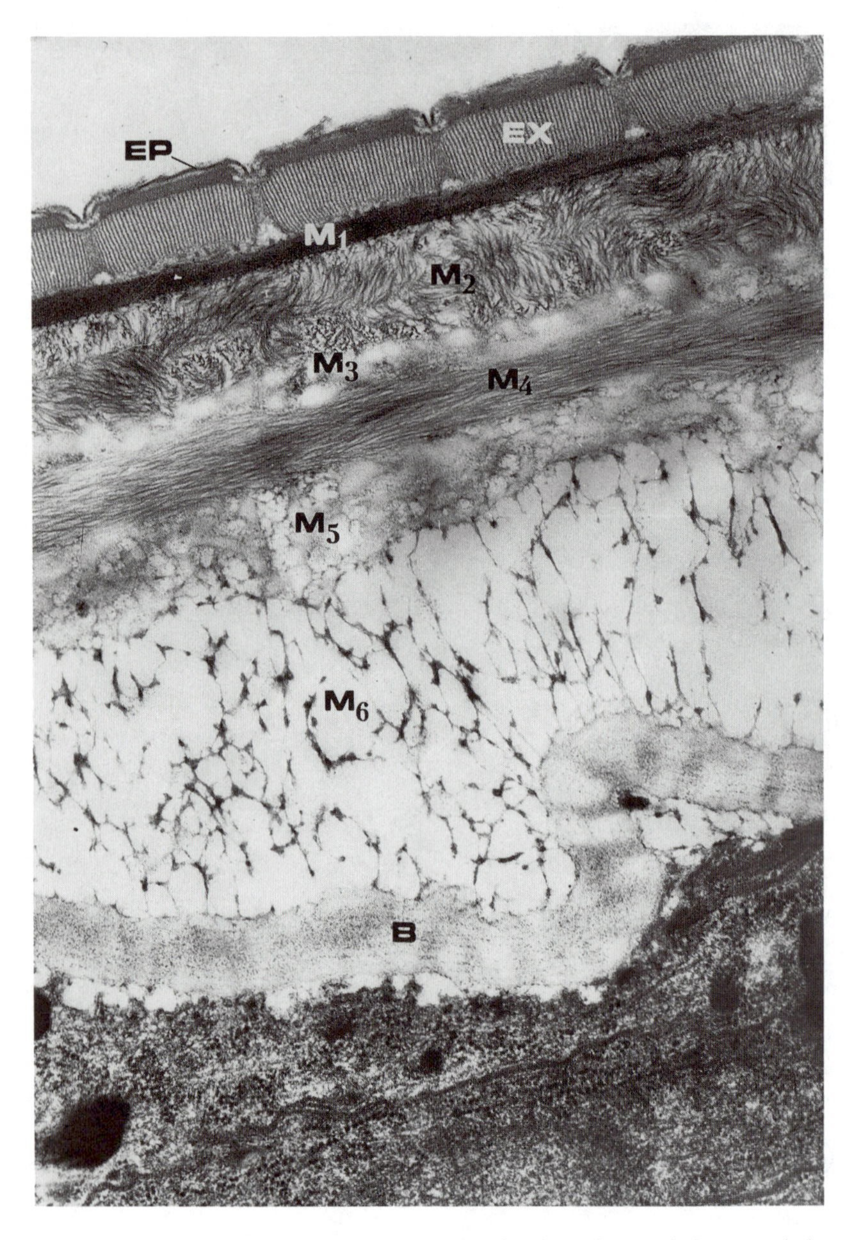

Figure 5. Longitudinal section of the cuticle of *Sphaerolaimus balticus*. Labels as in Figure 1. Magnification = 29,300.

striated layer is separated from the epicuticle by a thin layer of homogeneous material (Figure 5). The striated layer is absent among *Monoposthia* (order Desmodorida), investigated by the author, where the whole exocuticle is represented by homogeneous material and, probably, by mechanically durable material (Figure 7). Among Rhabditia, the striated layer is rather far from the epicuticle, and the interval is filled with homogeneous or longitudinally fibrous material (Eckert and Schwarz 1965; Wisse and Daems 1968; Wright 1968b; Bird 1971; Epstein et al. 1971; Byers and Anderson 1972; and others). In *Ascaris* and *Oxyuris*, the striated layer generally is absent (Hinz 1963; Watson 1965a; Anya 1966; Bird 1971). Generally, in Rhabditia there is a tendency toward thickening and complication of the exocuticle.

The mesocuticle is the layer of nematode cuticle with the widest and most varied component structures. Often within the boundaries of this layer is an intracuticular cavity filled with liquid or containing loose or porous material. In *Enoplus* (Figures 1–3), this cavity is intersected by columns of dense osmiophilic material supporting the upper layers of cuticle (epi- and exocuticle). In *Paracanthonchus* (Figure 8), an extensive internal cavity, constituting 75–80% of the cuticle's thickness, is also intersected by beams that support the epi- and exocuticle. The intracu-

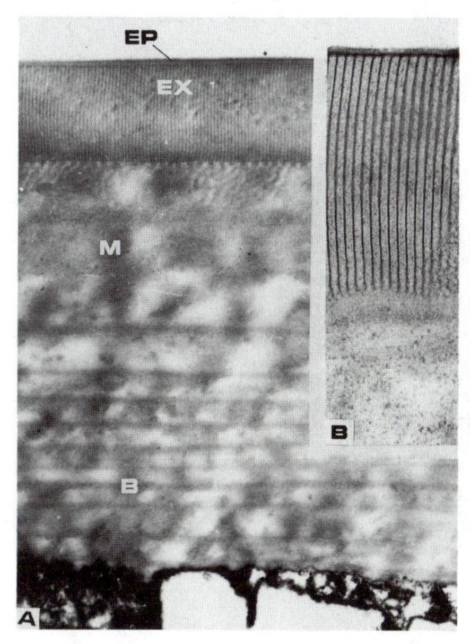

Figure 6. Transverse section of the cuticle of *Pontonema vulgare* (A) and an enlargement of its striated layer (B). Labels as in Figure 1. Magnifications: A = 18,800; B = 53,000.

ticular cavity of the *Sphaerolaimus* mesocuticle (Figure 5) is filled with coarsely alveolar osmiophilic mesh. An analogous internal cavity is characteristic of *Caenorhabditis,* where there are also columns that support the upper layers of cuticle (Epstein et al. 1971; Zuckerman et al. 1973). In many species, the intracuticular cavity is absent.

One of the characteristic components of mesocuticle consists of sublayers formed by obliquely oriented fibers. Usually there are two to three levels of such fibers, oriented at an angle of 60 to 75° from the longitudinal axis of the worm. These fibers envelop the body of the worm in a spiral in opposite intertwining directions. In chemical and diffraction properties, the fibers are close to a collagen, but they do not have striations (Fauré-Frémiet and Gerault 1944; Bird 1958, 1971). The thickness of this layer is from 1 to 2 μm in small forms and up to 25 μm in large parasites (Harris and Crofton 1957; Hinz 1963; Watson 1965a; Roggen et al. 1967). The layers of oblique fibers, as a rule, are situated in the outermost levels of the mesocuticle, adjoining the internal cavity or, if the latter is absent, adjoining the exocuticle directly. Sometimes, one of the layers of obliquely oriented fibers occupies the surface posi-

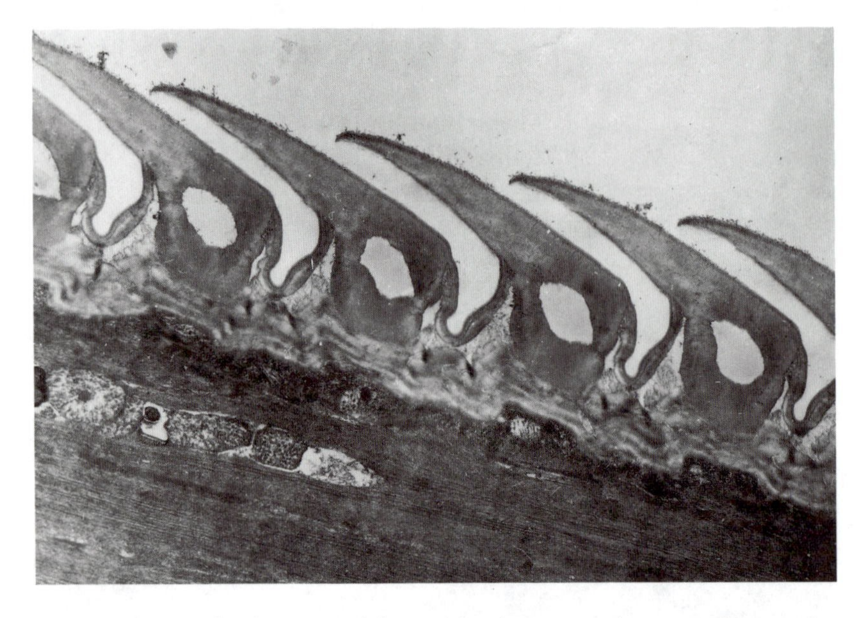

Figure 7. Longitudinal section of the cuticle of *Monoposthia costata.* Magnification = 13,500.

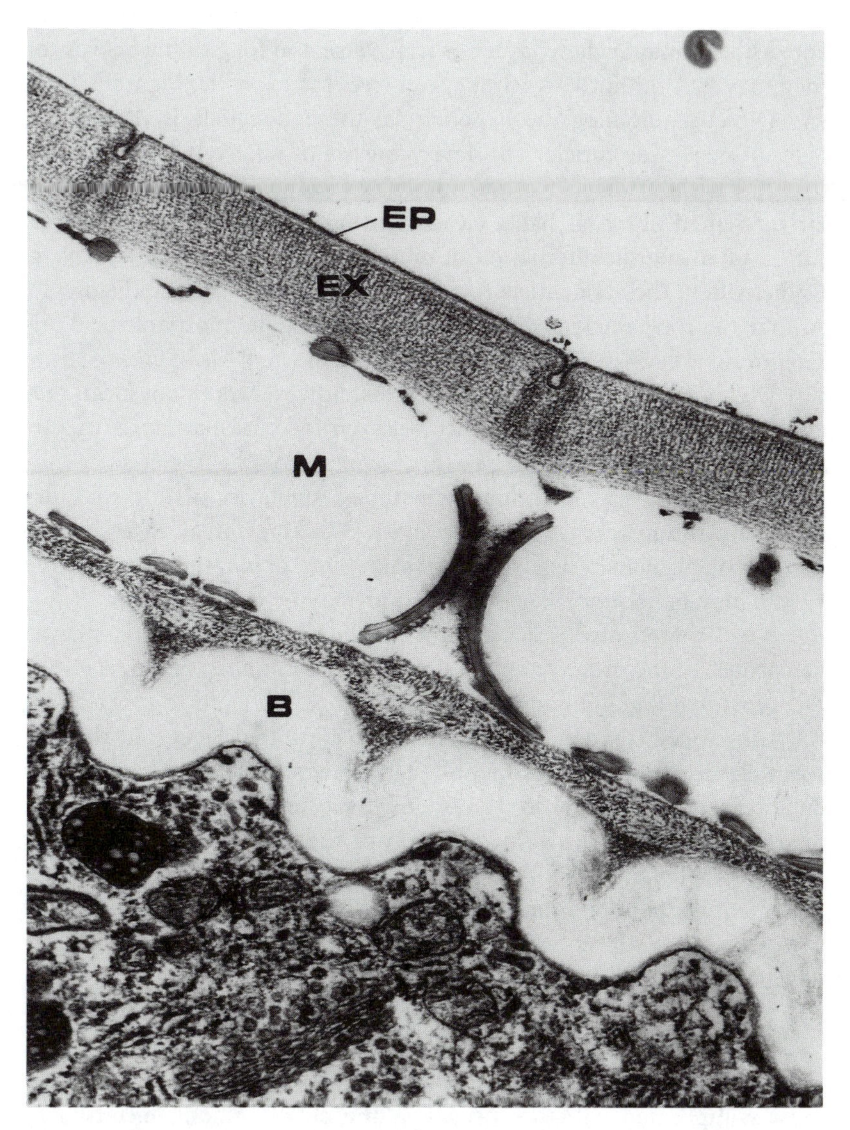

Figure 8. Longitudinal section of the cuticle of *Paracanthonchus macrodon*. Labels as in Figure 1. Magnification = 13,200.

tion in the mesocuticle, while the others are separated from it and adjoin the endocuticle.

The meoscuticle (usually deeper than the layers of obliquely oriented fibers) may include layers of transverse and longitudinal cords, or they may be thin-fibrillar, feltlike, etc. (see, for example, Figure 3).

Directly adjoining the hypodermis, the endocuticle is the deepest layer of nematode cuticle. This layer consists of relatively homogeneous material without distinct fibers, sometimes only with very thin, irregularly oriented fibrils. It has a clearly distinguishable secondary stratification involving the alternation of more dense and less dense sublayers. Rather often, the endocuticle forms a wavy, internal surface (Figure 1), so that the total thickness of the cuticle varies. The morphological criteria of homology cannot be fully applied in the analysis of cuticle structure (Remane 1956; Gilyarov 1964). This, however, does not mean that any morphological correspondence between the cuticular layers of various nematodes is impossible. Two principles should guide comparisons of the cuticle layers of various nematodes: similarity of infrastructure and the position relative to other layers. Looking for all layers in the cuticles of all nematodes is unnecessary. One or another layer in the cuticle may be more or less developed or may be entirely absent.

Most complicated is the structure of the cuticle in free-living marine nematodes of the order Enoplida, for example *Enoplus* (Figures 1–3) or *Deontostoma* (Siddiqui and Viglierchio 1977). Here, all four layers are fully developed, the exocuticle containing a striated layer and the mesocuticle developing layers of obliquely oriented fibers and sometimes an internal cavity (Figures 1–3). However, among enoplid groups, a tendency is observed toward reduction of the mesocuticle and a thickening of the endocuticle; this is distinctly observable in the series *Enoplus-Deontostoma-Pontonema-Mermis* (Figure 9). In *Mermis* (Lee 1970), only two layers of obliquely oriented fibers of mesocuticle are developed between the exocuticle and the multilayered endocuticle. In *Pontonema,* a weakly formed mesocuticle is also represented by a layer that is significantly inferior to the endocuticle (Figure 6).

Among nematodes of the Chromadorida-Monhysterida branch, the most complicated structure occurs in the cuticle among members of Monhysterida, for example *Sphaerolaimus.* All four layers are also present in members of Chromadorida (Figure 8). In the desmodorid genus *Monoposthia* (Figure 7), a dense, homogeneous exocuticle constitutes the foundation of the rigid framework of the cuticle. The mesocuticle is

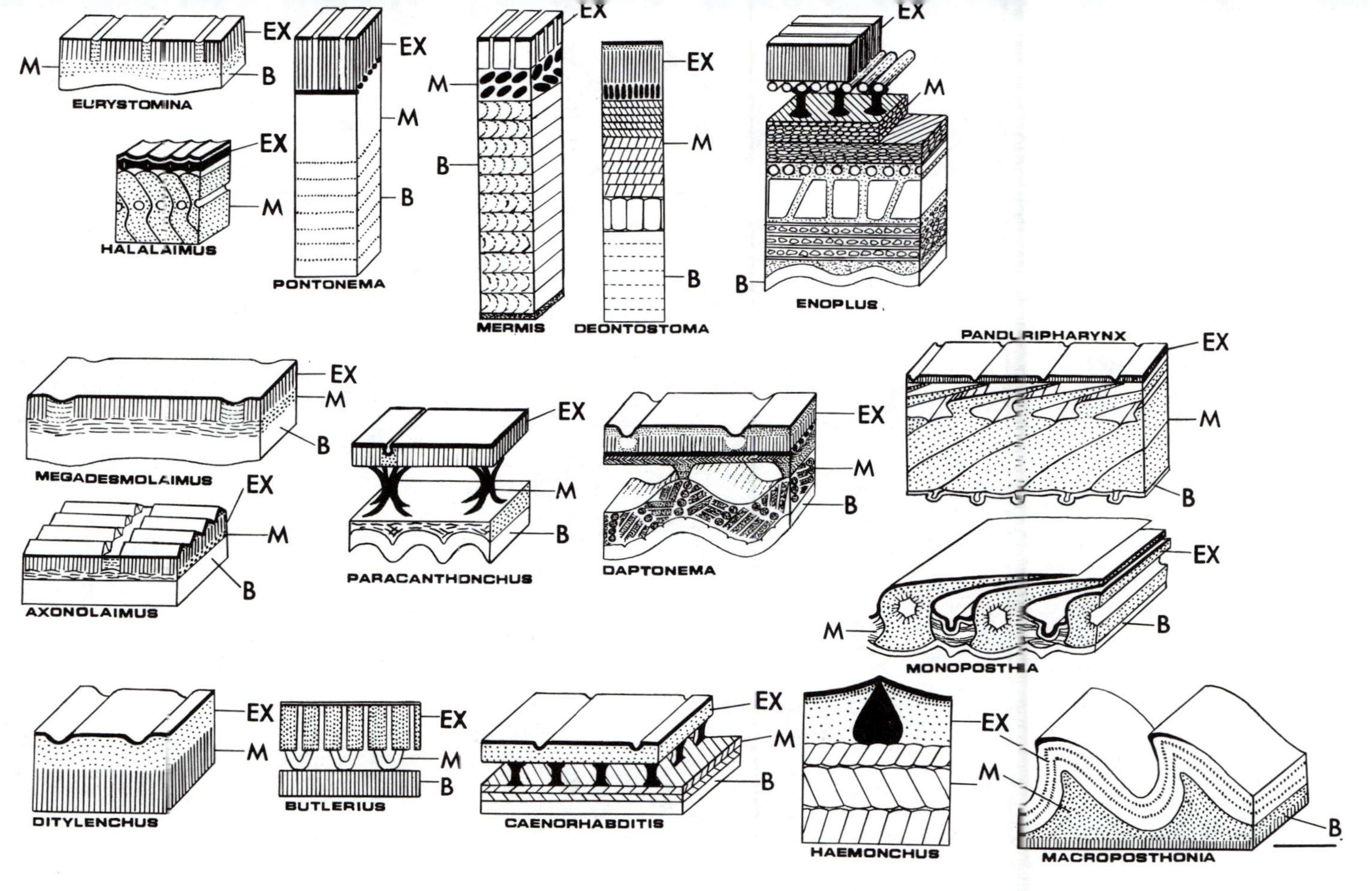

Figure 9. Diagrams of the structure of the cuticle of members of Enoplia (top row), Chromadoria (middle row), and Rhabditia (lower row). After Yushin 1988. Labels as in Figure 1.

reduced and is represented by minute spaces filled with fine fibrous material in the intervals between the plates.

A full four-layered structure is characteristic of the *Caenorhabditis briggsae* cuticle (Epstein et al. 1971; Zuckerman et al. 1973). Particularly characteristic of this species is the presence of an internal cavity in the mesocuticle, intersected by osmiophilic beams. In Maggenti's opinion (1979), a reduction of the endocuticle takes place in a number of Rhabditia, and, in many forms, in the mesocuticle as well. In *Ascaris, Oxyuris,* and other forms, there is a strongly developed mesocuticle containing several levels of obliquely oriented fibers directly adjoining the hypodermis (Bömmel 1894; Toldt 1899, 1904, 1905; Goldschmidt 1904, 1905; Bird 1958, 1971; Watson 1965a; Bogoyavlenskii 1973). Among small parasitic and saprobiotic members of Rhabditia, a thin cuticle contains only epi- and mesocuticular layers, so that the striated sublayer directly adjoins the hypodermis as, for example, in *Panagrellus* and various members of Tylenchida (Bird and Rogers 1965; Lee 1966a; Johnson et al. 1970; Bird 1971; Shepherd et al. 1972).

The data reviewed above pertain to the structure of the cuticle over most of the body other than the head end. For us, the organization of the cuticle at the head end of free-living marine nematodes has special interest. It was noted some time ago that at the head end of many nematodes the cuticle seems to split and its inner layer is attached to the stoma or to the upper end of the pharynx (de Man 1886; Stewart 1906; Filip'jev 1921; Kreis 1934; Inglis 1964b). Recently this phenomenon was, in essence, rediscovered by Belogurov and Belogurova (1975a), who employed the term "split cuticle." Electron microscope data have shown that at the head end of many nematodes there is an abrupt changeover in the cuticle, marking a more or less noticeable head seam (Figure 10). Further, there is a thickening of the compressible layer of the mesocuticle, and as a result, the dense layers of exo- and endocuticle separate (Figure 10). The sclerotized layer of the endocuticle is joined to the stoma walls, forming the internal skeleton of the head end—the endocupola. The middle compressible layer of the mesocuticle is observed with the light microscope as an interval between two layers of "split" cuticle. Thus, at the head end, there is not so much a splitting as a thickening of the cuticle. The middle compressible layer of the mesocuticle fills the role of a shock absorber or buffer, dissipating shocks to the anterior end. In combination with the solid endocupola, it creates an effective system for protecting the internal tissues of the head end.

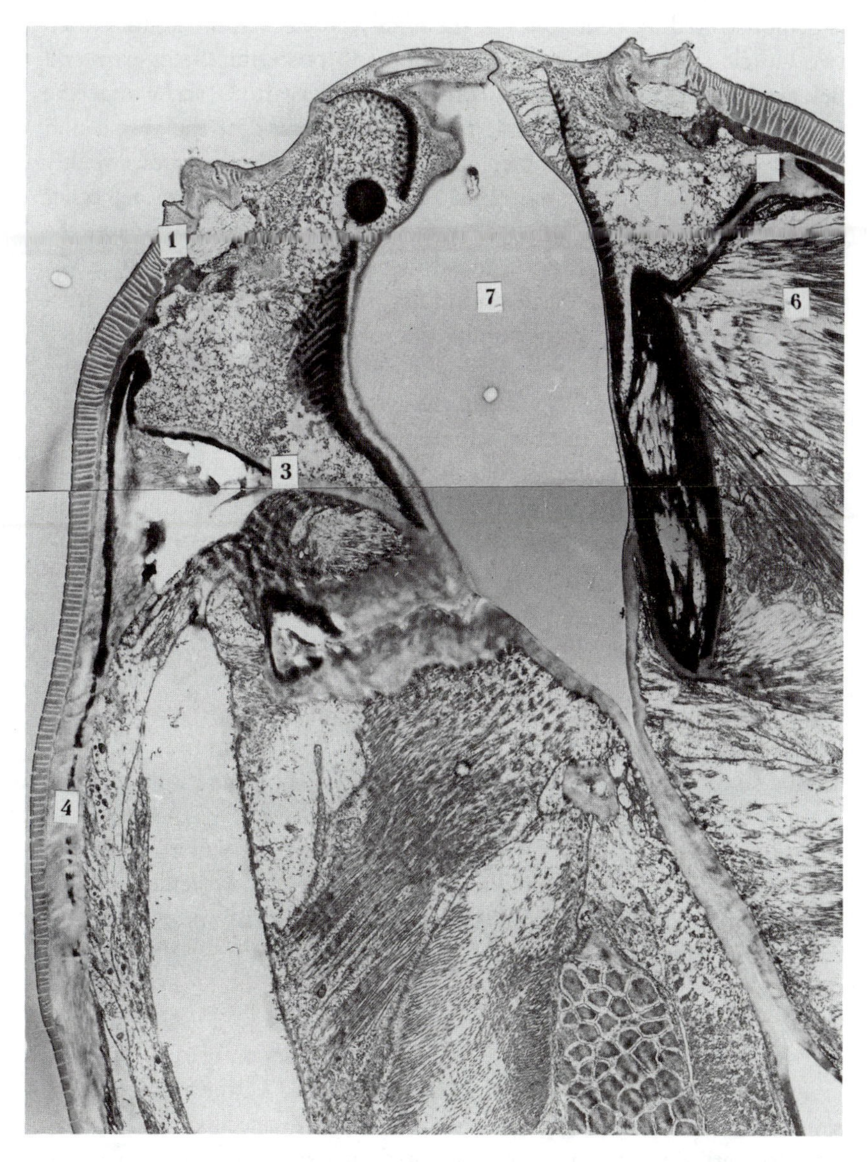

Figure 10. Longitudinal section of the head of the first-stage juvenile of *Enoplus demani*. Magnification = 21,700. 1, exocupola; 2, cephalic ventricle; 3, endocupola; 4, cephalic cuticle; 5, cephalic seam; 6, muscles; 7, mouth cavity.

An example of another organization of cephalic cuticular structure is demonstrated by nematodes of the family Sphaerolaimidae, for which two cuticle splits are described: anterior and posterior (Belogurov and Belogurova 1975a). Electron microscopic observations show that the posterior splitting of the cuticle represents a joining of the stoma and special connectors of the outer cuticle, and only the anterior, weakly noticeable "splitting" of the cuticle corresponds to this among other nematodes (Figure 11). Other examples of cuticle modifications at the head end in free-living marine nematodes are represented in Figure 12.

The cuticle also lines all ectodermal sections of the digestive tract (stoma, pharynx, and posterior intestine). The internal cuticle, which lines the lumen of portions of the digestive tract, is structurally more simple than the external cuticle. In the internal cuticle there is never the striated layer or layers of obliquely oriented fibers that are characteristic of exo- and mesocuticles. In the internal cuticle one can see a layer of epicuticle of more or less typical structure and an endocuticle either consisting of one homogeneous material or containing alternating bands of dark and light material.

Nematode growth is accompanied by molting; that is, the discarding of old cuticle and its replacement by new. Molting encompasses the external cuticle with all its derivatives (setae, papillae, amphid lining) and the internal cuticle of the stoma, pharynx, and posterior intestine. For all of the nematodes studied, four molts have been noted, although, among some very specialized parasitic species, one or two molts can take place within the ovum membrane.

Before molting, the hypodermis thickens, and in it there appears a large quantity of ribosomelike particles and multivesicular bodies. Chemical analysis shows an increase in protein content and RNA in the cytoplasm of hypodermal cells. The molting process also involves muscle cells in which, during molting, proteins and lipids are actively synthesized.

In the early molting stages, the hypodermis separates from the old cuticle and forms a new epicuticle on its surface. As the result of subsequent secretion, the deeper layers of cuticle are successively secreted. After secretion of the layers of the cuticle is complete, a sort of maturation takes place during which each cuticular layer gradually acquires its characteristic internal structure. The old cuticle may be simply discarded, as noted in many parasitic and free-living forms (Lee 1966b, 1970; Samoiloff and Pasternak 1969; Johnson et al. 1970). Among *Me-*

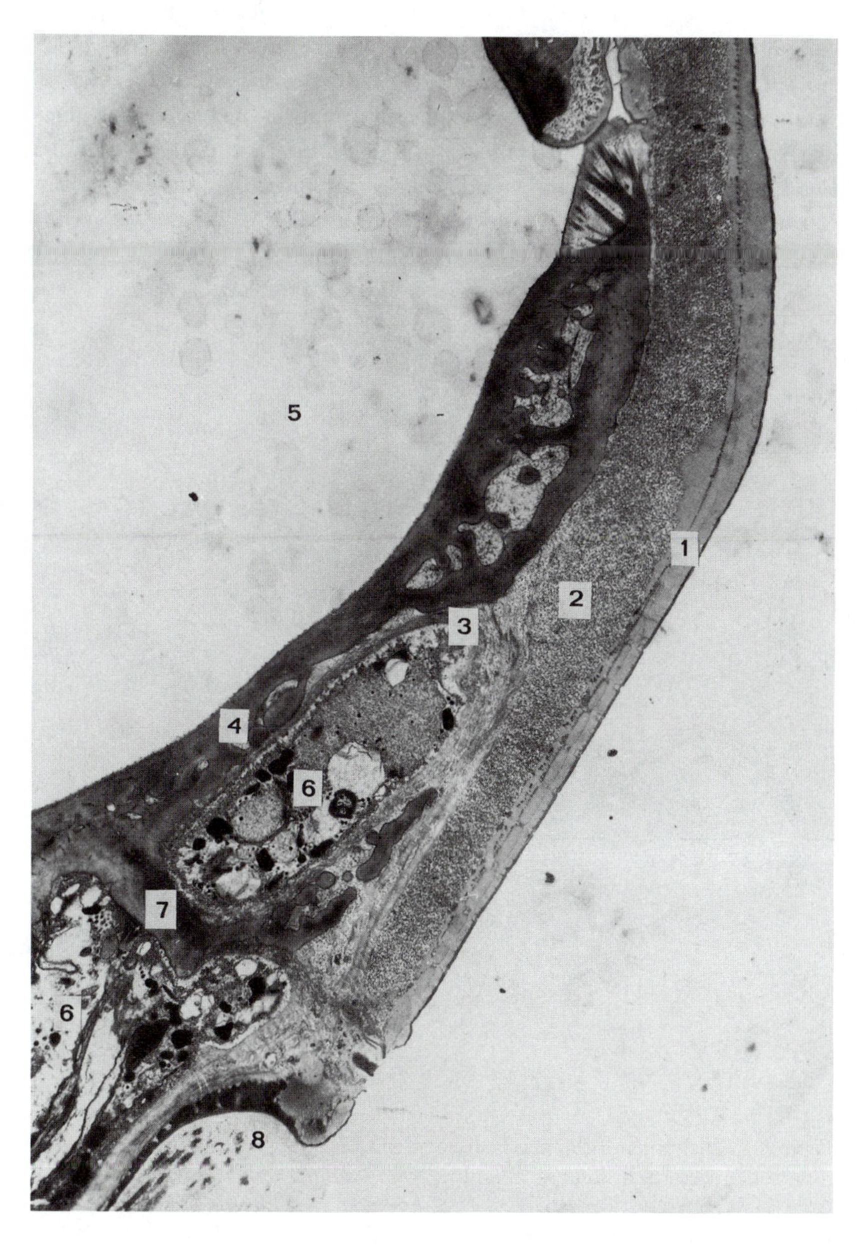

Figure 11. Longitudinal section of the head of *Sphaerolaimus balticus*. Magnification = 12,500. 1, exocupola; 2, cephalic ventricle; 3, endocupola; 4, stomatal cuticle; 5, mouth cavity; 6, hypodermis; 7, cephalic cuticular bridge.

loidogyne, Ascaris, and *Hemicycliophora,* a partial resorption of the old cuticular material is observed during the molting process (Bird and Rogers 1965; Thust 1968; Johnson et al. 1970). According to the data of some authors, the early stages of cuticle formation may occur intracellularly, as described for *Nematospiroides dubius* (Bonner et al. 1970). Later, a new cytoplasmic membrane is formed, dividing the newly formed cuticle from the hypodermis. This means of cuticle formation seems to be an exception; in most nematodes the cuticle is formed

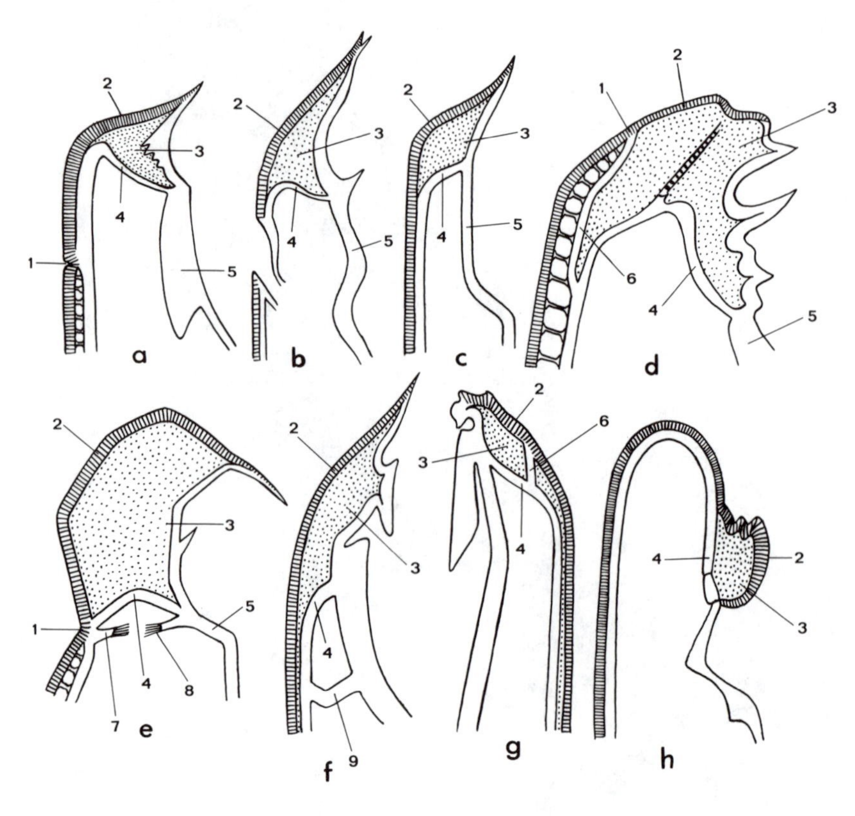

Figure 12. Diagrams of the head cuticular modifications in free-living marine nematodes (Figures a, b, d, e, g, and h from Yushin 1988; Figures c and f from Malakhov and Voronov 1981): a, *Enoplus;* b, *Bathyeurystomina;* c, *Pontonema;* d, *Halichoanolaimus;* e, *Daptonema;* f, *Sphaerolaimus;* g, *Axonolaimus;* h, *Megadesmolaimus.* 1, cephalic seam; 2, exocupola; 3, cephalic ventricle; 4, endocupola; 5, stomatal cuticle; 6, intracuticular septum; 7, rudimentary cuticular bridge; 8, tonofilaments; 9, cuticular bridge.

through secretion (Minier and Bonner 1975). Molting in nematodes occurs under the control of neurosecretory cells. In particular, for certain forms, it has been shown that under the effect of neurosecretory cells a secretion from the cervical gland of an enzyme of leucine-amino-peptide dissolves the old cuticle (Lee 1962, 1965).

Microvilli usually take part in the formation of invertebrate cuticles (Rieger 1984). In nematodes microvilli were found in embryogenesis of cuticle. Fifteen-day-old embryos of *Enoplus demani* have short microvilli filled with a dense substance (Figure 13a). Probably the material of which the cuticle is composed is excreted through the tips of the microvilli. The cuticle precursor looks like a thin granular membrane above

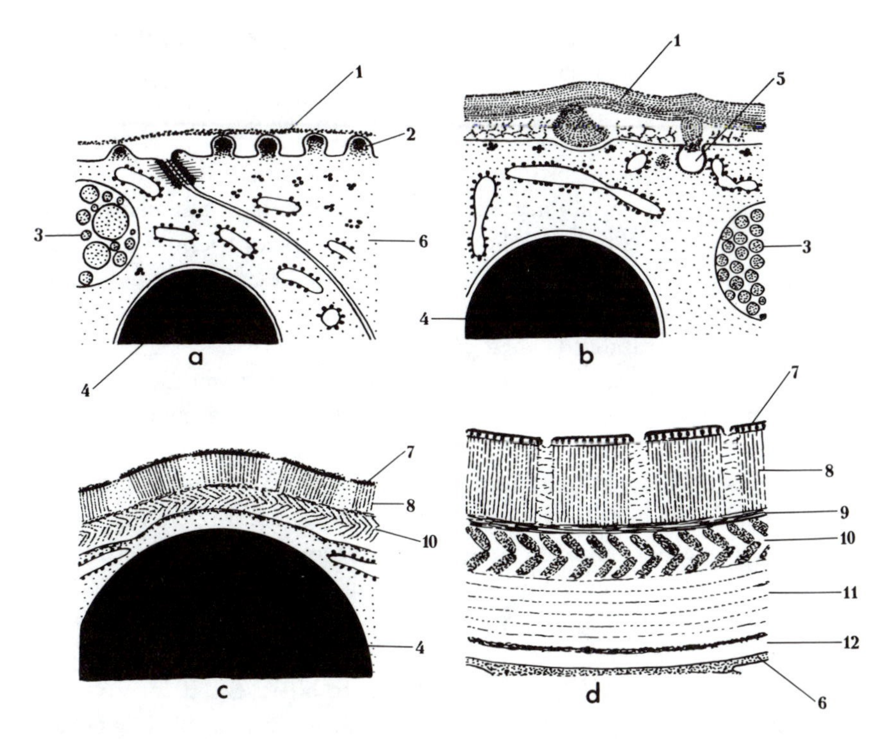

Figure 13. Embryogenesis of the cuticle of *Enoplus demani* (after Yushin and Malakhov 1989): a, 15 days old; b, 18 days old; c, 23 days old; d, first-stage juvenile. 1, cuticular precursor; 2, microvilli; 3, multivesicular bodies; 4, yolk bodies; 5, exocytosis of cuticular material; 6, hypodermis; 7, epicuticle; 8, striated exocuticle; 9, longitudinal fibers; 10, oblique fibers; 11, internal mesocuticular layers; 12, endocuticle.

the microvillar tips (Figure 13a). Microvilli disappear in the 18-day-old embryo stage, and the exocytosis of cuticular material may be observed (Figure 13b). The differentiation of the epicuticle, striated exocuticle, and rudimentary mesocuticle takes place in the 23-day-old embryo (Figure 13c). The cuticle of the first-stage juvenile of *E. demani* is essentially simpler than the adult cuticle (Figure 13d).

The nematode cuticle has three fundamental functions: permeability, that is, the capability for selective transport of matter; protection; and a dynamic function, that is, participation in the biomechanics of movement. Permeability is not a specific function of the cuticle. Animal integuments, whether cuticular or not, always facilitate the selective absorption of substances from the external environment. Undoubtedly, the fundamental role in the selective absorption of substances from the external environment is played by a layer of the ectodermal epithelium—the hypodermis. However, dense cuticle presents a significant barrier to penetration of substances and, evidently, possesses special mechanisms that facilitate transfer of some substances and inhibit the penetration of others. The semiporous properties of nematode cuticle are well known and are especially pronounced in parasitic forms. It is assumed that a layer of epicuticle is responsible for the selective properties of nematode cuticle (Bird 1980). Interestingly, among nematodes of Enoplida, the epicuticle layer is broken by ringed grooves or pores, which may relate to the high permeability of their cuticle.

The peculiarities of the nematodes' environments impose stringent requirements on the cuticle as a protective structure. Marine and freshwater nematodes inhabit interstitial spaces of benthic sediments; soil nematodes populate the capillary intervals between soil particles; phytonematodes live in narrow intercellular spaces; and gut-inhabiting, zooparasitic nematodes may be subjected to the mechanical properties of food masses. Living in "tight" environments, so typical of nematodes, would lead unavoidably to large or small damage to the body were it not for the cuticle serving as a reliable shield. In the external cuticle, the protective function belongs in some degree to all layers. Evidently, the protective function is fulfilled by various amorphous, marbled, lamellar layers, situated largely in the mesocuticle. The protective function is highly effective in the scleroticized endocuticle of the internal lining of the ectodermal sections of the digestive tract. The hardest layers of cuticle are those that, when observed with the light microscope, appear to strongly refract light and, under an electron microscope, are electron-lucid, structureless layers. According to their chemical nature, these layers consist of a special protein—scleroprotein, which is common in invertebrate cuticles.

The lacunar layers of nematode cuticle protect the organism from mechanical damage by absorbing shock from external sources. It is characteristic that strong development of the lacunar layer in nematode cuticle is detected at the anterior end of the body (the cuticle splitting in marine nematodes), which bears the greatest shock burden.

The cuticle also functions in the nematode's locomotory system. As is known, nematodes possess a very simple muscular system, principally a single layer of longitudinal muscles. The simplicity of the nematode muscular system contrasts sharply with the complexly organized musculo-cutaneous sac of other lower worms. Flatworms have strong layers of circular and longitudinal musculature and dorsoventral and oblique muscles. Lower Turbellaria have ciliary means of locomotion. Other flatworms, Nemertinea, and Trematoda move by peristalsis or by a type of pedal wave or other means requiring the interaction of a complex muscular system (Clark 1964). Flatworms and nemertines are able to alter sharply the shape of the body in the seizure of prey, copulation, and so forth. In contrast, common to nematodes are uniform movements consisting solely of bending the body. Nematodes are incapable of significantly changing the overall shape of the body in any way. Flexing the body is the basis for movement of almost all nematodes in all of their natural habitats, except for the aberrant means of movement among members of Desmoscolecida and Criconematidae (Stauffer 1924). The overwhelming majority of nematodes effectively move by undulation or serpentine creeping between particles of soil or sediment. These means of movement are similar to each other biomechanically (Gray and Lismann 1964; Gray 1968) and are characteristic of vertebrates. Nematodes are the only large group of lower Metazoa that travel in the same way.

Nematode movement is accomplished by the antagonistic interaction between longitudinal musculature and the cuticle, which is maintained in a stretched state by the high turgor pressure of internal tissues. The one-sided shortening of the musculature leads to the flexing of the body, and its extension takes place as a reaction of the stretched cuticle. The mechanism of nematode movement can be represented as the flexing of an elastic cylinder under the effect of two internal bands of muscles. The shape of the moving nematode fully satisfies this model and is approximated by an "elastic curve" (Calcoen and Roggen 1973). The complicated internal structure of the external cuticle, evidently, is largely explained by its dynamic functions. It is not by chance, therefore, that all the complicated structural components are encountered only in the outer cuticle. The cuticle of internal organs that do not participate in the biomechanics of movement is constructed significantly more simply than the outer cuticle and never contains, for example, a striated layer or a layer of oblique fibers. In juveniles of plant-parasitic nematodes that move freely in the soil, a relatively thin cuticle has a complicated internal structure and contains, specifically, a striated layer. After penetration into a host plant and transformation into an immobile parasitic form of life, the cuticle is much simplified, and the striated and other layers disappear (Bird and Rogers 1965; Lee 1966a; Bird 1971; Shepherd et al. 1972). It is interesting that among nematodes parasitic in animals the cuticle of intestinal helminths is much more complex in structure than that in pneumohelminths (Bogoyavlenskii 1973). These differences evidently are related to the greater mobility of nematodes inhabiting the intestinal lumen and having to withstand a constant flow of food masses, in comparison with immobile helminths in tissue.

The significance of separate layers relevant to the dynamic function of the cuticle largely remains unclear. An important biomechanical role is ascribed particularly to the striated layer or to the layer of oblique fibers, which, separately and sometimes together, are present in actively mobile forms (Figure 1). The striated layer presumably consists of regularly oriented protein molecules or groups of them. In the contractions of the longitudinal musculature on one side of the worm's body, a cuticle deformation occurs that leads to a convergence of the parallel components of the striated layer. The forces of resistance thus arising between the parallel protein molecules generate a resultant antideformation force, which returns the worm body to its original position.

The layer of obliquely oriented fibers that cover the nematode body in spirals can be compared with a system of spiral fibers in the basal membranes among members of Turbellaria and Nemertinea (Clark and Cowey 1958; Clark 1964; Inglis 1964a; Crofton 1966). Calculations show that such a system of fibrils has a maximum effectiveness with the angle of inclination from the axis of the body equal to 55° (Figure 14). Contraction of the longitudinal musculature increases the angle of the fibrils, and contraction of the circular musculature reduces the angle. The inclination angle of the spiral fibers in nematode cuticle is 60–75°.

The nematode body capacity precisely corresponds to the capacity of the fibril system at a given inclination angle. Any contraction of the longitudinal musculature causes an increase in the inclination angle of the fibrils and a decrease in the volume of the fibrils. This indicates an increase in the internal pressure (the liquids are incompressible) and causes a force that tends to return the system to its original state; that is, to straighten the nematode. Thus, at an inclination angle of the oblique fibers of more than 55°, longitudinal musculature is necessary and sufficient to ensure locomotion based on the bending of the body.

This concept neatly ties together the peculiarities of cuticle structure, the presence of only one layer of longitudinal musculature, and the existence of high internal turgor in nematodes. The only, and rather substantial, shortcoming of this theory is that the system of oblique fibers is far from being characteristic of all nematodes, although the locomotion of the overwhelming majority of the representatives of the class is based on the interaction of resilient cuticle, high internal stress, and longitudinal musculature.

By maintaining sufficient firmness, the cuticle should permit the nematode to flex within the necessary limits. This problem is resolved

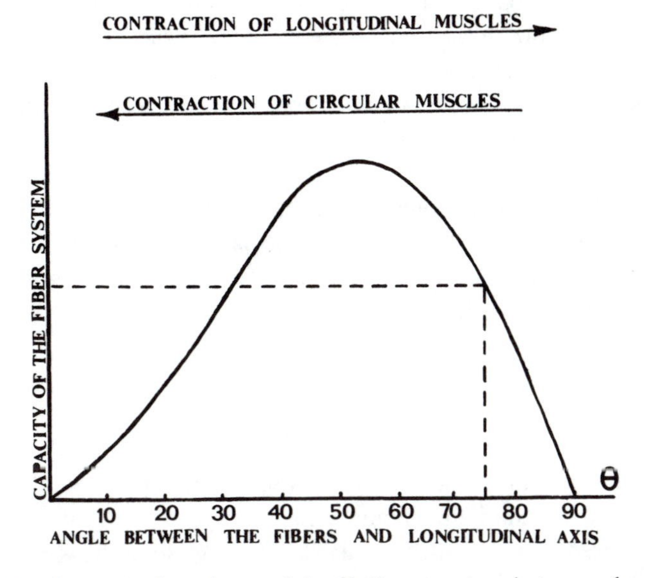

Figure 14. Change in the volume of the fibril system in relation to the angle of inclination from the longitudinal axis of the body (after Clark 1964).

by the creation of annulations that have a different nature, as we have seen. The peculiarities of nematode movements also specify the presence of a lateral differentiation of the cuticle. As is known, with movement, nematodes flex the body chiefly in the dorsoventral plane, and therefore the lateral sides of the body serve, in a way, as the nematodes' creeping surfaces. Many forms of the orders Enoplida, Monhysterida, and Desmodorida do not have lateral differentiation. But in many instances the first signs of its appearance can be observed. Thus, among Enoplida and some Dorylaimida (Chitwood and Chitwood 1950), the cuticle on the lateral sides of the body is thicker than on the dorsal and ventral sides. The highest development of lateral differentiation of cuticle is achieved in Chromadorida (Figure 15). Here, in a number of cases, there are real lateral cuticular alae, and the whole nematode appears to be flattened in the dorsoventral direction (Riemann 1976).

The acquisition of cuticle has played a major role in the formation of the whole organization of nematodes. Primitive ciliary epithelium directly in contact with the external environment is characteristic of lower free-living Bilateria. The acquisition of cuticle signifies an important step in the progressive evolution of integumentary tissues. Cuticular integument is also characteristic of many other lower non-coelomic Bilateria: Gastrotricha, Rotifera, Acanthocephala, Nematomorpha, Kinorhyncha, and Priapulida. But in not one of these groups is the cuticle as structurally complicated, nor does it play as significant a role in the biomechanics of movement, as in nematodes. Only in nematodes does the cuticle acquire varied functions that make it a major factor in forming the whole organization of this group. The dynamic function of the nematode cuticle, in addition to the protective function characteristic of all cuticular integuments in general, has permitted nematodes to develop extremely effective means for movement that have been partially responsible for the biological success of this group.

Hypodermis

The hypodermis, an epithelial tissue distinct from typical ectodermal epithelium, is the second chief component of the body wall of nematodes. It is composed of a thin subcuticular layer lying along the whole perimeter of the body between the cuticle and musculature, and of "longitudinal chords" penetrating deeply into the body. Each hypodermal cell in this connection consists of a thin flattened sheet extending between the cuticle and the muscle cells and a cytoplasmic protuberance bearing the nucleus and cytoplasmic organelles that contributes to the structure of the hypodermal chord.

The ultrastructure of the hypodermis indicates extensive physiological activity in this tissue. That part of each hypodermal cell that lies within the chord has a large nucleus with a large nucleolus, an irregular endoplasmic reticulum, elements of a Golgi lamellar complex, and large

Figure 15. Lateral differentiation of the cuticle in *Neochromadora.*

vacuoles with homogeneous contents. The subcuticular portion of the cell is characterized by an impoverishment of organelles, but there are large vacuoles and vesicles that create an abundance of membrane structures. The hypodermis is isolated from the cuticle by a clear cytoplasmic membrane. Connection of the hypodermis to the cuticle is accomplished by a system of hemidesmosomes, from which tonofibrils branch out, penetrating the subcuticle. The interchordal hypodermis is attached to the musculature through a layer of intercellular amorphous substance. The total thickness of the interchordal hypodermis among free-living forms is 0.1–0.8 μm in all. Yet the nerve fibers of the nerve commissures still pass between the cuticle and the musculature.

In parasitic forms, the subcuticular layer reaches significant thickness (up to 30 μm), and a strong system of various types of supporting fibers is developed in the hypodermis (Martini 1906, 1907, 1908, 1909; Chitwood and Chitwood 1950; Byers and Anderson 1972; Bogoyavlenskii 1973).

The nematode hypodermis is a tissue with few cells. The arrangement of hypodermal cells is orderly, and their boundaries form a more or less regular hypodermal mosaic (Figure 16). Because of this, the architectonic principles of hypodermal organization are manifested even in the arrangement of individual cells. Hypodermal cells are arranged in regular longitudinal rows varying from five to twelve in different species. With change in the number of rows of hypodermal cells, the number and degree of development of hypodermal chords also changes as does the symmetry of the whole body wall (Figures 16 and 17).

In forms with a five-row hypodermis (*Oxystomina*), the hypodermal mosaic conforms to bilateral symmetry; there are, in all, two basic chords of hypodermis (lateral), and if the sequence of nuclei in the dorsal row is disregarded, the body wall tends toward biradial symmetry (Figure 17).

The six-row hypodermis is less widespread and is noted only in *Dolicholaimus* (order Enoplida) and in *Camacolaimus* (order Plectida). In this instance, both the hypodermal mosaic and the body wall tend toward biradial symmetry (Figure 17). We encounter special forms of six-row hypodermis in *Metalinhomoeus* and *Daptonema* of the order Monhysterida, where there is also a hypodermal mosaic, and the body wall conforms to bilateral symmetry (Figure 17). These types can be regarded as transitional to the more widespread seven-row type of hypodermal structure.

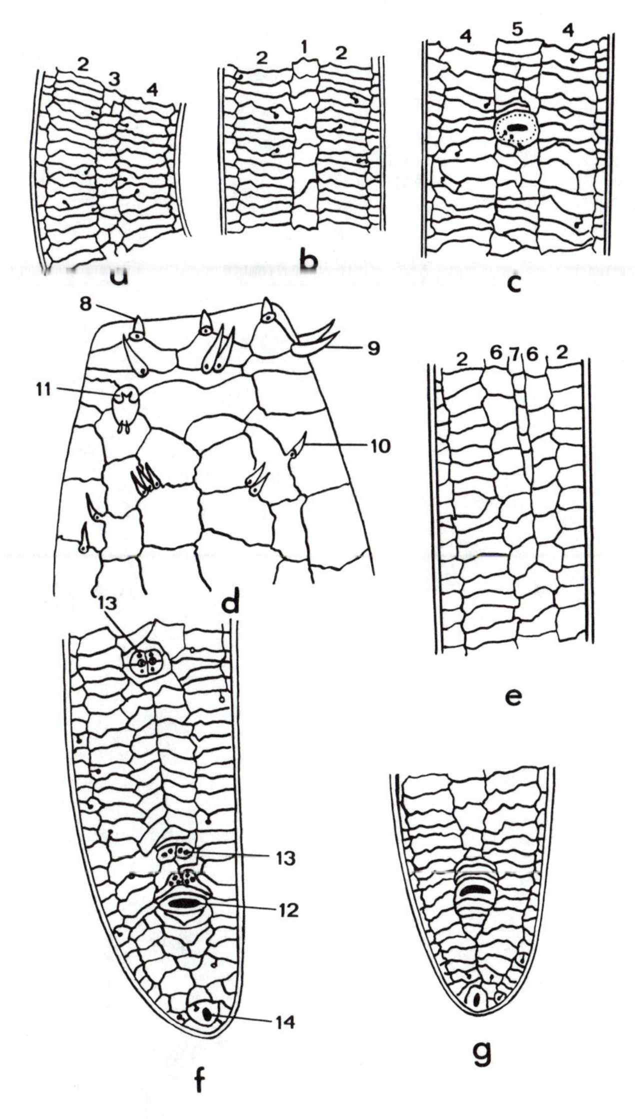

Figure 16. The hypodermal mosaic of *Pontonema vulgare:* a, lateral view of the midbody region; b, dorsal view of the midbody region; c, ventral view of the vulvar region; d, head end; e, beginning of additional rows of cells in the region of the nerve ring; f, ventral view of the male tail; g, ventral view of the tail of a female. 1, dorsal; 2, subdorsal; 3, lateral; 4, subventral; 5, ventral; 6, cephalic subdorsal; 7, cephalic dorsal; 8, labial papillae; 9, cephalic setae; 10, cervical setae; 11, amphid; 12, anus; 13, genital papillae; 14, caudal gland pore.

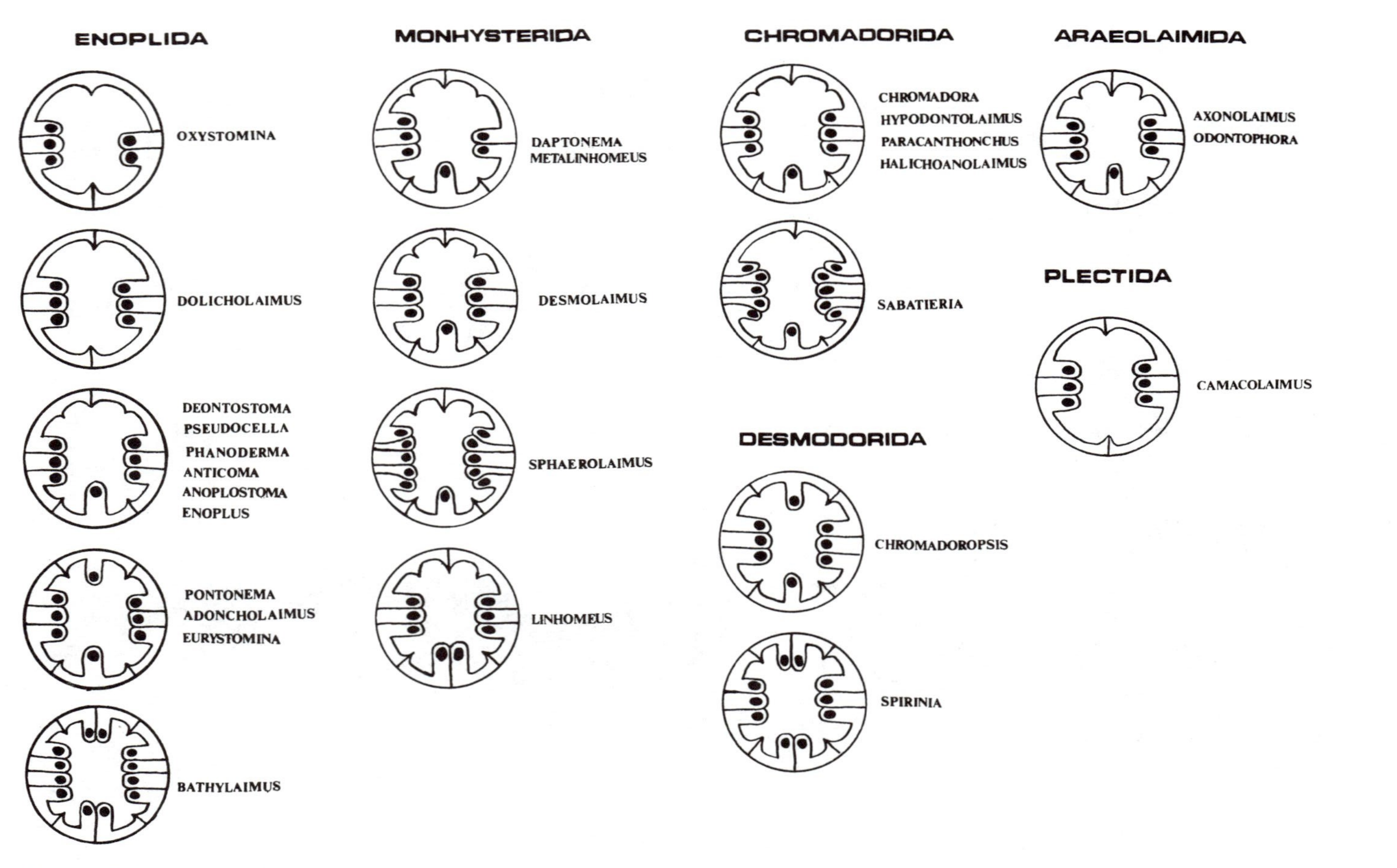

Figure 17. Diagram of the hypodermal structure in various groups of free-living nematodes.

The seven-row hypodermis is the common type among the investigated species of nematodes. In this case, the hypodermal mosaic and body wall as a whole conform to bilateral symmetry (Figure 17). Owing to protuberances of hypodermal cells, three basic chords of hypodermis are formed: a pair of lateral chords, each of which contains nuclei of three rows of cells, and an unpaired ventral chord that contains the nuclei of the unpaired ventral row. This type of structure has been noted among *Deontostoma, Pseudocella, Anticoma, Anoplostoma, Enoplus, Mesacanthion, Chromadora, Hypodontolaimus, Prochromadorella, Paracanthonchus, Halichoanolaimus,* and *Axonolaimus.*

An eight-row hypodermis is very characteristic of representatives of the families Oncholaimidae and Enchelidiidae in the order Enoplida and of *Chromadoropsis* in the order Desmodorida. In this case, the hypodermal mosaic and the body wall as a whole acquire a complete biradial symmetry (Figure 17). The twelve-row hypodermis of *Bathylaimus* is derived easily from the eight-row type by augmenting the rows of hypodermal cells without changing the general symmetry of the hypodermis. In the same way, the ten-row hypodermis of *Spirina* can be derived from the eight-row hypodermis of *Chromadoropsis* (Figure 17).

By doubling the cells of the ventral row of the seven-row hypodermis of the *Desmolaimus* type, the nine-row hypodermis of *Linhomoeus* can be derived. The unusual types of hypodermal structure in *Sabatieria* of the order Chromadorida and *Sphaerolaimus* of the order Monhysterida can also be derived from the simple seven-row prototype if the possibility of adding special paralateral rows of cells is assumed (Figure 17).

Although the data available on hypodermal structure among various free-living nematodes are at this time incomplete, they still portray the scale of variation in the organization of this structure among nematodes. The presence of such variation is a rather unexpected fact that results in important consequences for comparative anatomy.

First, it is interesting to try to establish the direction in which the evolution of hypodermal structure took place in nematodes. With more or less completeness, we have obtained in each order the morphological series of hypodermal types, but these series can be read in two directions: from the forms with the least number of hypodermal cells to multirow types of hypodermal structure and vice versa (Figure 17).

In this connection, it is interesting that, at least for a number of forms possessing a seven- or eight-row hypodermis in the mature state, it has been shown that juveniles hatching from an ovum have five rows of hypodermal cells in a total of two hypodermal chords (Figure 18). Unfortunately, it still remains unknown how, during the process of postembryonic development from a five-cell-row hypodermis, the multicell-row hypodermis of adults is formed. Nevertheless, the presence of a five-cell-row hypodermis in the first postembryonic stages is very indicative. It is also striking that among mature forms we find species with a five- or six-cell-row hypodermis and, among species related to them, forms with significantly more rows.

For nematodes of the subclass Rhabditia, a syncytial hypodermis is very characteristic, although among a number of forms of the orders Rhabditida and Tylenchida a cellular hypodermis has been noted (Wright 1965, 1968b; Roggen et al. 1967; Yuen 1967; Bird 1971). It is significant, however, that juveniles of parasitic nematodes have a cellular hypodermis containing five rows of cells arranged as shown above for juveniles of free-living forms (Martini 1906; Thust 1968). The hypodermis changes into the syncytial state only in juveniles at stage III. Cellular hypodermal structure in juveniles of parasitic nematodes evidently recapitulates the original ancestral state. Among mature forms of the sub-

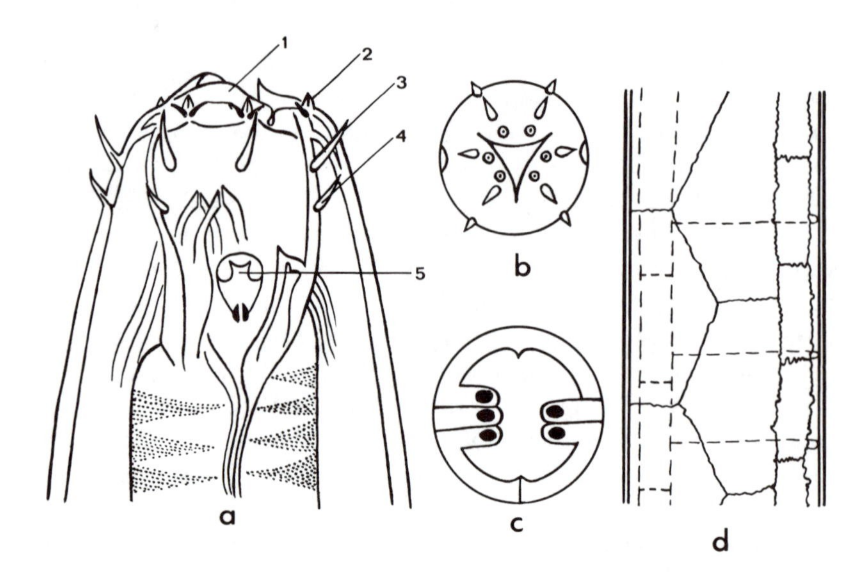

Figure 18. Structure of the first-stage larva of *Pontonema vulgare:* a, head end; b, diagram of cephalic sensilla; c, diagram of hypodermal structure; d, hypodermal mosaic in midbody region. 1, labia; 2, labial papillae; 3, cephalic sensilla of anterior circle; 4, cephalic sensilla of posterior circle; 5, amphid.

class Rhabditia, the nuclei are arranged largely in lateral chords of hypodermis. Only in nematodes of the order Rhabditida are the nuclei found in the dorsal and ventral chords of the hypodermis (Chitwood and Chitwood 1950; Bogoyavlenskii 1973). Among large parasitic nematodes (Ascaridida and some Oxyurida), there is an expulsion of the hypodermal nuclei from the longitudinal chords to the subcuticle, which is very thick among these forms (Hinz 1963; Watson 1965a; Bogoyavlenskii 1973). The hypodermis of parasitic nematodes is extremely complicated in its internal structure. It is abundantly penetrated by supporting fibers and tonofibrils and is subdivided into several layers, etc. (Martini 1906, 1907, 1908, 1909; Hinz 1963; Bird 1971; Bogoyavlenskii 1973).

The initially small number of hypodermal cells and their regular arrangement are evidently related to the sharp reduction of body size in nematodes (more precisely, to reduction in body diameter). As we will see, reduction in body size has been determined by the evolutionary trend of many nematode organs toward a small number of cells and has predetermined a tendency toward producing constancy in cell composition.

The hypodermis represents a variety of embedded epithelium with a specific internal structure. Both the peculiarities of cell differentiation and the general architecture of the hypodermal layer to a significant degree are determined by the peculiarities of the biomechanics of nematode movement. The nematode hypodermis is sharply differentiated into a very thin layer of subcuticle and conspicuous hypodermal folds or chords. This differentiation is the consequence of the unusual locomotory mechanism of nematodes, based on the interaction of cuticle and musculature. The biomechanics of nematode movements require a close mechanical relationship between the cuticle and the muscles. Each soft interlayer inevitably weakens the interaction between these two constituent elements of the locomotory mechanism. Yet, between the cuticle and the musculature inevitably appears an interlayer of covering epithelium. The resolution of this contradiction consists of the division of the hypodermis into a thin subcuticle, which provides a close relationship of cuticle and musculature, and hypodermal chords, in which a large part of the cell organelles are collected.

The characteristics of the nematode locomotory system are also determined by the general architecture of the hypodermis. Undulating movement requires an antagonistic system of muscles, the contraction of which causes a wave to pass through the animal body, providing forward movement of the organism. In vertebrates, the muscles of the right and left sides are the antagonists. In nematodes, the muscles of the dorsal and ventral sides are the antagonists. Hypodermal chords interrupt the layer of longitudinal musculature, dividing it into a number of muscular bands. With all the variation in types of hypodermal organization among nematodes, there is one general pattern: the lateral chords are much broader than the others, and in this connection the dorsal and ventral sides of the body are richer in musculature than the lateral sides. The flexing of the nematode body occurs in the dorsoventral plane, and with movement through the substratum, the lateral sides of the body serve as creeping surfaces. The correspondence of the wide lateral chords of the hypodermis to the creeping surfaces of the nematode body, along with the lateral differentiation of the cuticle, underscores the effects that locomotion has on the whole organization of the integument of these animals. In many nematodes, there is a real lateral differentiation of the hypodermis in that the cell borders of lateral rows of cells are strongly coupled with adjacent cells (Figure 19).

The supporting and mechanical function of the hypodermis is undoubtedly one of the most important and largely determines its structure. However, the hypodermis also has a number of other no less important functions. The secretion of new cuticle in the molting period and the maintenance and growth of the cuticle during the intermolt period are accomplished through hypodermal activity. As indicated by numerous researchers (Bird and Rogers 1965; Thust 1968; Samoiloff and Pasternak 1969; Bonner et al. 1970; Johnson et al. 1970; Lee 1970; Bonner and Weinstein 1972; Grotaert and Lippens 1974; Minier and Bonner 1975; and others), the hypodermis undergoes significant changes before and during molting. The hypodermis thickens, the dimensions of nucleoli in the nuclei increase, the endoplasmic reticulum develops hypertrophically, and a large quantity of free ribosomes appear. These changes are related to the biosynthesis of protein components of the new cuticle. The new cuticular material accumulates in the surface layers of the cytoplasm of hypodermal cells, being concentrated partially in the multivesicular corpuscles. Although a majority of authors describe the formation of new cuticle as a process of secretion, some researchers treat the cuticle as a kind of rebirth of the surface layers of the cytoplasm of hypodermal cells (Hinz 1963; Bonner et al. 1970).

It is important to point out that the synthetic activity of the hypodermis is rather extensive even during the intermolt period. It has been shown that even in the periods between moltings inclusions of new components occur in the cuticle. The synthesis takes place in the hypodermis (Samoiloff 1973).

The hypodermis is the most important barrier tissue in nematodes. Although the transport of substances through the cuticle is acknowledged, the basic function is selective absorption of substances. The hypodermis is also one of nematodes' main storage tissues, in which fat and, especially, glycogen are deposited (Lee 1965; Bird 1971).

In conclusion, the hope can be expressed that study of the hypodermis, which, up to now, has been given undeservedly little attention, will open new perspectives for nematology, not only for knowledge of this structure in itself, but also for the purposes of taxonomy and the improvement of the classification. Thus, study of the hypodermal mosaic among representatives of various groups of free-living nematodes shows a great variation in the types of hypodermal structure within the class boundaries. The method for demonstrating the hypodermal mosaic is

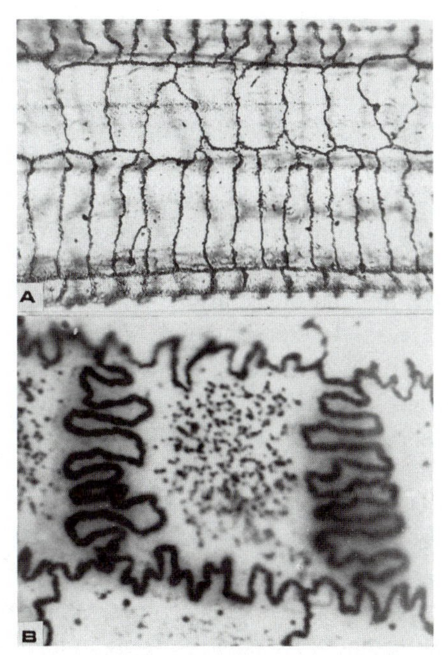

Figure 19. Hypodermal mosaic in *Pontonema vulgare* (A) and cell margins in the lateral row of hypodermal cells of *Metaparoncholaimus longispiculum* (B). Magnifications: A = 160; B = 300.

very simple and accessible to systematists. Considering the great difficulties experienced in nematode systematics in connection with the known morphological anatomy of nematodes, careful study of the hypodermal mosaic for purposes of systematics is recommended.

Somatic Musculature

The musculature of the nematode body is represented overall by a single layer of longitudinally oriented muscle cells. In comparison with the complex system of muscles making up the body wall in other lower worms, the nematode musculature looks impoverished and rather reduced. As we have seen, the presence of a single layer of longitudinal muscle cells is related to the specific manner of nematode movement. However, this means, evidently, that a significant force from muscular contractions is provided in nematodes by just one layer of muscle cells. One would therefore expect definite structural specializations of the muscle cells themselves, related to the intensification of their contractile function.

For a long time the belief was held that nematodes have the same smooth muscle cells that occur among other lower worms. Only relatively recently was it shown that the muscle fibers of nematode are obliquely striated (Rosenbluth 1963, 1965a, b, 1967). Obliquely striated muscle cells are found also among many other groups of invertebrates, including annelids (Kawaguti and Ikemoto 1959; Kawaguti 1962; Lindner and Fischer 1964; Heumann and Zebe 1967; Rosenbluth 1968; and others) and molluscs (Kawaguti and Ikemoto 1960; Hanson and Lowy 1961; Hoyle 1964; Kalamkarova and Kryukova 1966; Twarog 1967; Kryukova 1972). Obliquely striated musculature is also found among the Gastrotricha (Teuchert 1974) and among Priapulida (Zavarzin 1976).

Obliquely striated musculature has much in common with the transversely striated musculature of vertebrates and arthropods. However, in contrast with transversely striated musculature, the planes of z-bundles in obliquely striated muscle cells are arranged at an acute angle to the longitudinal axis of the cell (Figure 20). With contractile filaments of actin and myosin arranged parallel to one another, but staggered relative to one another, it is possible to distinguish H, A, I, and Z zones. Unlike transversely striated muscle cells, these zones are visible not only in lon-

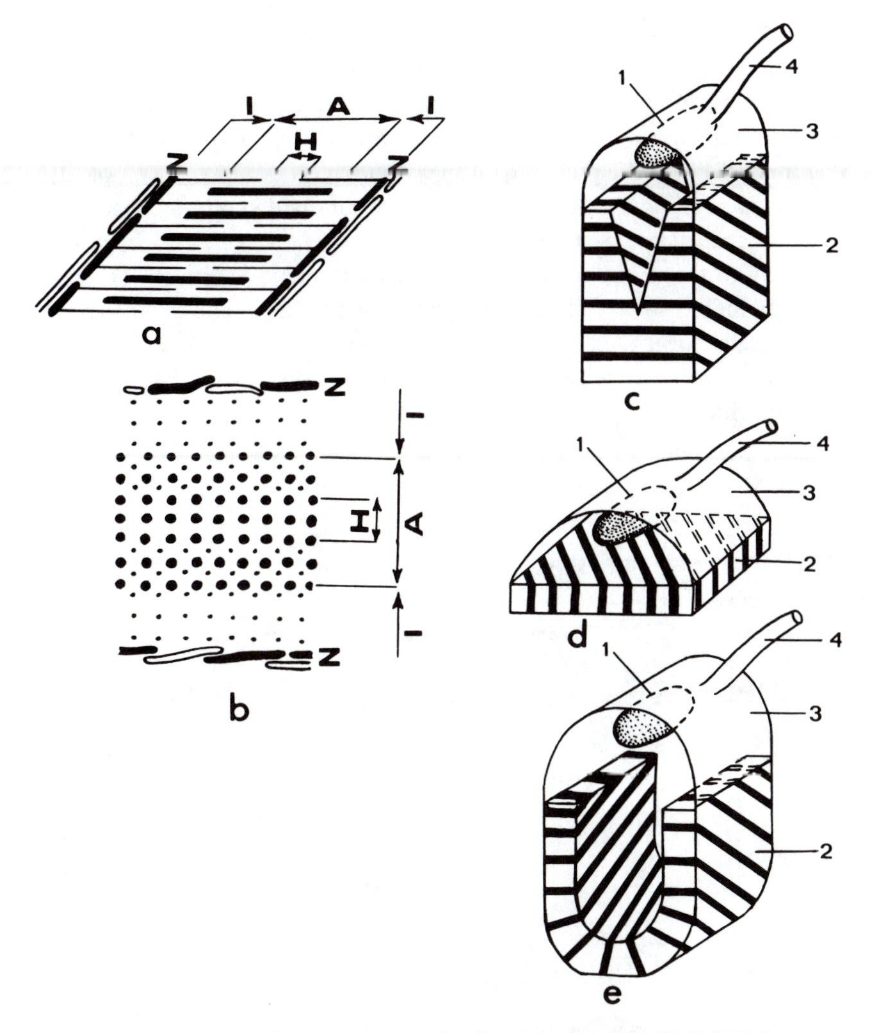

Figure 20. Structure of nematode muscle, showing zones A, H, I, Z (after Hope 1969, with changes): a, longitudinal section of a sarcomere; b, transverse section of a sarcomere; c, primary coelomyarian cell; d, platymyarian cell; e, secondary coelomyarian cell. 1, nucleus; 2, contractile part; 3, plasmic part; 4, innervation process.

gitudinal but in transverse section as well (Figure 20). These zones are more clearly distinguished when the inclination angles of the z-bundles are not too acute, such as in slack fibers. In contracted muscles with a very acute inclination angle, the H, A, and I zones are poorly defined.

Let us examine the overall morphology of nematode muscle cells by using as an example the muscles of free-living marine nematodes of the order Enoplida. Somatic muscle cells as a rule have a spindlelike form (Figure 21), but those that fulfill a special function can have a complexly irregular form (Figure 21). Muscle cells are subdivided into a contractile zone, a cytoplasmic part, and innervation processes, which go from the cell body to the nerve trunk. Each cell, as a rule, sends out several processes, which also may branch (Figures 20 and 21). The processes of muscle cells form a complex spatial network in the pseudocoel. The processes of several cells are united in tufts, approach the hypodermal chords, which support the nerve trunks, and make contact with nerve

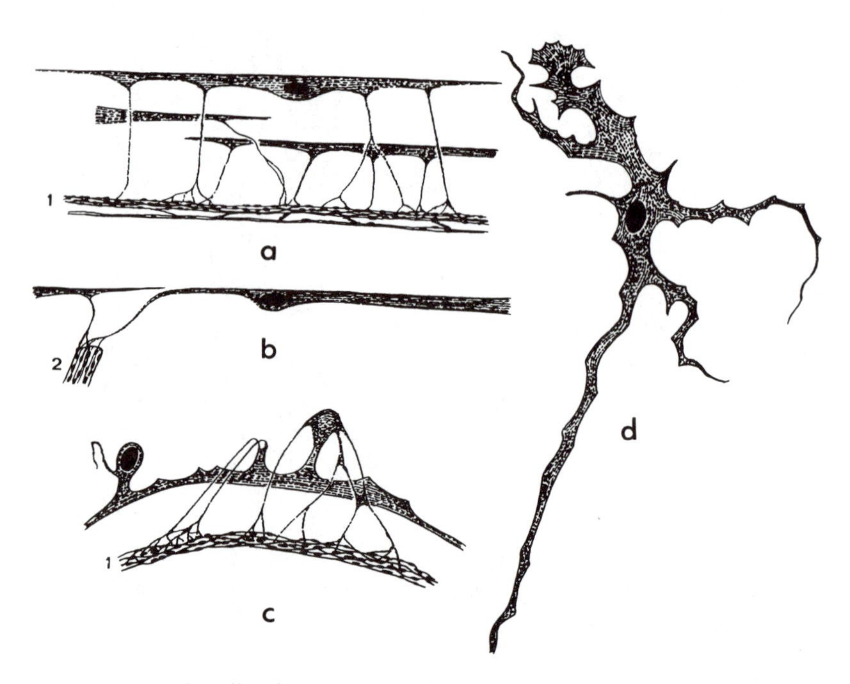

Figure 21. Muscle cells of *Pontonema vulgare* in the midregion of the body (a), the anterior region (b), the posterior region (c), and the area of the vulva (d). 1, ventral nerve trunk; 2, circumpharyngeal nerve ring.

fibers. Muscle cells situated in front of the nerve ring send processes directly into the nerve ring (Figure 21). In the cytoplasmic zone of the muscle cells are arranged the nucleus, mitochondria, various supporting fibers, granules of glycogen, and other organelles. In large nematode species, the cytoplasmic zone in the cells is rather broad, and in many small forms there is scarcely room for the nucleus (Figure 22).

The contractile zone of the cell, when viewed in transverse section, is observed to have thick (20–23 nm) myosin filaments and around each are arranged ten to twelve thin (about 6 nm) actin filaments (Figure 23). The inclination angle of the contractile filaments to the z-bundle is 14–17° (Figure 23). With such inclination angles, zones H, A, and I are visible in transverse sections, but not always clearly so (Figure 23). The z-bundles consist of electron-dense amorphous substance, and with them are associated cisternae and vesicles of sarcoplasmic reticulum (Figure 23).

The contractile zone in muscle cells among members of Enoplida lies basally, and in the more apical part of the cell it spreads into two parts in the cytoplasmic zone (Figure 24). The z-bundles in the transverse sections are parallel to the base of the cell and, in the basal region, stretch from one side of the fiber to the other (Figure 24). The muscle cells structured in this manner can be called primary coelomyarian. They are characteristic of representatives of the subclass Enoplia (Hope 1969).

Nematodes of the subclass Chromadoria are characterized by platymyarian muscle cells (Figures 20 and 25). In this case, the contractile zone also lies in the basal part of the cell, but z-bundles of dense material are perpendicular to the base of the cell and at an acute angle to the long axis of the cell. The contractile zone in cells of this type can curve up along the sides of the cell and, at the ends of the cell, can come together in a ring. In sections, therefore, it is possible to see both typical platymyarian cells and cells in which the contractile zone is situated along the periphery in the shape of a ring.

Platymyarian muscle cells are also characteristic of representatives of the subclass Rhabditia, particularly of nematodes of the orders Rhabditida, Tylenchida, many Strongylida, and Oxyurida (Wright 1964; Jamuar 1966; Lee 1966b; Yuen 1967; Wisse and Daems 1968; Bird 1971; Bogoyavlenskii 1973).

Among large nematodes, namely nematode parasites of animals, is found a coelomyarian musculature, which was used as the example for the first studies of this structure in nematode muscle cells (Hinz 1963;

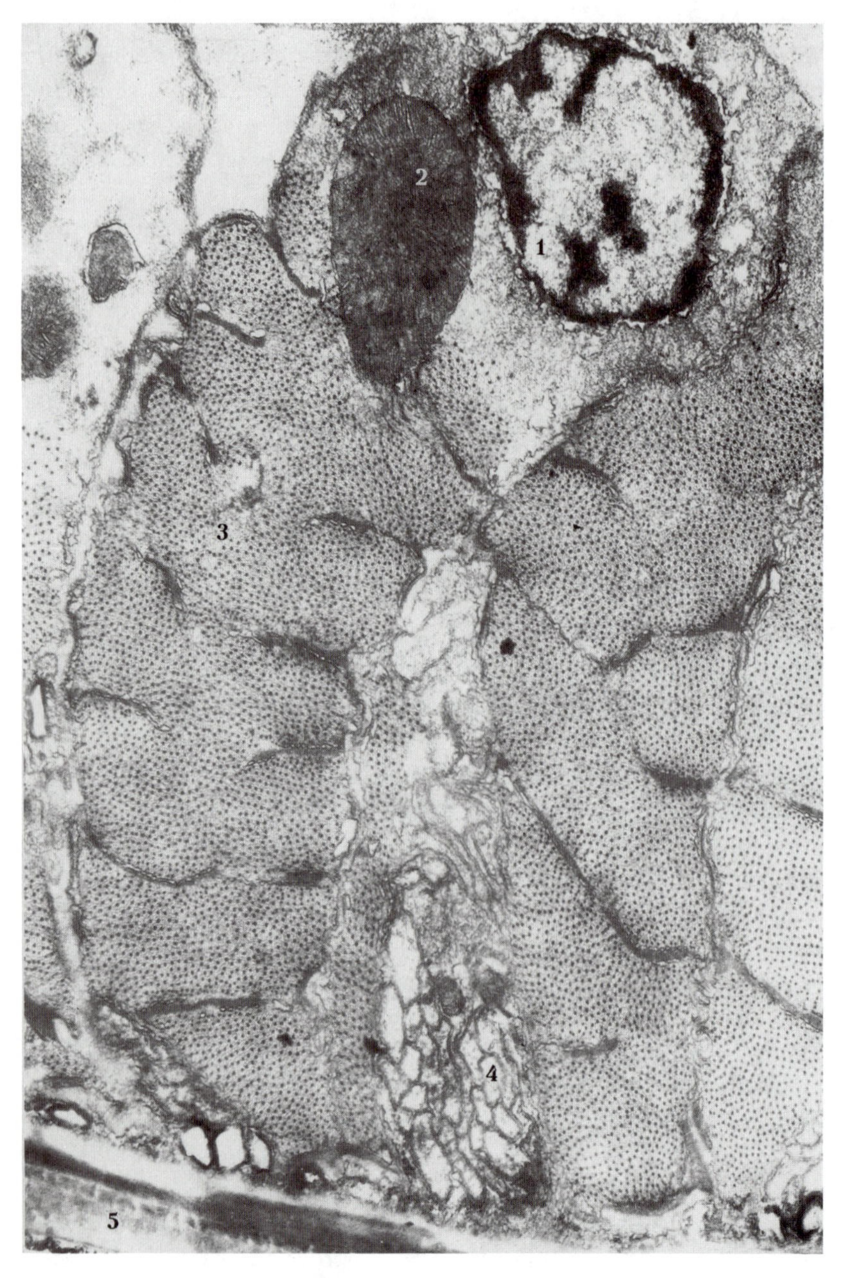

Figure 22. Muscle cells of *Anoplostoma rectospiculum.* Magnification = 24,300. 1, nucleus; 2, mitochondria; 3, contractile zone; 4, dorsal nerve; 5, cuticle.

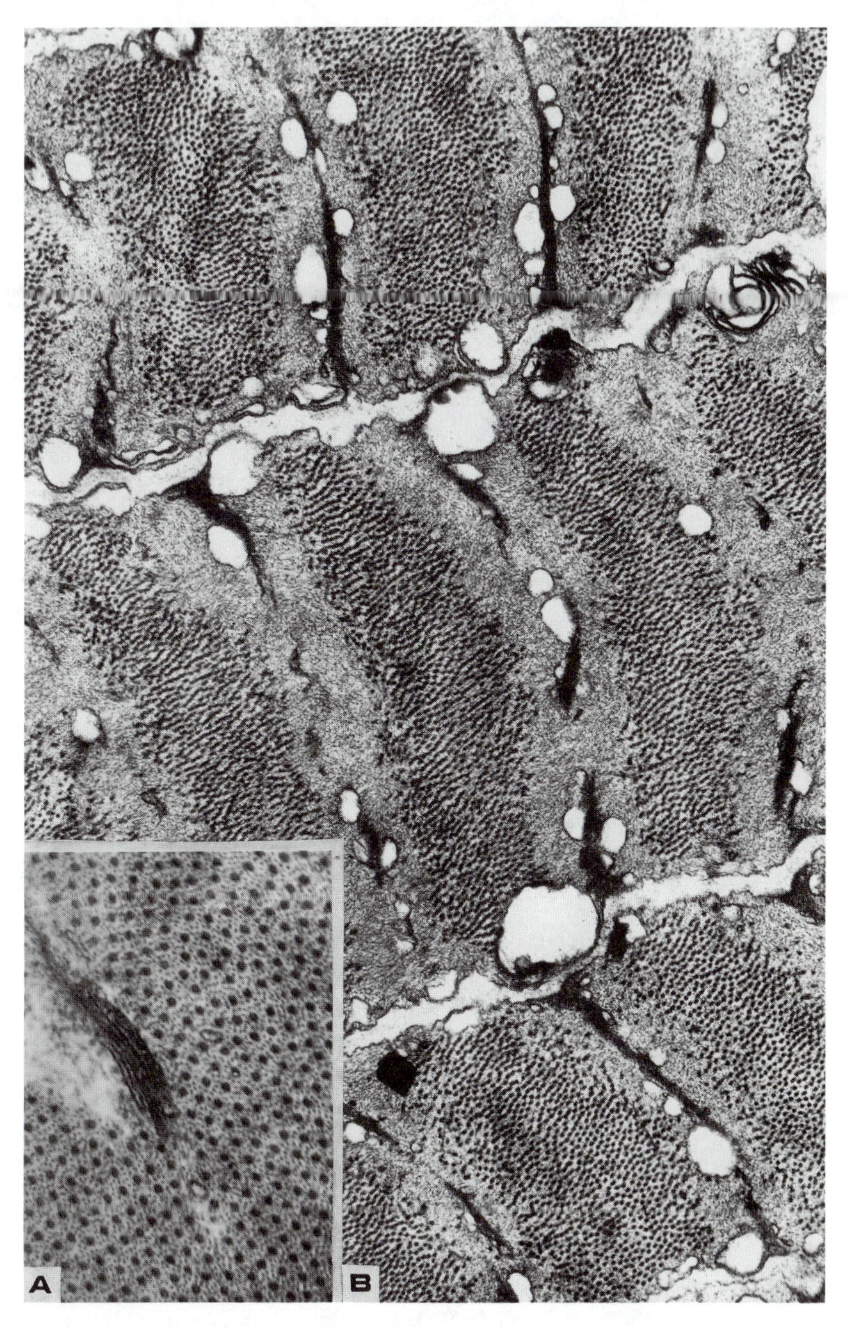

Figure 23. Actin and myosin filaments of *Enoplus* (A) and various zones in the sarcomeres (B). Magnifications: A = 42,200; B = 23,100.

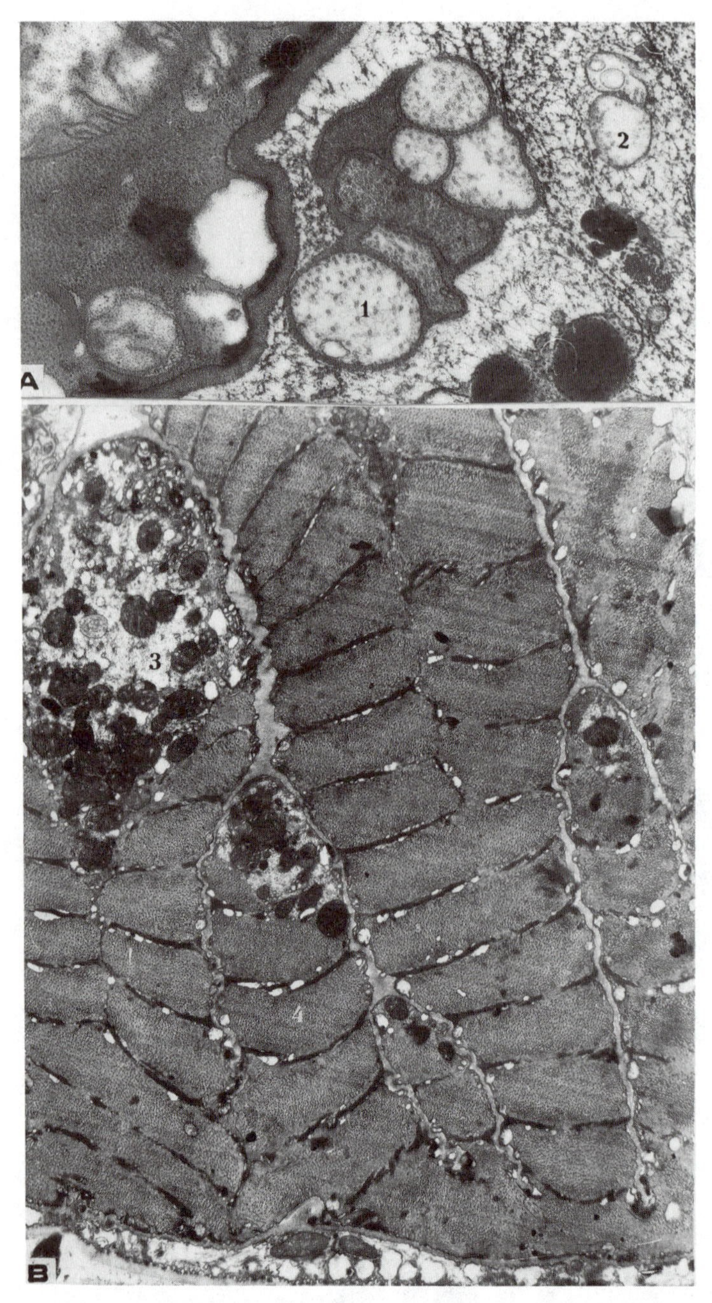

Figure 24. Transverse sections of a group of innervation processes of muscle cells in the pseudocoelom of *Pontonema* (A), and muscle cells of *Enoplus* (B). Magnifications: A = 13,000; B = 92,000. 1, innervation processes of muscle cells; 2, basic substance of pseudocoelom; 3, sarcoplasmic part of muscle cell; 4, contractile part of muscle cell.

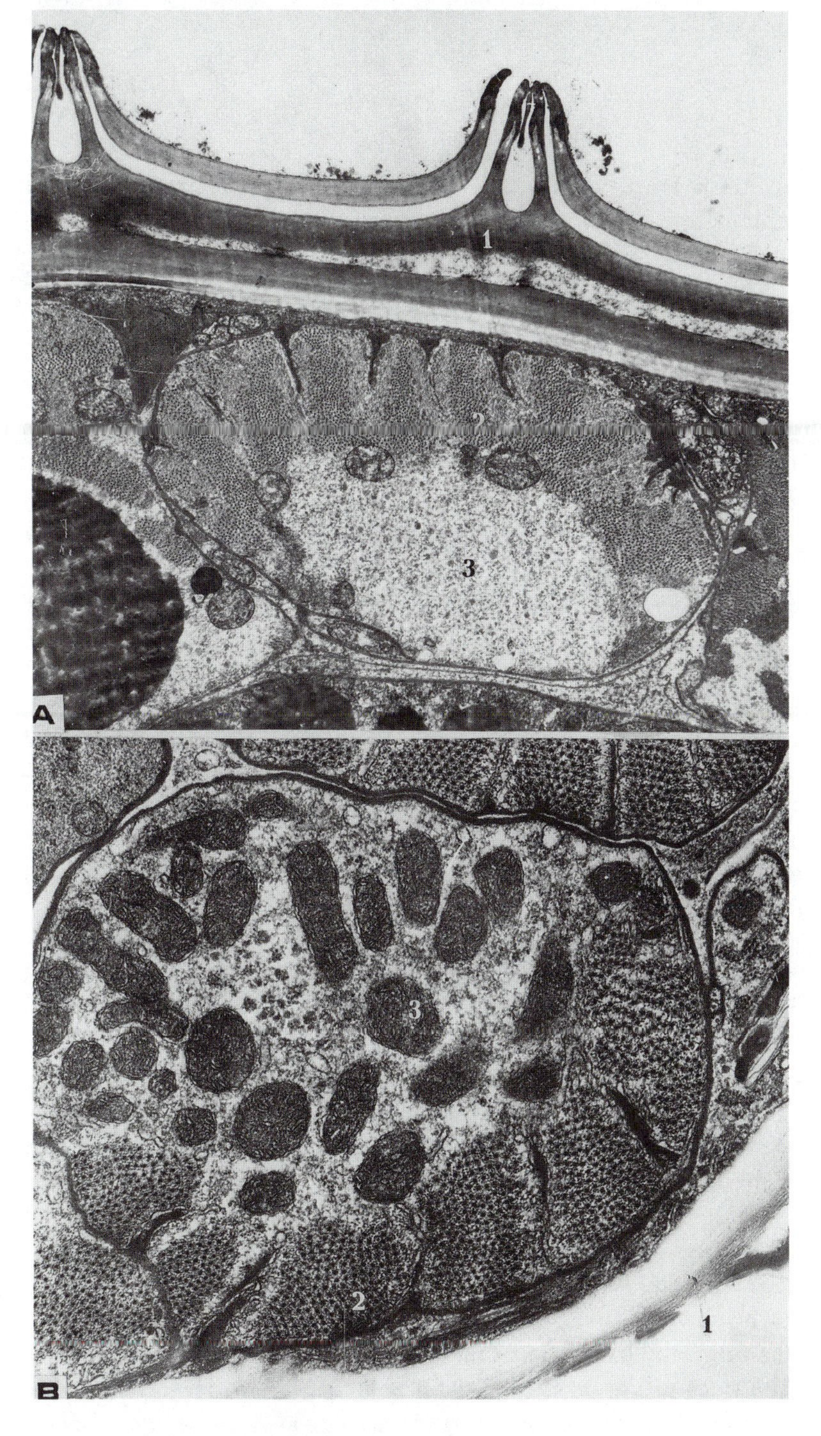

Figure 25. Platymyarian muscle cells of *Monoposthia costata* (A) and *Paracanthonchus macrodon* (B). Magnifications: A = 10,000; B = 16,800. 1, cuticle; 2, contractile zone; 3, plasmic zone.

Rosenbluth 1963, 1965a, b, 1967, 1968; Auber-Thomay 1964). The contractile zone in these cells, when viewed in transverse section, is U-shaped and the z-planes look like portions of a spiral (Figure 20).

Muscle cells of this type can be distinguished from the platymyarian cells in that these cells in cross section have assumed a U-shaped contractile zone. And, actually, in the ontogeny of nematodes with coelomyarian muscle cells of this type, the latter are formed from platymyarian muscle cells. The juveniles of large parasitic nematodes have typical platymyarian muscle cells that subsequently are transformed into coelomyarian (Martini 1906, 1907, 1908; Thust 1968). Because of this, the muscle cells characteristic of large parasitic Rhabditia can be called secondary coelomyarian.

It is interesting that among some groups of Rhabditia (Filariina, for example), the contractile zone in the muscle cells comes together in a ring along the whole length of the cell. Thus, among these nematodes, there are real circomyarian cells. Some authors (Bogoyavlenskii 1973, for example) were inclined to see in this the final stages of the process of the transformation of epithelial-muscle cells into actual muscle cells. Actually, nematode musculature is constructed from real muscle cells with complete organization of the contractile elements and does not have anything in common with primitive epithelial-muscle cells of Cnidaria and certain lower Turbellaria. But in nematode evolution, the process of transformation of some types of muscle cells into others is actually observed. Thus, from platymyarian muscle cells, in evolution (and this is affirmed by data on the ontogenetic development of the muscle cells) there is a formation of secondary coelomyarian cells, and from the latter are formed circomyarian cells.

The reason for the emergence of secondary coelomyarian musculature can be seen in the following. The free-living precursors of modern large parasitic nematodes were small forms like a majority of the presently living Chromadoria and certain free-living Rhabditia, and they had platymyarian musculature. The increase in body size of large animal parasitic nematodes also required an increase in the contractile strength of each individual muscle cell. This was accomplished by the spreading of the originally basal contractile zone and the basal zone then curving up the sides of the muscle cell into what appears to be the form of a horseshoe when the muscle cell is viewed in transverse section; this led to the formation of secondary coelomyarian cells. The secondary coelomyarian cells of parasitic forms possess large sarcoplasmic sacs pierced by a system of supporting fibrils, and the borders of the cells are strengthened by a system of border membranes (Bogoyavlenskii 1973).

The musculature of Enoplia cannot be derived from platymyarian muscles, because in this case the z-zones go in a completely different direction and the plasmic part seems to cut the single contractile zone in two halves (Figure 20). Thus, the muscle cells of Enoplia represent an independent type of musculature and can be called primary coelomyarian. The muscle cells of Rhabditia, as already indicated, should be termed secondary coelomyarian.

Analogous considerations can also explain the evolution in the number of muscle cells. Schneider (1866, 1869) divided all nematodes according to the number of muscle cells into polymyarian (tens and hundreds of muscle cells) and meromyarian (usually with a total of eight cells in transverse section). In addition, meromyarian forms possess constancy of cell composition (Martini 1906, 1907, 1908, 1909, 1912; Pai 1928; Schönberg 1944; Chitwood and Chitwood 1950; Wessing 1953). The Schneider classification, of course, does not correspond to the classification of the class Nematoda. Platymyarian nematode forms of various groups, as a rule, possess a small number of muscle cells and, consequently, are meromyarian. Meromyarian musculature is characteristic, as a rule, of small forms. A condition close to meromyarian is characteristic of a majority of Chromodoria. Among Rhabditia, real meromyarianism is characteristic of Rhabditida, Tylenchida, and a number of zooparasitic nematodes of various orders (for a detailed review, see Bogoyavlenskii 1973). Large parasitic nematodes possess polymyarian musculature. However, juveniles of these forms are meromyarian (Martini 1906, 1907, 1908, 1909; Thust 1968). These data indicate the secondary polymyarianism of large parasitic nematodes.

All representatives of the order Enoplida are polymyarian. Polymyarian forms are also prevalent among other nematode orders. Evidently, in this case, we have original, primary polymyarianism. In connection with progressive decrease in body size, there is a gradual decrease in the number of muscle cells to six to eight in each quadrant (in Chromadoria) and, further, to two with actual meromyarianism. Large parasitic nematodes descending, evidently, from small forms again went into polymyarianism. This was related to the necessity for increasing muscular strength. Thus, the polymyarianism of large parasitic nematodes, as in the case of the coelomyarianism of muscle cells, is a secondary phenomenon. Let us note that with secondary polymyarianism the number of muscle cells significantly exceeds that in primary polymyarian nematodes. The number of muscle cells in transverse section of marine members of Enoplida do not exceed several tens, while in *Ascaris,* the number of cells reaches 600 (Bogoyavlenskii 1973).

The distinctiveness of nematode musculature lies in the peculiarities of its innervation. The question of the innervation of nematode muscle cells has a long history. Schneider (1866) early established that in nematodes the processes of muscle cells go up to the nerves, and his observation was supported by later researchers (Bütschli 1874; Rohde 1892; Goldschmidt 1908, 1909, 1910). Other authors believed that the innervation of nematode musculature did not differ from that of vertebrates (Deyneka 1912). At the present time, the questions of the innervation of nematode muscle cells can be considered settled. The processes of nematode muscle cells penetrate the nerve trunk and only here do nerve-muscle synapses take place (Hinz 1963; Auber-Thomay 1964; Rosenbluth 1965a, b). The further propagation of a pulse from the nerve trunks to the body of muscle cells proceeds along the sarcoplasmic processes of the muscle cells themselves. The sarcoplasmic processes of the muscle cells form a complex network where stimulation is transferred from one muscle cell to another through a system of "electric synapses" that are present in the sarcoplasmic processes (de Bell et al. 1963). Owing to the conduction of the stimulus through the system of sarcoplasmic processes, the nematode musculature acquires the properties of a "physiological syncytium," all the cells lying to one side of the body (dorsal or ventral) acting as one muscle (Jarman 1959; de Bell et al. 1963).

Thus, in nematodes we have a situation where the nerves do not go to the muscles but, in contrast, the muscle cells send processes to the nerves. At the present time, it has been shown that similar relationships are characteristic of Coelenterata (Gulyayev 1976), Annelida (Rosenbluth 1968), Echinodermata (Cobb 1967), and even lancelets (Flood 1966). But nowhere in the animal kingdom will we encounter such a complicated system of sarcoplasmic processes as those characteristic of nematodes. Essentially, in nematodes we have a partial substitution of motor sections of the nervous system with sarcoplasmic processes of muscle cells. Thus, the nematode musculature, besides the contractile function, partially takes on the function of conducting stimulation and possesses special morphological structures (the system of sarcoplasmic processes) that facilitate this function.

With respect to comparative anatomy, as already indicated, the nematode musculature seems partially reduced and simplified. In Turbellaria, for example, the cutaneous-muscular envelope contains a complicated system of muscles represented by circular and longitudinal muscles, and oblique and dorsoventral muscles. With respect to histology, muscle cells of platyhelminths remain basically at the level of smooth muscle. The musculature of small Turbellaria still has no locomotor value and movement is accomplished by cilia of the covering epithelium. The musculature plays an auxiliary role, serving in the seizure of prey, in copulation, and other such functions.

Nematode locomotion is based exclusively on muscular contraction. Therefore, nematode muscle cells have achieved a high level of perfection in the contractile function; with respect to morphology they are fully comparable to muscle elements of coelomate worms. On the other hand, the peculiarities of nematode locomotion with resilient cuticle and the uniformity of movements limited to body flexing have made having a complicated system of muscles unnecessary. The evolution of nematode musculature has proceeded, therefore, not toward anatomical complication but toward specialization in the structure of the muscle cells themselves.

The originally small sizes of the primitive representatives of this group left a certain imprint on the structure of the nematode muscular system. The number of muscular cells arranged over the perimeter of a body section in free-living forms never exceeds a few tens. Muscle cells always lie in a single row. These are all consequences of the small body diameter of free-living nematodes.

The overall architectonics of the nematode body wall has been examined. The undulating type of nematode movement requires strong contractions in one plane. In other words, it is necessary to have two antagonistic muscles contracting in counterphase. In nematodes these muscle systems are the muscles of the dorsal and ventral sides of the body. Even in cases where the ventral hypodermal chord is stronger than the dorsal, the total number of muscle cells on the dorsal and ventral sides is even. In other words, even in the case of bilateral symmetry in the body wall as a whole, the nematode musculature is subject to biradial symmetry.

Nervous System

The presentation of the nematode nervous system is based upon studies of extremely specialized parasitic species. The technically unsurpassed work of Goldschmidt (1903, 1905, 1906, 1908, 1909, 1910), Deyneka (1912), and other authors up to the present provides the basis for sections on the nematode nervous system in all contemporary textbooks and references. Having paid due respect to the precision and detail of these works, let us note that they were done on such highly specialized parasites as human or horse ascarids. It is known that the nervous system undergoes extensive changes with transition to the parasitic form of life. Therefore, from the point of view of comparative anatomy, it is necessary to know the structure of the nervous system of free-living primitive forms. Only these kinds of data will enable us to construct comparative anatomical series and to judge the basic design of the nematode nervous system as a whole.

Data on the structure of the nervous system of free-living nematodes, particularly the marine representatives, are rather scarce. Some information can be drawn from works on the microscopic anatomy of these animals (Türk 1903; zur Strassen 1904; Stewart 1906; Kreis 1934; Timm 1953). Much more complete data are provided by works accomplished by applying special neurological methods, for example, using methylene blue dye (Filip'jev 1912; Malakhov 1978a, b, 1982).

With respect to morphology, the most complex nervous system is characteristic of free-living marine nematodes of the order Enoplida. In the middle part of the body there are ten longitudinal nerves: ventral, dorsal, paired subdorsal, subventral, ventrolateral, and dorsolateral. The most highly developed of these is the ventral nerve trunk, with a large portion of the fibers passing through the ventral hypodermal chord; other fibers are subcuticular. The ventral nerve trunk contains numerous bi- and unipolar neurons. Other longitudinal nerves occupy a subcuticular position, and their fibers are in the surface layer of the hypodermis. All other longitudinal nerves (besides the ventral nerve trunk) are thin and consist of fibers alone. The longitudinal nerves are linked by semicircular commissures and individual nerve fibers. The arrangement of the longitudinal nerves of marine Enoplida is tied to the epidermal mosaic and the arrangement of hypodermal chords (Figure 26).

A characteristic feature of the marine enoplid nervous system is the strong development of the lateral sensory plexus (Figure 27), composed of neurons of several types. The sensory neurons of the lateral plexus innervate the numerous somatic sensory setae. In the female vulvar region, the lateral plexus becomes more complicated, and its neurons probably coordinate the activity of this organ. In the posterior part of the male body, the lateral plexus also becomes more complicated, and this evidently can be explained by the richness of sensory structures and the complex movements of the tail section during copulation.

The ventral nerve trunk is infused by paired components in the circumpharyngeal nerve ring. Other longitudinal nerves may or may not be connected to the nerve ring. Around the nerve ring are arranged neurons of several types. Neurons extending anteriorly from the nerve ring go to the head as basic sensory nerves. These nerves lie deep in the body of the nematode, adjacent to the pharyngeal wall. The subdorsal and subventral nerves each innervate one labial papilla and a pair of cephalic setae. Lateral nerves each innervate a labial papilla, a lateral cephalic seta, and the amphid. Processes of numerous sensory neurons situated

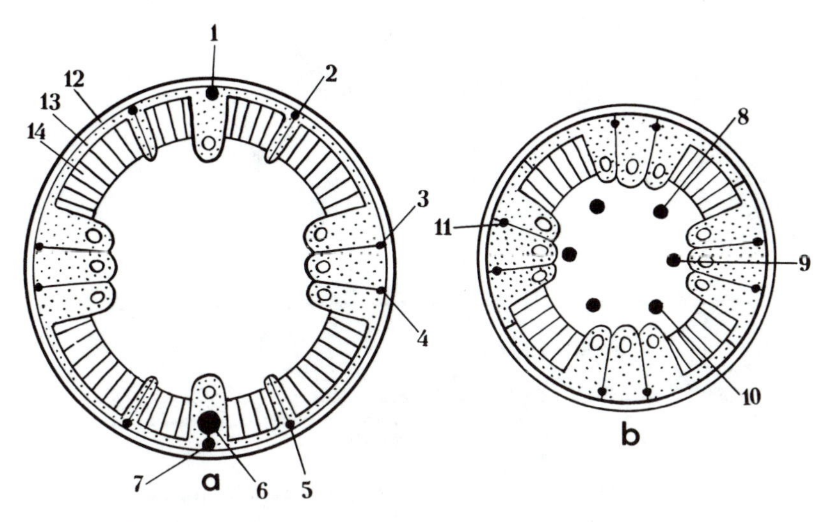

Figure 26. Diagram of the arrangement of longitudinal nerves of *Pontonema vulgare* in the midregion of the body (a) and anterior to the nerve ring (b). 1, dorsal nerve; 2, subdorsal nerves; 3, dorsolateral nerves; 4, ventrolateral nerves; 5, subventral nerves; 6, ventral nerve trunk; 7, surface branch of the ventral nerve trunk; 8, subdorsal cephalic nerves; 9, lateral cephalic nerves; 10, subventral cephalic nerves; 11, surface cephalic nerves; 12, cuticle; 13, hypodermis; 14, musculature.

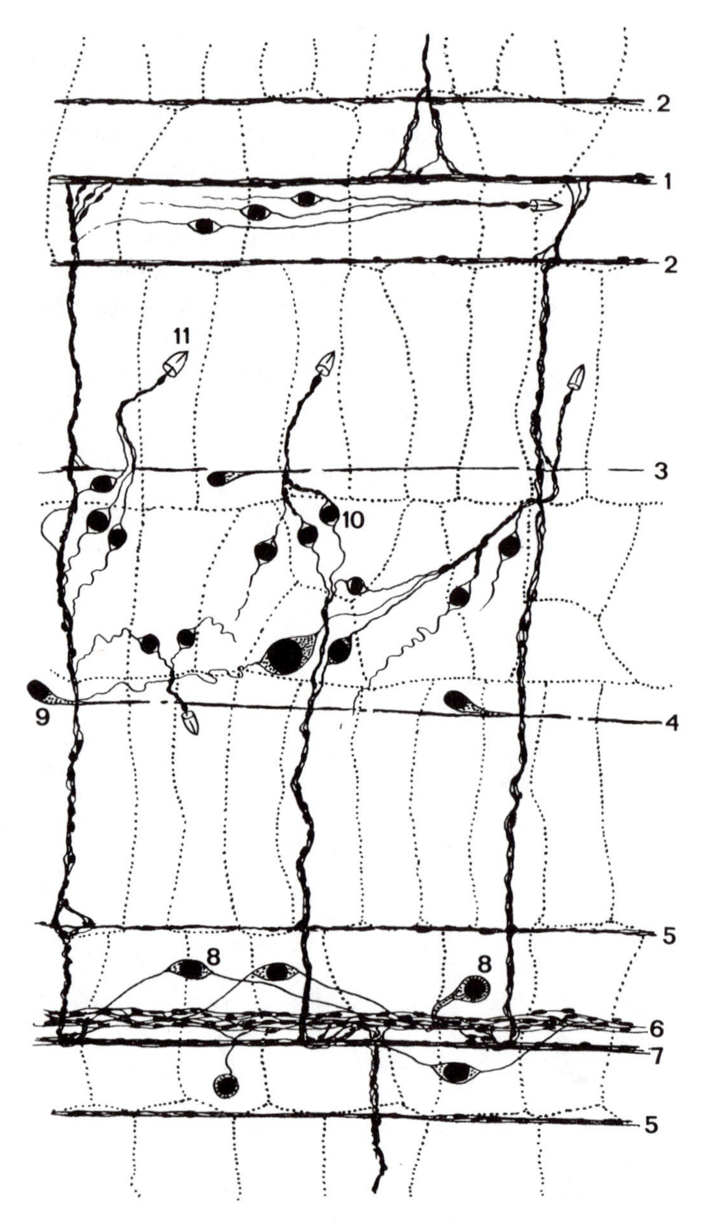

Figure 27. Part of the nervous system in the midbody region of *Pontonema vulgare*. Dotted lines represent borders of hypodermal cells. 1–7, as in Figure 26; 8, neurons of the ventral trunk; 9, unipolar neurons of lateral nerves; 10, sensory bipolar neurons of lateral plexus; 11, somatic setae.

both anterior and posterior to the nerve ring proceed within eight thin peripheral nerves to the cervical setae. The arrangement of the peripheral cephalic nerves is determined by the hypodermal mosaic of the anterior end of the body. The morphological continuity of the longitudinal nerves of the trunk and the surface sensory nerves of the head end are related to the continuity in the structure of the hypodermis (Figure 28).

The neurocytes that innervate the amphids lie laterally posterior to the nerve ring (Figure 29). Their dendrites are part of the lateral, deep sensory nerves, and the axons pass into the nerve ring through the subventral rootlets. The processes of numerous interneurons, situated pos-

Figure 28. The innervation of the sensory organs of the head end of *Pontonema vulgare*. 1, lateral cephalic nerves; 2, dorsolateral cephalic nerves; 3, subdorsal cephalic nerves; 4, dorsal surface nerves; 5, labial papillae; 6, cephalic setae; 7, amphids; 8, cervical setae.

terior to the nerve ring (Figure 29), also pass into the ring through the subventral rootlets.

In marine enoplids there are no real nerve ganglia. Neurons associated with the nerve ring are collected into four weakly formed groups, the arrangement of which coincides with the position of the basic chords of the hypodermis. The largest number of cells is collected in the lateral groups.

At the posterior end of the body, the ventral nerve trunk divides into several symmetrical branches (Figure 30). Part of the fibers of the ventral trunk form a thin circumanal commissure, and others go to the area of the lateral chords of the hypodermis. Other longitudinal nerves thin out and disappear. Special sensory neurons innervate the genital papillae, the spicules, and the gubernaculum of the male (Figure 31). The dorsal submedian nerves and part of the fibers of the ventral nerve trunk are motor nerve processes; to them go the innervation processes of the muscle cells. The lateral plexus and, evidently, the lateral nerves

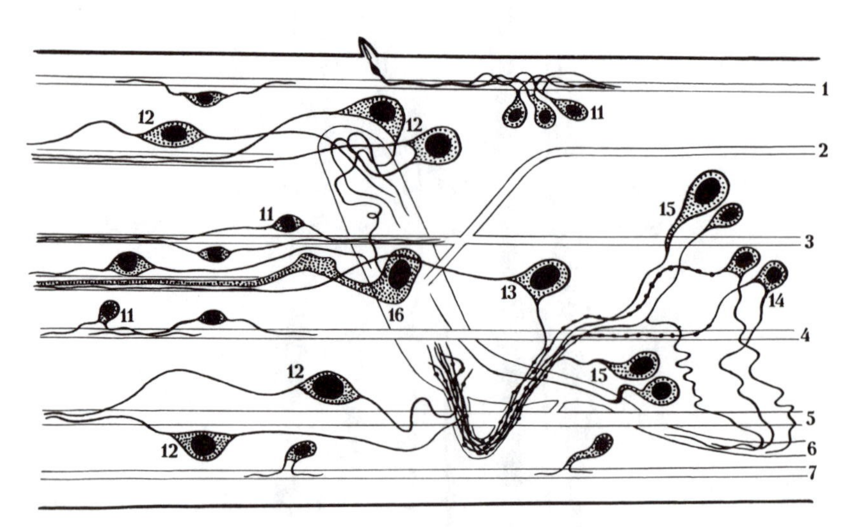

Figure 29. Arrangement of basic types of nerve cells in the region of the nerve ring of *Pontonema vulgare*. 1–7, as in Figures 26 and 27; 11, small sensory neurons of cervical setae; 12, large sensory neurons of cephalic sensory organs; 13, amphidial neurons; 14, bipolar neurons, the processes of which enter the nerve ring; 15, unipolar neurons connected to the nerve ring; 16, amphidial glands.

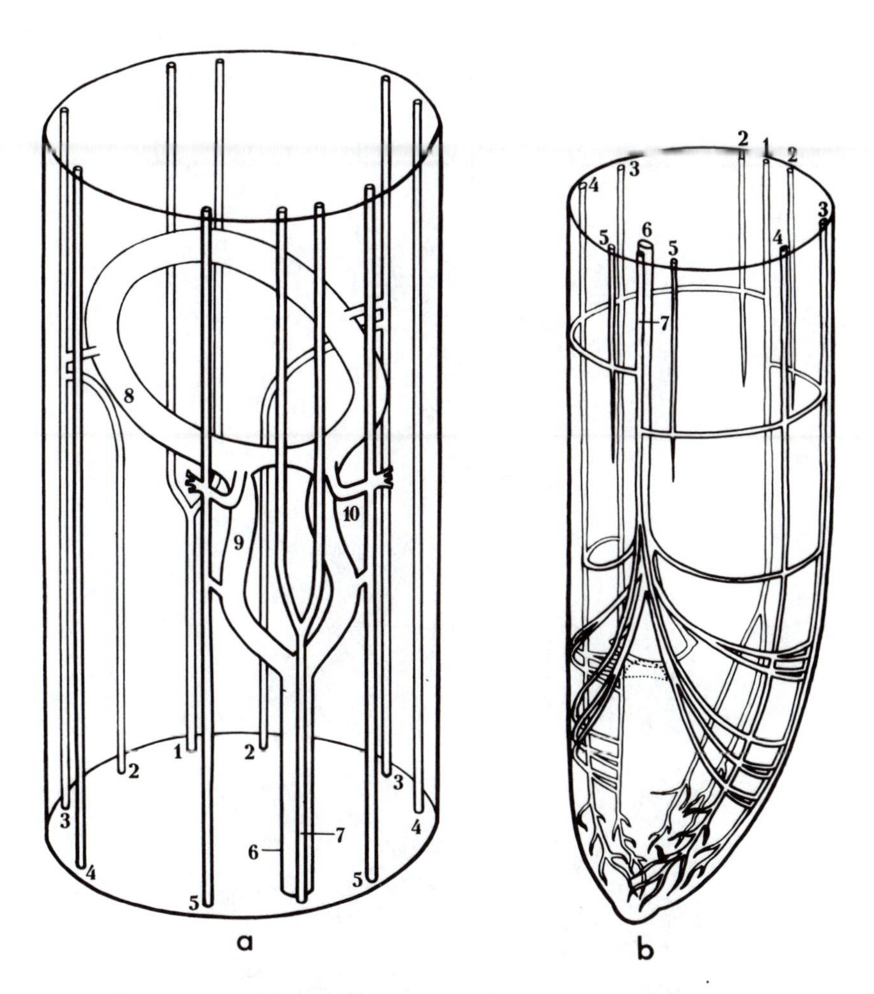

Figure 30. Diagram of longitudinal nerves of *Pontonema vulgare* in the region of the nerve ring (a) and in the region of the posterior end of the body (b). 1–7, as in Figure 26; 8, nerve ring; 9, rootlets of the ventral trunk; 10, subdorsal rootlets.

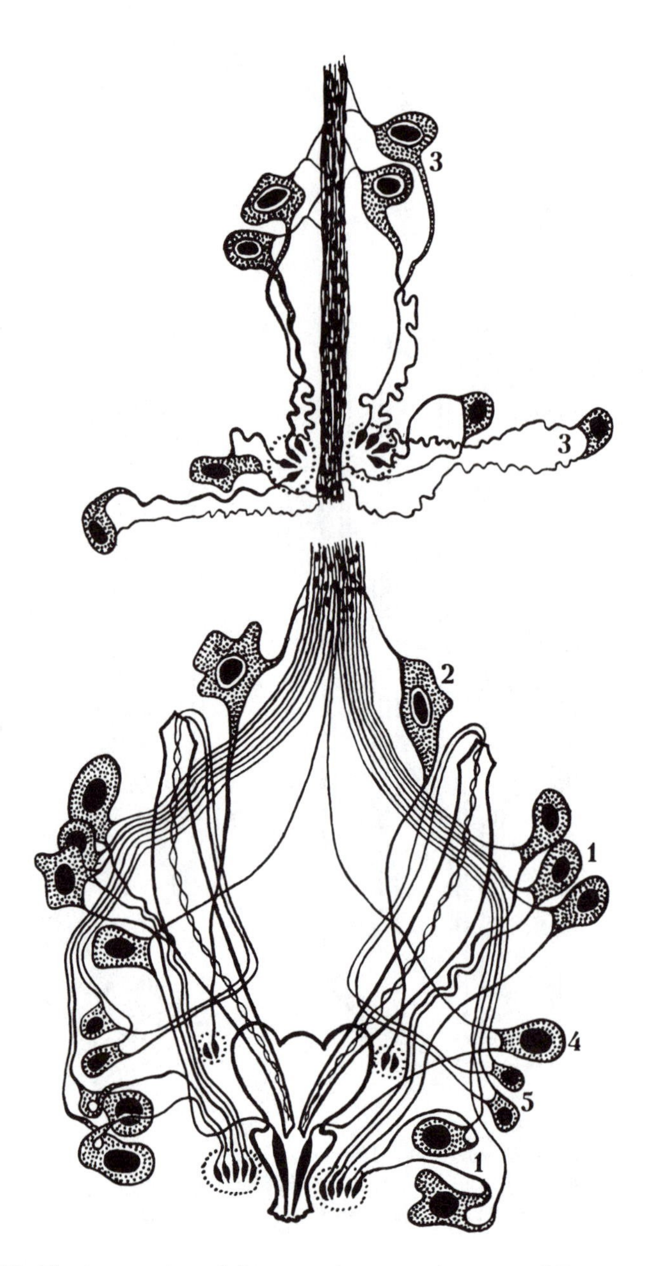

Figure 31. The innervation of the secondary sexual organs of *Pontonema vulgare.* 1, neurons innervating first pair of papillae; 2, neurons innervating second pair of papillae; 3, neurons innervating third pair of papillae; 4, neurons innervating gubernaculum; 5, neurons innervating spicula.

belong to the sensory section of the nervous system. The head nerves undoubtedly also belong to the sensory part of the nervous system.

The data presented above undoubtedly reflect one of the most primitive of nervous systems within the class Nematoda. In the evolutionarily more advanced nematodes of the subclass Rhabditia, the nervous system has undergone definite changes. First of all, the number of neurons of the nervous system decreases and becomes constant. In marine nematodes of the order Enoplida, the total number of nerve cells is several thousand, and it is not constant (because of the variability in the number and position of the somatic setae, as a consequence of which the lateral plexus is different among various specimens). In Rhabditia the total number of neurons is strictly constant within the boundaries of the species and is about 150–300 cells. It is interesting that among small members of Rhabditia, in connection with the overall small-cell nature of the organism, nerve tissue composes up to 40% of the total number of cells that construct the body (Pai 1928; Schönberg 1944; Wessing 1953). Perhaps in the ratio of the number of nerve cells to the total number of cells in the organism nematodes hold first place in the animal kingdom. Large parasitic nematodes evidently have evolved from small saprobiotic forms and preserve the very small and constant number of nerve cells (Goldschmidt 1908, 1909, 1910). The decreased total number of neurons among members of Rhabditia is the consequence of the simplification of the environment it inhabits, a reduction of sensory structures related to this, and the loss of complex forms of behavior that are especially obvious in the parasitic form of life.

The nervous system of all nematodes is closely related to the hypodermis, and this in itself is a very primitive trait. However, in marine nematodes, the longitudinal nerves lie almost directly adjoining the cuticle. A more superficial arrangement of the nervous system has never been encountered in the animal kingdom. Among parasitic Rhabditia the hypodermal chords run deeply into the body and with them also go the nerve trunks.

The close relationship of the hypodermis and nervous system is stressed not only by the coincidence of architectonics, but also, in the literal sense, by the "close" relationship on a cytological level. The longitudinal nerves proceed into the hypodermal chords, and the nerve tissue and hypodermal tissue are separated only by cytoplasmic membranes. The morphological unity of the hypodermis and nerve tissue is underscored by the fact that both of them are surrounded by an overall sheath—a layer of amorphous substance separating them from adjoining tissues and enclosing the body cavity. Within the interchordal hypodermis proceed nerve fibers of semicircular commissures uniting the longitudinal nerves. The close relationship of the hypodermal and nerve tissues suggests that the hypodermis has the function of a kind of glial tissue supporting the nerve elements of the body.

The general small-cell nature of the nervous system among members of Rhabditia is reflected in the structure of the nerve trunks. Thus, in *Ascaris,* close to the place of departure of the longitudinal nerves from the nerve ring, the ventral nerve trunk contains a total of fifty-five fibers; the dorsal trunk, three; and the submedian and lateral trunks, four each (Goldschmidt 1908, 1909, 1910). To compensate, the thickness of individual fibers reaches up to 40 μm (Deyneka 1912; Hinz 1963).

The total number of longitudinal nerves in nematodes is eight to twelve. There is always a ventral nerve trunk and a dorsal nerve. Several forms have one or two pairs of lateral nerves. The number of submedian nerves varies from two to four pairs. The longitudinal nerves of nematodes are far from uniform. The ventral nerve trunk occupies a special place in the nematode nervous system and is incomparably better developed than all other longitudinal nerves. In it are many nerve cells (bi- and unipolar). In function, the neurons of the ventral trunk are motor neurons and interneurons (White et al. 1976). To the ventral trunk, as we have seen, also come the processes of the neurons of the lateral sensory plexus. Thus, the ventral nerve trunk is a multifunctional formation containing nerve fibers of various types. The dominance of the ventral trunk is the general rule for all nematodes (Bütschli 1874; Brandes 1892, 1899; Türk 1903; zur Strassen 1904; Looss 1905; Rauther 1906; Stewart 1906; Goldschmidt 1908, 1909, 1910; Deyneka 1912; Filip'jev 1912, 1921; Martini 1916; Pai 1928; Chitwood and Chitwood 1933, 1950; Chitwood and Wehr 1934; Paramonov 1962; Bogoyavlenskii et al. 1974; White et al. 1976; and others).

The ventral nerve trunk in section looks like a single formation, but there is indication that in its origin it is a paired structure. In the anterior part of the body, the ventral trunk is formed by the merging of well-developed rootlets or, as they are still called, "subventral nerves." At the posterior end, the ventral trunk again bifurcates before the anus, forming a circumanal ring and several branches that go into the lateral chords of the hypodermis.

The neuropil of the ventral trunk goes directly across to the neuropil of the circumpharyngeal ring through strong rootlets. The way in which the ventral trunk is connected to the nerve ring is the same for all nematodes: at some distance from the ring, the ventral trunk splits into two branches ("subventral nerves"), which pass to the nerve ring and merge into its lower half. In Rhabditia, the place of bifurcation of the ventral trunk corresponds to the position of the excretory pore. In many Enoplida, the excretory pore is found far forward of the nerve ring, and the bifurcation of the ventral trunk is not related to it.

The circumpharyngeal nerve ring in Enoplida occupies a slanting position: its dorsal edge is moved anteriorly and the ventral edge posteriorly. In many Rhabditia the situation is reversed. The circumpharyngeal nerve ring is represented by a bundle of nerve fibers embedded in the common stroma formed by four to eight glial cells. In addition, a body of four to seven commissural neurons lies in the ring. The basic mass of the ring is composed of a neuropil distinguished by the great complexity of its external structure. Neuron processes of various types enter the nerve ring: sensory, motor, internuncial, and also innervation processes of the muscle cells of the anterior part of the body. Numerous synapses between neuron fibers of various types are noted in the nerve ring. Ultrastructural investigations reveal several sorts of chemical and electric synapses (Ware et al. 1975). Even direct sensory-motor synapses are found between processes of sensory neurons and processes of muscle cells (Ware et al. 1975). In *Parascaris,* several zones of nerve plexuses differing in types of contacting fibers were distinguished (Deyneka 1912). All this shows that the nerve ring cannot be regarded as a simple commissure. The nerve ring is a central organ in which various types of information processing take place.

The complex neuropil of the nerve ring is formed from fibers coming from the ventral trunk and from processes of cells situated around the ring. In the free-living marine Enoplida, the body of neurons is arranged as a simple layer around the pharynx, as noted by authors who have investigated the microscopic anatomy of these forms (de Man 1886; Jägerskiöld 1901; Türk 1903; Stewart 1906; Timm 1953). One can speak only conditionally of four groups of cells. In large parasitic members of Rhabditia, it is customary to speak of nerve ganglia around the ring, but, in reality, there are no real ganglia, even here. In this case, we have simply more or less formed groups in which are gathered bodies of neurons, the processes of which extend to the nerve ring. The division into cell layer and neuropil, typical for ganglia, is not present here naturally. These "ganglia" do not have membranes isolating them from surrounding tissue. There is no glial tissue forming a stroma of ganglia, and there are only individual supporting cells. An exception is the predatory marine nematode of the order Monhysterida, *Siphonolaimus,* in which the "ganglia" are clothed in a special membrane (zur Strassen 1904).

The difference between the so-called ganglia of nematodes and the real ganglia of other animals has long been stressed by a number of authors (Filip'jev 1921; Chitwood and Chitwood 1950; Schuurmans-Stekhoven 1959). It is not just a matter of a formal, terminological disparity. The usual ganglion represents a morphologically unified formation in which bodies of cells lie closer to the periphery, and the central zone is occupied by neuropil. In nematodes, the structures corresponding to the neuropil and cell layer are spatially separated and do not form a morphological unit. The part corresponding to a neuropil forms the nerve ring in which the processing of information also is accomplished. The part corresponding to the cell layer is carried beyond the boundaries of the nerve ring and provides several more or less separated groups—"ganglia." From the functional point of view, in nematodes, the whole system, the nerve ring plus the so-called ganglia, corresponds to the single ganglion in other animals.

All of the above shows that in nematodes we encounter a rather low level of morphological integration of the nervous system, even in its central sections. The evolution of the central nervous system in nematodes has proceeded distinctively, unlike those in other taxa of the animal kingdom. In nematodes, also, no real ganglia developed. The central sections of the nervous system apparently represent systems that are sufficiently complete functionally but poorly integrated morphologically.

The total number of nerve cells related to the nerve ring in Rhabditia is small and strictly constant. The better developed are the lateral "ganglia," each containing from thirty to fifty-four cells. The ventral "ganglion" unites the bodies of sixteen to thirty-three neurons. The dorsal "ganglion" has one to seven, or is entirely absent, as in *Oxyuris* (Martini 1916). Neuron processes of the ventral "ganglion" enter the ring through rootlets of the ventral trunk. Neurons of the lateral "ganglia" send processes through the lateroventral commissures into the rootlets of the ventral trunk, from which they extend to the nerve ring. Thus, in the arrangement of the nerve ring, the cells associated with it, and their connections, there is bilateral symmetry.

The significant development of the ventral nerve trunk and the circumpharyngeal nerve ring and their associated nerve cells permits them to be considered the central structures of the nematode nervous system.

Some authors consider the ventral nerve trunk and nerve ring as the nematode central nervous system (Chitwood and Wehr 1934; Chitwood and Chitwood 1950; Ansari and Basir 1964), believing that they are analogous in some degree to the central nervous systems of higher animals. Chitwood and Chitwood (1950) even look upon the nematode ventral nerve trunk as a nerve ganglion extended in length. In ontogeny the nematode nervous system develops from two basic groups of rudiments: groups of neuroblasts in the anterior part of the body provide the nerve ring and the cells related to it, but in addition, along the ventral side of the animal is laid a chain of neuroblasts that ultimately provides the ventral nerve trunk (Sulston 1976; Sulston and Horvitz 1977; Ehrenstein and Schierenberg 1980).

Other longitudinal nerves of the nematode body are formed largely from processes that enter them from the ventral trunk and the nerve ring. The most prominent of these, as a rule, is the dorsal nerve. As special ultrastructural investigations have shown (White et al. 1976), there is full symmetry in the quantity and quality of motor innervation between the dorsal and ventral sides. This provides equal strength in contracting the dorsal and ventral muscular bands, which is necessary for accomplishing the undulating mechanism for movement. However, the bodies of motor neurons lie in the ventral trunk and their processes go out from it, reaching the dorsal nerve by way of semicircular commissures. Thus, the dorsal nerve, to a significant degree, is composed of fibers coming from the ventral trunk (White et al. 1976). This can also be seen in the example of *Ascaris*. Immediately upon leaving the nerve ring, the dorsal nerve contains a total of three fibers; after the first semicircular commissure there are six; after the second, there are nine; and, in the middle of the body, the dorsal trunk contains thirteen fibers (Goldschmidt 1908). The dorsal and submedian nerves, as a rule, are devoid of neurons. Only in *Ascaris* and forms close to it have rare individual nerve cells been found in the dorsal nerve (Goldschmidt 1908, 1910; Deyneka 1912; Bogoyavlenskii et al. 1974). The dorsal nerve is considered to be motor in its function. A purely motor function is also characteristic of the submedian nerves.

The sensory sections of the nervous system are situated laterally. In free-living members of Enoplida, we encounter a broad lateral plexus that innervates the sensory setae and the lateral nerves relating to them. For all Rhabditia and for parasitic forms in particular, a significant reduction of the sensory sections is characteristic. The lateral sensory plexus is absent. Only the lateral sensory nerves (one or two pairs) are preserved. It is interesting that in *Oxyuris equi* (Martini 1916) and *Ascaridia galli* (Bogoyavlenskii et al. 1974) bipolar neurons connected to lateral nerves have been found. In *O. equi,* there are three of them and in *A. galli* five to seven. It is clear that in this case we have the last remnants of a lateral plexus. Interestingly, in certain Rhabditia postdeirids have been noted, which are minute papillae innervated from neurons lying along the path of the lateral nerves (Hesse 1892).

Another remnant of a lateral plexus is the lumbar ganglia lying in the posterior part of the body. In some cases (in *Oxyuris*), they have the appearance of a nerve plexus (Martini 1916). Lumbar ganglia innervate phasmids, which are probably analogs of lateral sensory setae. On the lateral sides of the Rhabditia body are situated pre- and postanal sensory ganglia, which innervate the genital papillae of males.

The overall reduction of the sensory sections of the nervous system also affects the anterior end of the body. In Rhabditia, in most cases, we do not find a complex system of surface sensory nerves that innervate cervical setae. The only analogs of cervical setae are the deirids—small papillae in the anterior part of the body. The deirids are innervated from the lateral nerves or from lateral "ganglia" (Bütschli 1874; Goldschmidt 1910; Ansari and Basir 1964). The internal sensory nerves extending to the cephalic sensory organs in Rhabditia are more differentiated: the amphidial nerves are always clearly isolated from the lateral papillary nerves.

The innervation of nematode internal organs has not been studied sufficiently. The pharynx is innervated only by longitudinal nerves (a dorsal and a pair of subventral) connected by circular commissures at the base and in the anterior part of the pharynx, and also by sector commissures (Goldschmidt 1910; Martini 1916; Hsu 1929; Chitwood and Chitwood 1933, 1950; Frenzen 1954; Bogoyavlenskii et al. 1974). In *Caenorhabditis,* the pharynx consists of thirty-four muscle, nine marginal, nine epithelial, five glandular, and twenty nerve cells (Albertson and Thomson 1976). Among these neurons, there are sensory neurons, interneurons, and motor neurons. In the wall of the middle intestine of parasitic nematodes have been found longitudinal nerves and a sensory nerve plexus (Nurseitov 1972). A nerve plexus is described on the wall of the *Parascaris* gonoducts (Zacharias 1913). Along the path of the canals of the excretory system of the parasitic nematodes, bipolar neurons have been described, which, it is assumed, participate in the innervation of the excretory system (Bogoyavlenskii et al. 1974).

For some time, thoughts have been expressed as to the cholinergic nature of nerve impulse transmission in nematodes (Lee 1962, 1965; Castillo et al. 1963; de Bell et al. 1963). It is known that nematode muscles possess a sensitivity to acetylcholine, the most sensitive being the portions of muscle cell membrane in the area of contact between their sarcoplasmic processes and nerve fibers (de Bell et al. 1963). In this sense, it is very demonstrable that the cholinesterase activity in nematodes is found largely in motor nerves: in the ventral, dorsal, and submedian nerves (see, for example, Lee 1965; Shishov 1971; Shishov et al. 1973; Malakhov and Belova 1977; Zhuchkova and Shishov 1979). Thus, it can be assumed that the transmission of stimulation from nerve fibers to muscle cell processes takes place with the aid of a substance close to acetylcholine.

In recent years the method of demonstrating cholinesterase activity often is used to determine the overall topography of an invertebrate nervous system (Kotikova 1967, 1973; Boguta 1976; Ospovat 1978; and others). In nematodes, the distribution of cholinesterase activity does not provide a complete representation of the topography of the nervous system. Cholinesterase activity in nematodes is related primarily to motor nerves and therefore is keyed to the nerves of the ventral and dorsal sides of the body. It is easy to guess that this is related to the concentration of muscle elements on the dorsal and ventral sides of the body. The nerve structures in the middle part of the body, defined by the determination of cholinesterase activity, have uniform or nearly uniform development both on the dorsal and ventral sides of the body (Figure 32). This is related to the fact that ventral and dorsal muscle bands are equally developed and have equal innervation. The nerve structures in the central part of the body, again as revealed by cholinesterase activity, are subject to the same biradial symmetry, as is the whole body wall.

Cholinesterase activity is found also in certain sensory structures of nematodes (amphids, phasmids, and genital and head sensilla). However, acetylcholine is less characteristic than catecholamine for the sensory regions of the nematode nervous system. In free-living nematodes, catecholamines are found in the sensory plexus and also in the circumpharyngeal nerve ring and related nerve cells and cephalic sensory nerves

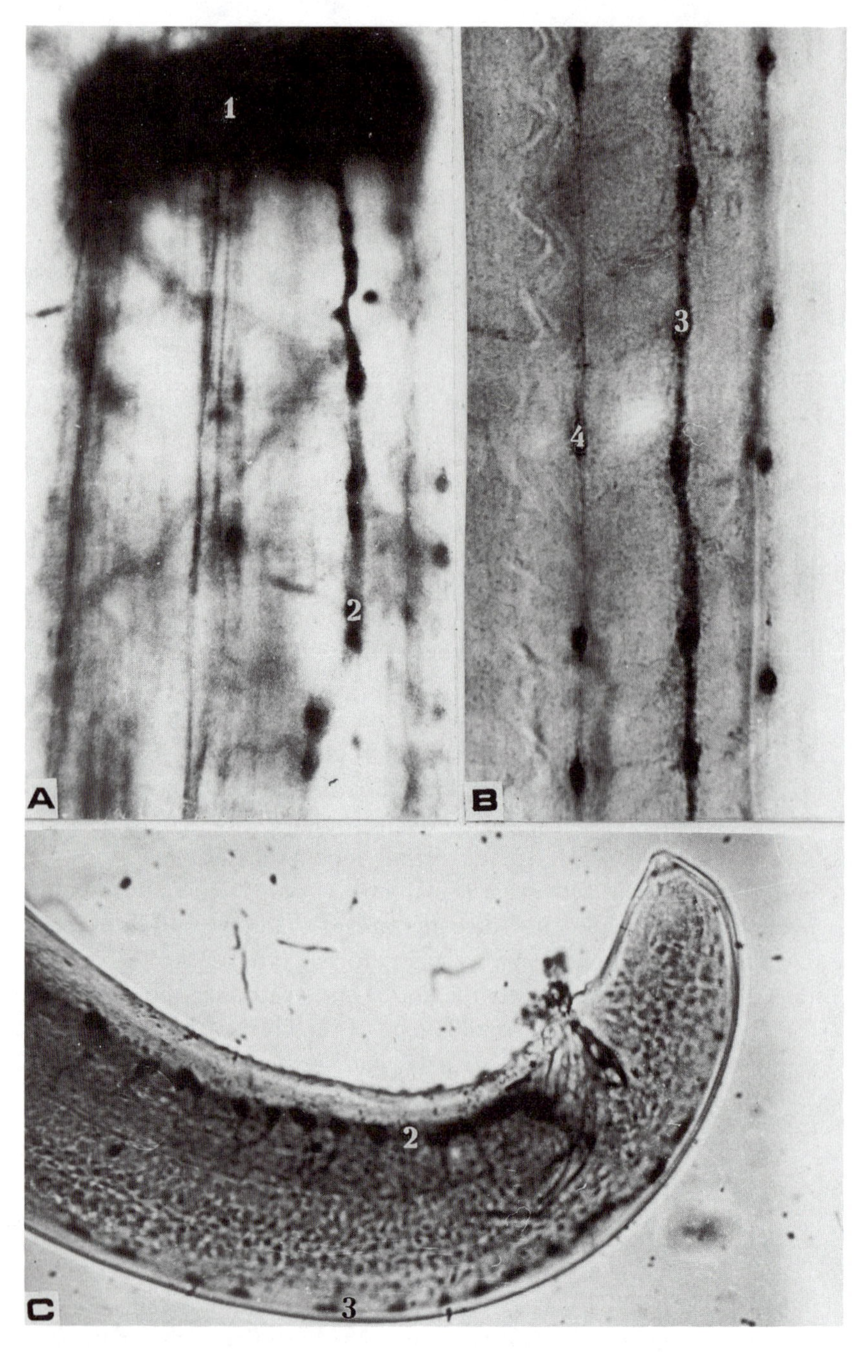

Figure 32. Cholinesterases in the region of the nerve ring (A), in the dorsal and subdorsal nerves (B), and in the posterior part of the male body (C) of *Pontonema vulgare*. Magnifications: A and B = 310; C = 240. 1, nerve ring; 2, ventral nerve trunk; 3, dorsal nerve; 4, subdorsal nerves.

and in the ventral nerve chord (Figures 33 and 34). The distribution of catecholamines is subject to bilateral symmetry.

In saprobiotic and parasitic nematodes, the sensory sections of the nervous system are significantly reduced. In these forms only four catecholamine neurons remain. They innervate the cephalic papillae of the outer circle, the male genital papillae, and one or two pairs of neurons on the lateral sides of the body, and all are remnants of the catecholamine sensory plexus of free-living forms (Sulston et al. 1975; Zhuchkova and Shishov 1979).

In the nematode nervous system an odd combination of various elements of symmetry is observed. The cephalic nerves generally are subject to hexaradiate symmetry, reflecting the general triradiate symmetry of the anterior end of nematodes, which traces back to the triradiate symmetry of the pharynx. The motor cholinergic sections of the nematode nervous system are subject to the biradial symmetry of the body wall. However, bilateral symmetry is undoubtedly dominant. It appears first in the sharp dominance of the ventral trunk. But this dominance is not only quantitative. The ventral trunk is rich in neurons of various types, whereas other longitudinal nerves consist predominantly of single fibers. Fibers coming from the ventral trunk to a significant degree form other longitudinal nerves. The ventral trunk is closely connected to the nerve ring, and together they represent a kind of nematode "central nervous system." The overall bilateral plan of the nematode nervous system clearly appears in the arrangement of the catecholaminergic elements.

The preservation of the overall bilateral symmetry of the nematode nervous system contrasts sharply with the radiality of many organ systems and, above all, of the sensory organs and body wall. The bilateral symmetry of the nematode nervous system, preserved despite the radiality of many other organ systems, cannot be viewed as other than an ancestral trait—a legacy of the bilateral organization of its ancestors.

Sensory Organs and Glands

The dense cuticle that covers the whole nematode body reliably protects the organism from the injurious effects of inhospitable elements in the external environment but also prevents the entry of sensory information. As has been noted, nematodes are practically devoid of free sensory endings in the body wall. All sensory endings in nematodes are connected

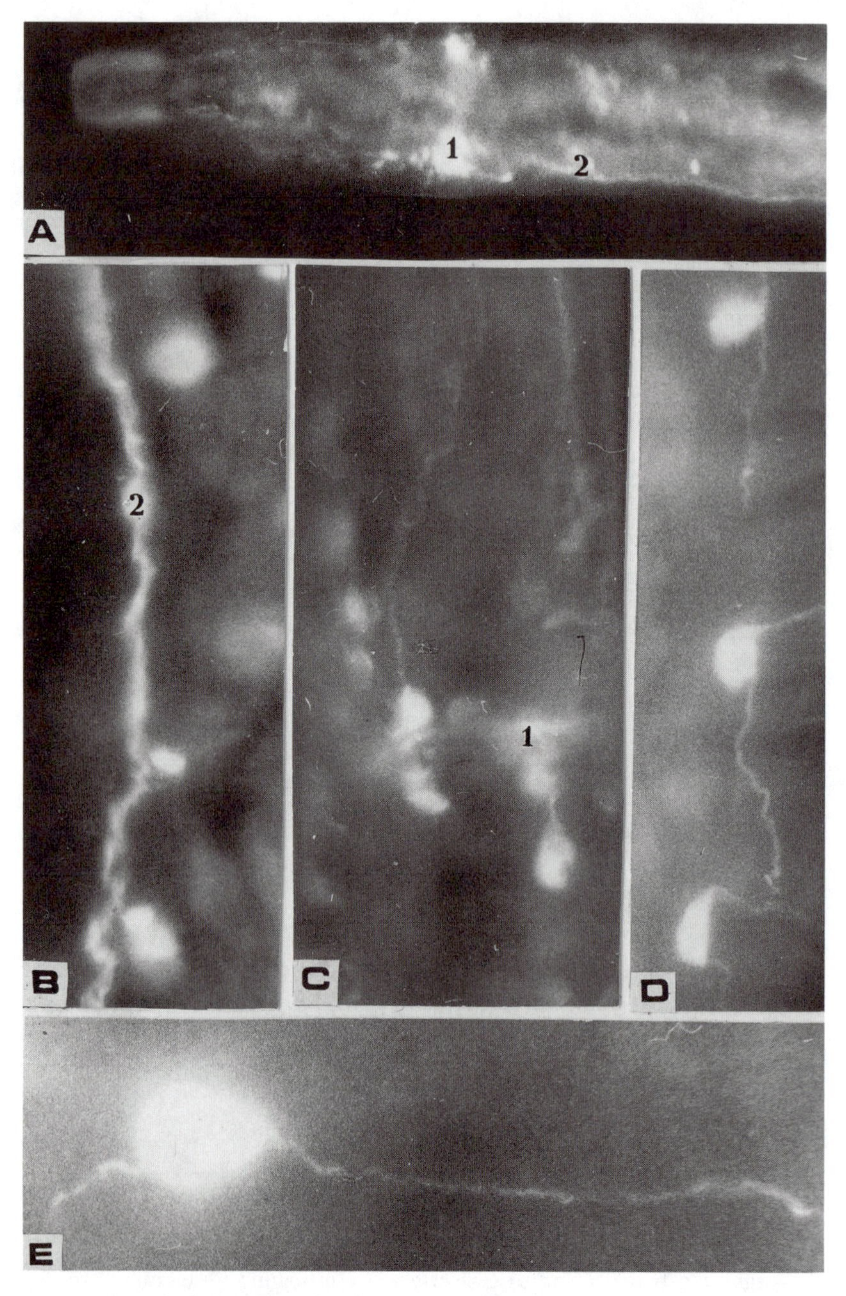

Figure 33. Catecholamines of the nervous system in the anterior part of the body (A); in the ventral nerve trunk (B); in the area of the nerve ring (C); in the neurons of the lateral plexus (D); and in one neuron of the lateral plexus (E) of *Pontonema vulgare*. 1, nerve ring; 2, ventral nerve trunk.

to special organs: setae, papillae, etc. Essentially, nematodes do not sense touch at any place on the body, except at papillae or setae.

Any sensory organ in nematodes contains three components: cuticular structures, nerve cell processes, and accompanying cells (Figure 35). Cuticular structures play an important role in providing specificity of reception for a sensory organ, facilitating the reception of stimuli of a definite type to sensory processes of nerve cells and hindering the penetration of stimuli of other types. The structure of the cuticular components of a sensory receptor often serves as important indirect evidence of the type of reception by that sensory organ. Thus, a sensory receptor with an opening that directly connects the nerve endings to the outer environment is considered to have chemical sensitivity. For mechanoreceptors, on the other hand, it is characteristic to have some modification of cuticle facilitating the entry of mechanical stimuli to the nerve endings. Of course, judgments as to the type of reception based on the structure of only the cuticular elements will not be definitive by any means.

The cell body of neurons that innervate sensory receptors lies rather far, as a rule, from the sensory organ itself. Thus, the bodies of nerve cells innervating the sensory organs of the head end are situated near

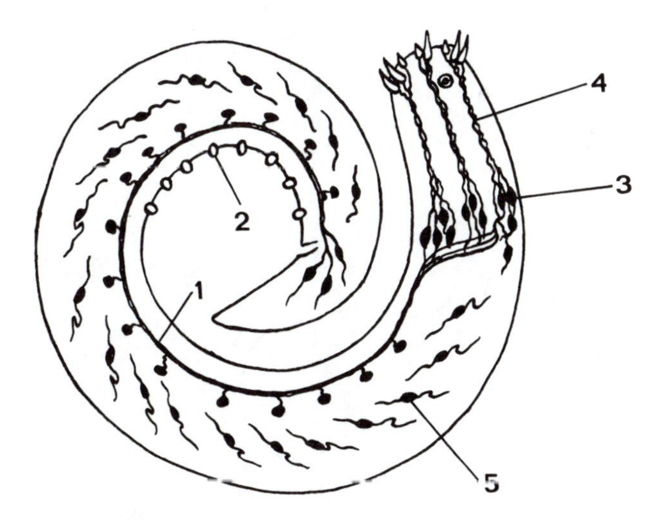

Figure 34. Diagram of the catecholamine-containing elements of the nervous system in free-living nematodes. 1, abdominal nerve trunk; 2, supplementary organs; 3, sensory neurons near the nerve ring; 4, cephalic sensory nerves; 5, peripheral sensory neurons.

the circumpharyngeal nerve ring. The tip of the dendrite innervating a sensory organ forms a swelling from which extends one or more dendritic branches. Dendritic branches usually have the structure of more or less modified cilia. Near the base, the dendritic branches contain nine doublets of microtubules and one or more individual central microtubules. At the base of the dendritic branches, the microtubules can form a structure similar to a basal body. Similarity to a normal cilium is confirmed by the presence of a striated rootlet extending into the depth

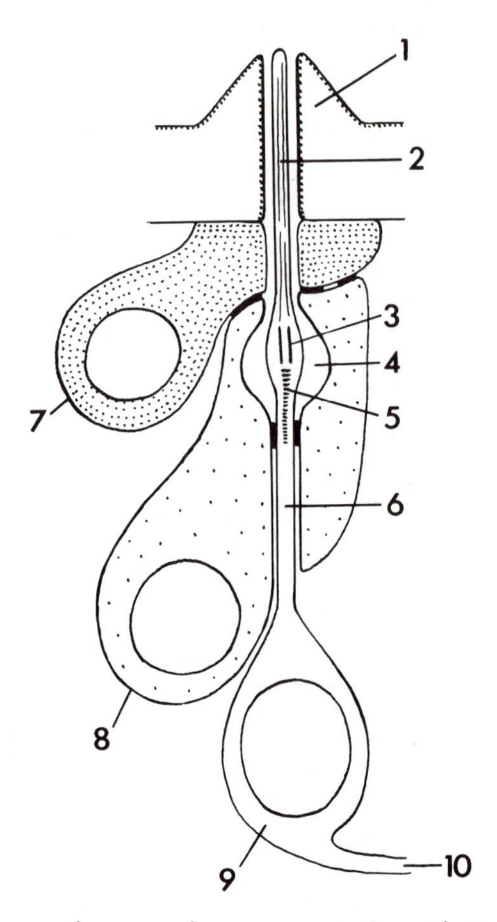

Figure 35. Diagram of a nematode sensory organ. 1. cuticle; 2, dendritic branch; 3, rudiment of a basal body; 4, receptor cavity; 5, striated rootlet; 6, dendrite; 7, socket cell; 8, sheath cell; 9, sensory neuron; 10, axon.

of the dendrite. The regular arrangement of microtubules is preserved only near the base of the dendritic branch, and this arrangement disappears distally. Dendritic branches can expand, and the number of microtubules sometimes increases to eighty or 100. Often, vacuoles are encountered in dendritic branches, and sometimes an accumulation of dark material can be seen between microtubules. The presence of such material is usually regarded, by analogy with members of Arthropoda, as evidence of a mechanoreceptor function for the dendritic branch. On the other hand, thin dendritic branches with a small number of microtubules without accumulations of dark material, if they end near the pore, usually are treated as chemoreceptors.

Associated with sensory organs are special accompanying cells: sheath cells and socket cells. Sheath cells enclose the dendrites for some length and form a so-called receptor cavity at the level of "basal bodies" of dendritic branches. The sheath cell is characterized by relatively light cytoplasm and frequently has a glandular nature. Probably this cell is responsible for the secretion that surrounds the dendritic branches. The socket cell envelops the distal end of the canal through which the dendritic branches pass, and it directly adjoins the cuticle. The socket cell is considered responsible for forming the cuticular structures of the sensory organ.

Somatic Sensory Organs

Somatic sensory organs are well developed in free-living nematodes of the subclasses Enoplia and Chromadoria and are represented by cuticular pores, by setae, or by both. Despite the fact that somatic sensory organs are numerous and densely strewn over the body of free-living nematodes, they remain very poorly studied, and we have only a few examples to explain the structure of these formations.

In *Paracanthonchus,* somatic sensory organs are represented by cuticular pores and setae. The pores open at the bottom of cuticular depressions that extend in rows over the whole body. A canal to the receptor cavity extends from the pore through the cuticle. The cuticular pore is innervated by one or two dendrites that have dendritic branches (Figure 36a). The sheath and socket cells are represented here by hypodermal cells that are comparatively little changed. The socket cell has dense cytoplasm, and the sheath cell contains large vacuoles and has a glandular nature (Malakhov and Yushin 1984). The numerous somatic

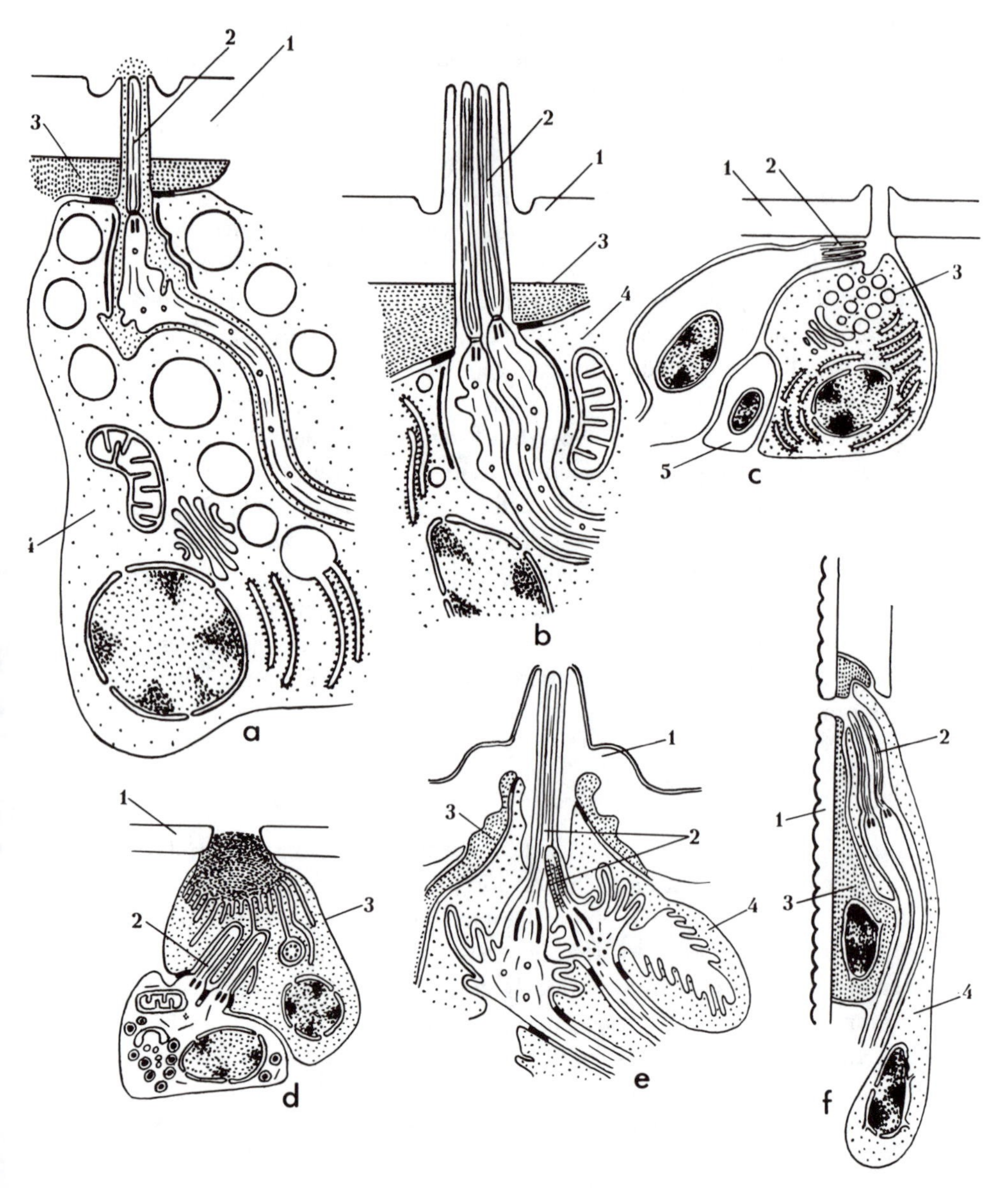

Figure 36. Somatic sensory organs in nematodes (Figures a and b are original; figures c–f are after Wright 1980): a, somatic pore in *Paracanthonchus macrodon;* b, somatic seta in *Paracanthonchus macrodon;* c, somatic pore in *Chromadorina germanica;* d, bacillary band of *Capillaria;* e, cervical papilla of *Heterakis gallinarum;* f, somatic pore in *Xiphinema americanum.* 1, cuticle; 2, dendritic branch; 3, socket cell; 4, sheath cell; 5, hypodermal cell.

setae of *Paracanthonchus* are represented by hollow processes of cuticle innervated by five or six dendritic branches (Figure 36b). At the tip of the setae is a wide opening. The sheath and socket cells here are relatively poorly modified hypodermal cells (Malakhov and Yushin 1984). A similar structure of cuticular pores has also been noted in *Chromadorina germanica*, where the hypodermal cell associated with the sensory cell (sheath cell?) has a glandular nature (Lippens 1974b). In *Deontostoma californicum*, somatic setae are represented by conical processes of cuticle with an opening at the tip, to which extend several dendritic branches and a glandular (sheath?) cell (Maggenti 1964; Siddiqui and Viglierchio 1977).

In *Xiphinema americanum*, a member of the order Dorylaimida, cuticular pores are represented by openings in the cuticle to which dendritic branches extend (Figure 36f). The sheath and socket cells are also present here (Wright and Carter 1980). In parasitic nematodes of the order Trichocephalida, there are pores that open on the ventral side in the pharyngeal region. As in Dorylaimida, dendritic branches are connected to these pores (McLaren 1976; Wright 1980). The order Trichocephalida has unusual formations—bacillary bands. The bacillary bands are rows of pores in the cuticle to which are connected glandular cells of the lateral chords of the hypodermis (Figure 36d). With them are associated neuronlike cells having four to six dendritic branches that take root in the glandular (sheath) cells (Wright 1968a, b, 1980). The neuronlike cells do not have axons but in the basal parts contain a large quantity of neuro-secretory vesicles.

The somatic sensory organs in Rhabditia are significantly reduced. They have deirids and post-deirids. In *Caenorhabditis* a deirid is represented by a cuticular papilla without an opening and innervated by a single dendritic branch containing electron-dense material (Ward et al. 1975). In *Heterakis gallinarum* and *Ascaridia galli* there are postlabial papillae represented by cuticular processes with an opening at the tip and innervated by two dendritic branches (Wright 1980). One of these branches goes to the pore at the tip of the papilla, and the other ends at the papilla base and contains electron-dense material (Figure 36e).

The functions of the somatic sensory organs are varied. The somatic pores of Chromadorida and Dorylaimida usually represent chemoreceptors. Indirect support for this belief is the presence of the glands associated with these organs that, as a rule, accompany the chemoreceptor organs. Somatic setae are usually regarded as mechanoreceptors (Croll and Smith 1974). However, the fact that at the tip of the seta there is an opening and associated glands suggests that there is a chemoreceptor function along with the mechanoreceptor function. The bacillary band organs of Trichocephalida are thought to be organs of osmotic or ionic regulation (Wright 1968a, b, 1980). Postlabial papillae in *A. galli* and *H. gallinarum* look like organs of combined reception: chemical (accomplished by a dendritic branch extending to the pore at the tip of the papilla) and mechanical (accomplished by a short dendritic branch containing dense material). Deirids and post-deirids in members of Rhabditia probably represent purely mechanical receptors.

Cephalic Sensory Organs

Sensory organs of the head end are represented by labial papillae, cephalic papillae or setae, and amphids. At the head end are also situated cervical papillae or setae, which are somatic sensory organs that have moved forward to the anterior end.

In free-living marine nematodes of the subclass Enoplia, labial papillae and cephalic setae are conical processes of the cuticle, surrounded at the base by cuticular folds (Figure 37A and B). Cephalic setae differ from labial papillae only by greater length. The papillae and setae are penetrated by a canal that opens at the tip. The opening is clearly visible with scanning electron microscopic examination (Figure 37A and B).

Among representatives of *Pontonema vulgare,* free-living marine nematodes of the order Enoplida, the head has six labial papillae, ten cephalic setae, and a pair of pocket-shaped amphids (Figure 38a). In *Pontonema* each labial papilla or cephalic seta is innervated by nine to twelve dendritic branches (Figure 38b, c, and d). Most dendritic branches are threadlike structures that contain irregularly scattered, primarily single microtubules (Figure 39C). Inside the threadlike dendritic branches, near the tip of a papilla or seta, are found osmiophilic vesicles (Figure 39A). One of the dendritic branches widens and contains a vacuole, and the number of microtubules increases to 80–100 (Figure 39C). Electron-dense material occurs between the microtubules (Figure 39C). Near the tip of the labial papilla or seta, the branch thins out, and the osmiophilic vesicles are absent (Figure 39A).

The amphidial nerve opens at the bottom of the pocket-shaped amphids of *P. vulgare* (Figure 39). The amphidial receptor apparatus is represented by seventeen dendritic branches of which one is widened and

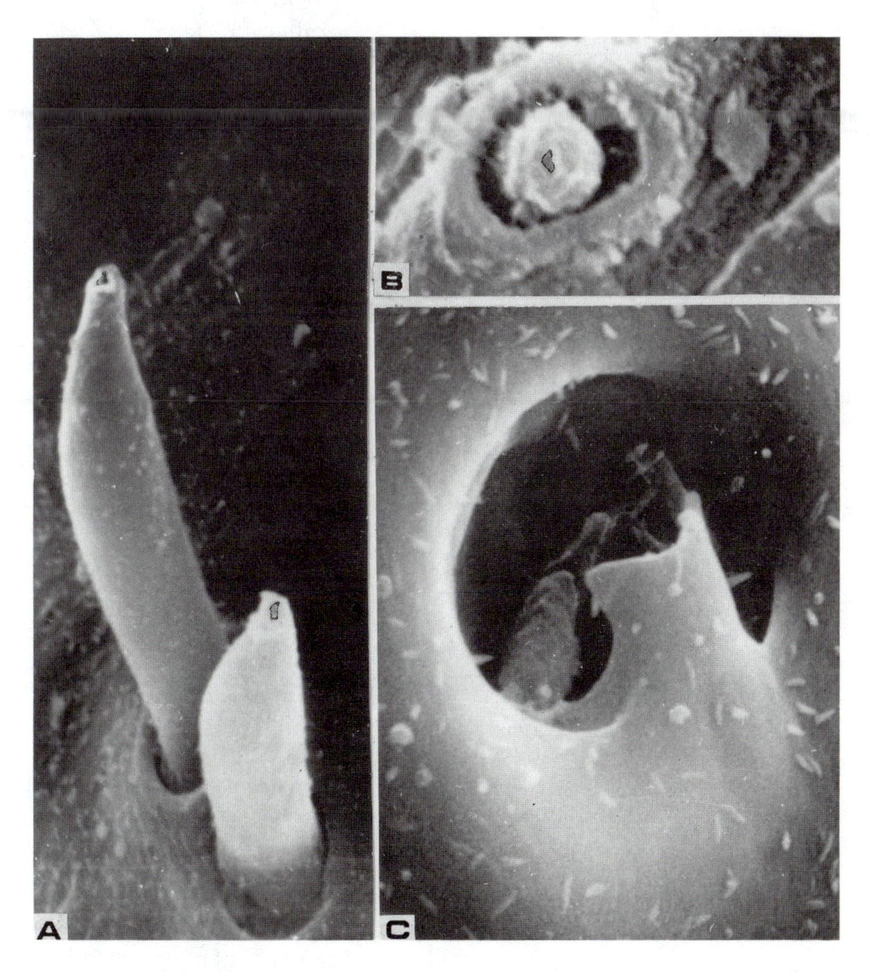

Figure 37. Fine structure of sensory organs in *Pontonema vulgare*. Magnifications: A = 5,500; B = 23,400; C = 15,600. A, cephalic setae; B, labial papillae; C, amphid.

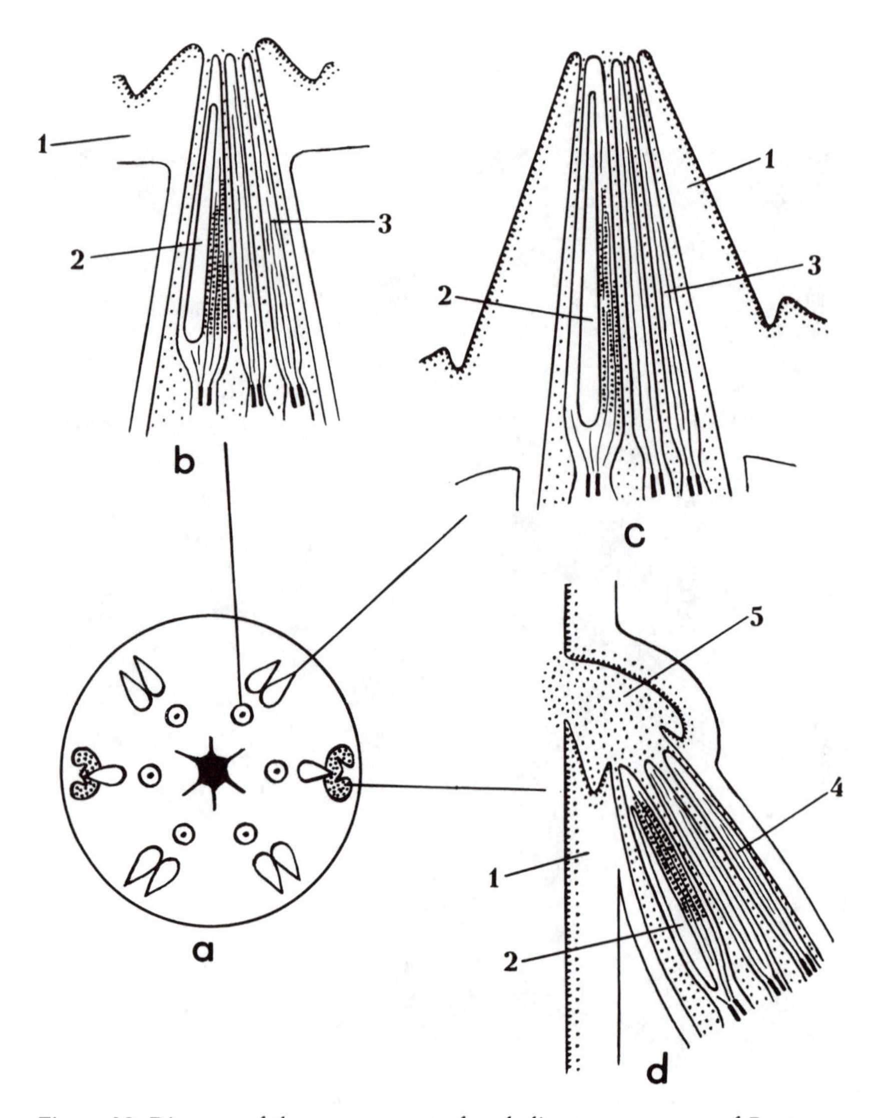

Figure 38. Diagram of the arrangement of cephalic sensory organs of *Pontonema vulgare* (a) and diagrams of the infrastructure of a labial papilla (b), cephalic seta (c), and amphids (d). 1, cuticle; 2, widened dendritic branch; 3, dendritic branch with alveolar structures; 4, threadlike dendritic processes without alveolar structures; 5, amphidial pocket.

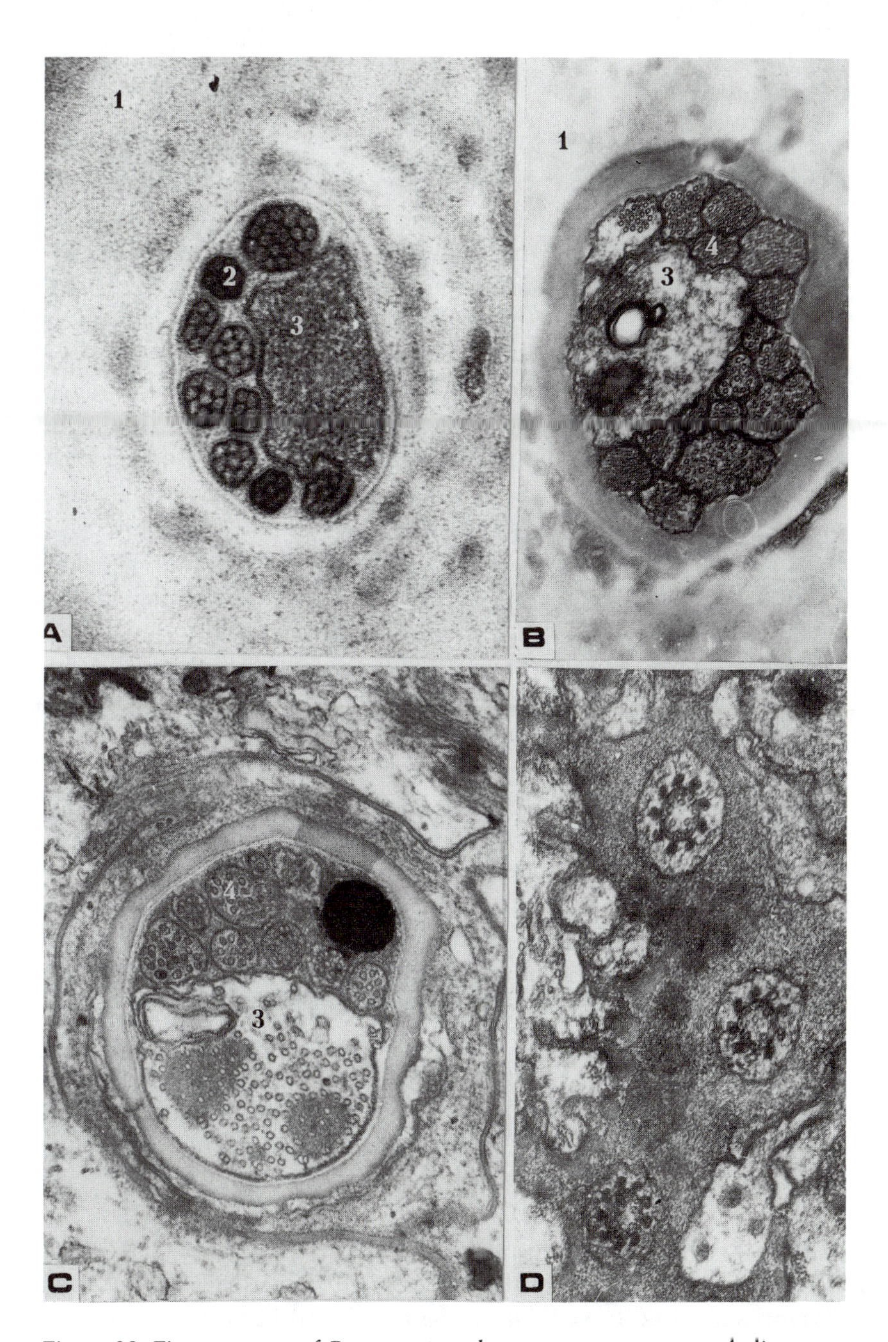

Figure 39. Fine structure of *Pontonema vulgare* sensory organs: cephalic setae in cross section (A); cross section of pore at posterior end of amphidial pocket (B); cross section of nerve innervating labial papilla (C); cross section of basal body of dendritic branch of amphidial nerve (D). Magnifications: A = 16,800; B = 25,400; C = 35,000; D = 58,500. 1, cuticle; 2, alveolar dendritic branch; 3, swollen dendritic branch; 4, threadlike dendritic branch.

contains a vacuole and an accumulation of dense material (Figure 39B). The remaining dendritic branches have the usual threadlike form and contain a small number of microtubule singlets (Figure 39B). The osmiophilic vesicles are not found in the dendritic branches of the *P. vulgare* amphid. The dendritic branches of the *P. vulgare* amphid end at the level of the amphidial nerve pore and do not penetrate into the cavity of the amphidial fovea (Figure 38). The amphidial nerve canal, through which pass dendritic branches, and the amphidial fovea are filled with granular material, the secretion of the amphidial gland. The gland itself lies far from the head end and is embedded in the stroma of the nerve ring.

In *Xiphinema* of the order Dorylaimida, there are six labial papillae, six cephalic papillae in an inner circle, four cephalic papillae in an outer circle, and a pair of pocket-shaped amphids (Wright and Carter 1980). The labial and cephalic papillae of the inner circle are represented by pores in the cuticle to each of which extend four dendritic branches. Papillae of the outer circle contain three dendritic branches each. Dark inclusions, vacuoles, and so forth are not contained in the dendritic branches. However, one of the branches in each papilla is thicker than the rest. The amphid is innervated by dendrites of fourteen neurons that result in nineteen dendritic branches (five dendrites have two branches each). One of the dendrites innervating the amphid lies outside the amphidial nerve, passes through the sheath cell, and enters the amphidial nerve canal. The dendritic branch of this dendrite is very short and does not reach the pore of the amphidial nerve. The remaining branches go through the amphidial nerve canal, which is surrounded by a glandular sheath cell, and go out through a pore into the cavity of the amphidial fovea. One of the dendritic branches is twice as thick as the others. The dendritic branches originating from dendrites having two branches contain accumulations of dark material.

The amphids of *Oncholaimus vesicarius* of the order Enoplida are innervated by thirty-eight dendritic branches coming from three dendrites (Figure 40b). Some of these dendritic branches have darker cytoplasm than others. In addition to the three dendrites mentioned, there is a dendrite in the amphidial nerve that has ten dendritic branches that enter a depression inside the sheath cell (Figure 40b). They lie opposite the area of accumulation of dark granules—"eyespots"—in the pharyngeal tissue.

In *Tobrilus aberrans* (order Enoplida) the pocket-shaped amphid is innervated by sixteen dendrites, each of which has one dendritic branch (Storch and Riemann 1973). In free-living juveniles of *Gastromermis boophthorae* (order Mermithida) the pocket-shaped amphid is innervated by eighteen thin dendritic branches going out through the pore into the fovea of the amphid (Figure 40a). The glandular sheath cell forms numerous processes extending into the receptor cavity.

Among representatives of *Capillaria hepatica* (order Trichocephalida), a highly specialized parasitic member of Enoplia, the head end has six labial papillae, six cephalic papillae in an inner circle, four cephalic papillae in an outer circle, and a pair of porelike amphids (Figure 41). All of these sensory receptors have the appearance of pores in the cuticle. Dorso- and ventrolateral papillae are innervated by two thin dendritic branches extending to the pore itself (Figure 41b). The lateral labial papillae contain three dendritic branches: two are threadlike and do not

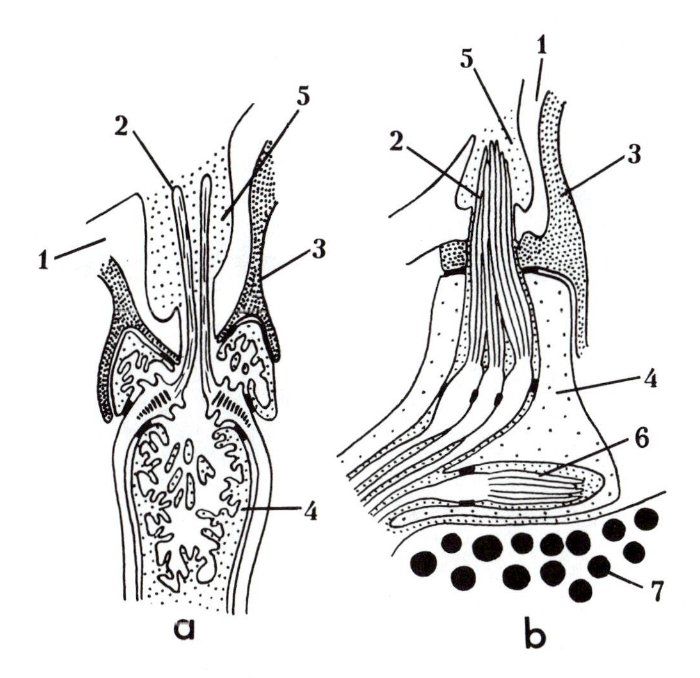

Figure 40. Diagram of the amphid of *Mermis* (a) and *Oncholaimus vesicarius* (b). 1, cuticle; 2, dendritic branches; 3, socket cell; 4, sheath cell; 5, secretion filling amphidial canal; 6, dendritic branch of a proprioreceptor; 7, pigmented granules in pharyngeal tissue.

extend to the pore, and one widens and contains dark material (Figure 41c). Lateral cephalic papillae in the outer circle each contain two dendritic branches that do not go to the pore itself (Figure 41d). The rest of the cephalic papillae are similarly constructed, but each contains three dendritic branches (Figure 41f). The amphids contain ten dendritic branches each that end at various levels inside the amphidial canal (Figure 41e).

Among representatives of the subclass Chromodoria, the free-living marine nematode *Sphaerolaimus balticus*, there are six labial papillae and ten cephalic setae, and eight groups of cervical setae (Figure 42).

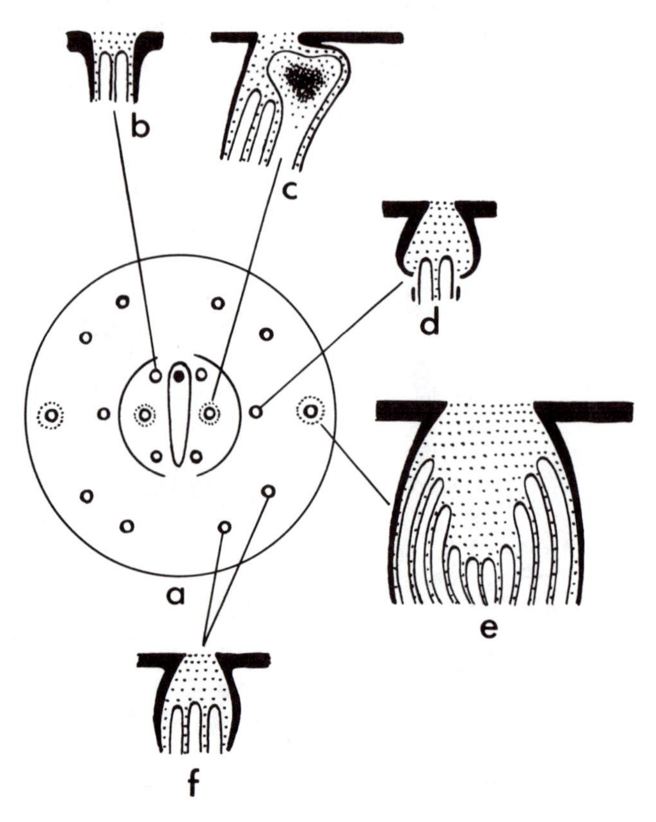

Figure 41. Arrangement and structure of the cephalic sensory organs of *Capillaria hepatica* (after Wright 1974b): a, arrangement of cephalic sensory organs; b, submedian papilla; c, lateral labial papilla; d, lateral cephalic papilla; e, amphid; f, submedian cephalic papilla.

Each papilla or seta is innervated by a single dendritic branch originating from a dendrite deep in the cephalic tissues. The dendritic branches attain a length of 25 m (Malakhov and Ovchinnikov 1980). A dendritic branch goes up to the pore at the tip of the papilla or seta (Figures 42 and 43A). In a cross section of a dendritic branch is visible an irregular arrangement of microtubules (Figure 43B). Near the tip, the dendritic branch contains dense material (Figure 43A).

The amphids of *S. balticus* have a circular form. Each amphid is innervated by threadlike dendritic branches that extend through a pore to the cavity of the amphidial fovea and there make several turns (Figures 42 and 43). One of the dendritic branches goes into the sheath cell and in it forms a descending spiral, enveloping the canal of the amphidial nerve (Figures 42 and 43). This last dendritic branch is characterized by significant thickness, translucent content, and the presence of vacuoles (Figure 43D).

In *Paracanthonchus*, at the head end, there is an arrangement of six labial papillae, ten cephalic setae, and a pair of amphids that have a spiral form (Figure 44). Labial papillae and cephalic setae are similar in structure to the somatic setae of this species but differ in having thicker cuticular walls. Each labial papilla is innervated by five dendritic branches of identical structure. Each of the six short cephalic setae contain five dendritic branches and each of the four long setae contain four each (Malakhov and Yushin 1984). The amphid is represented by a spiral trough on the surface of the cuticle that is filled with a secretion from the amphidial gland (Figure 44). Into the midst of this secretion extend dendritic branches that penetrate through the pore of the amphid (Figure 44). Each amphid is innervated by eighteen dendritic branches, one of which branches into the sheath cell (Figure 44).

In the saprobiotic nematode *Caenorhabditis elegans,* the arrangement of cephalic sensory receptors is six labial papillae, six cephalic papillae in the inner circle, and four cephalic papillae in the outer circle (Figure 45). The amphids of *C. elegans,* as in all Rhabditia, are porelike. Labial papillae are represented by small raised places in the cuticle with an opening at the top (Ward et al. 1975; Ware et al. 1975). Each labial papilla is innervated by two dendritic branches. One of them goes to the pore of the papilla and the other, containing an accumulation of dark material, ends in the depth of the papilla (Figure 45b). Subdorsal and subventral cephalic papillae of the inner circle, together with the corresponding papillae of the outer circle, form doubled submedial papillae.

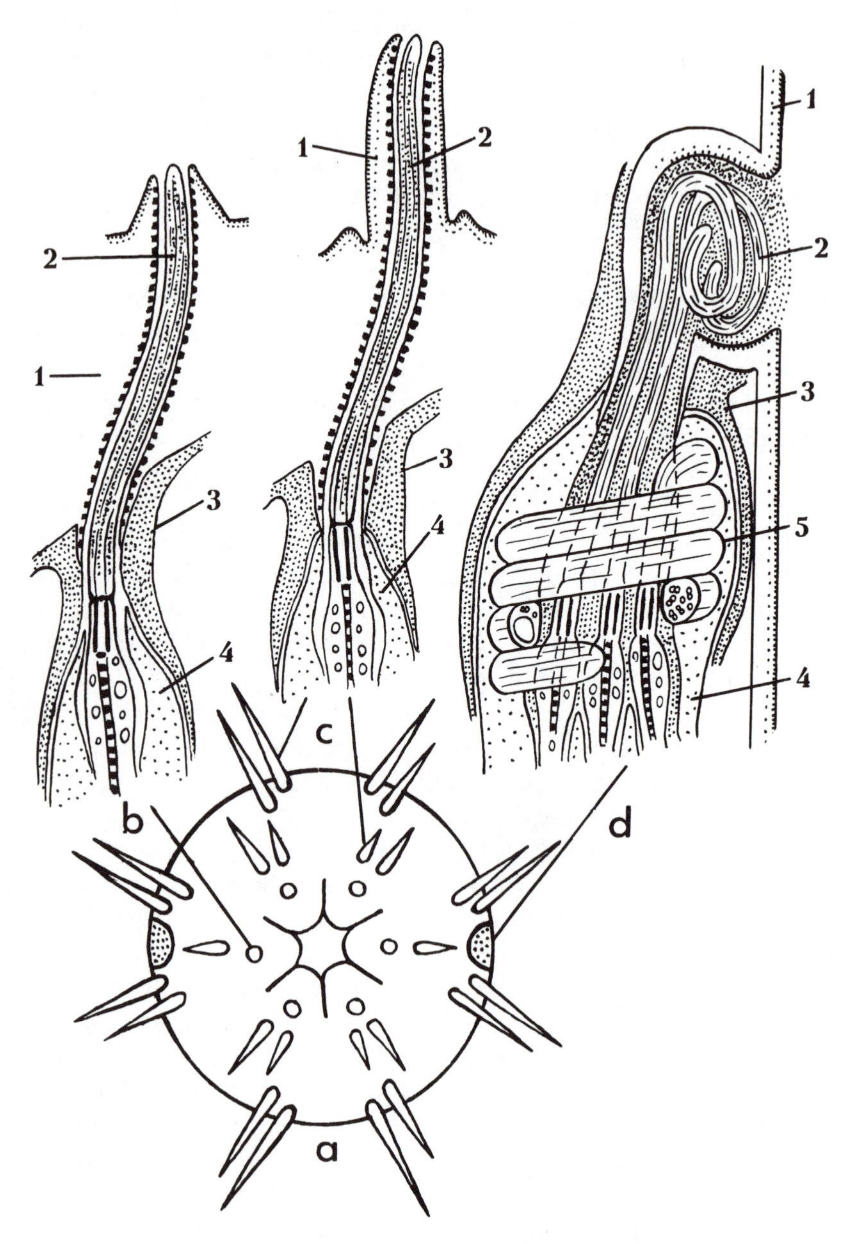

Figure 42. Diagram of the arrangement (a) and structure of the labial papilla (b), cephalic and cervical setae (c), and amphids (d) of *Sphaerolaimus balticus*. Labels as in Figure 36 with the following exception: 5, dendritic branch within the sheath cell.

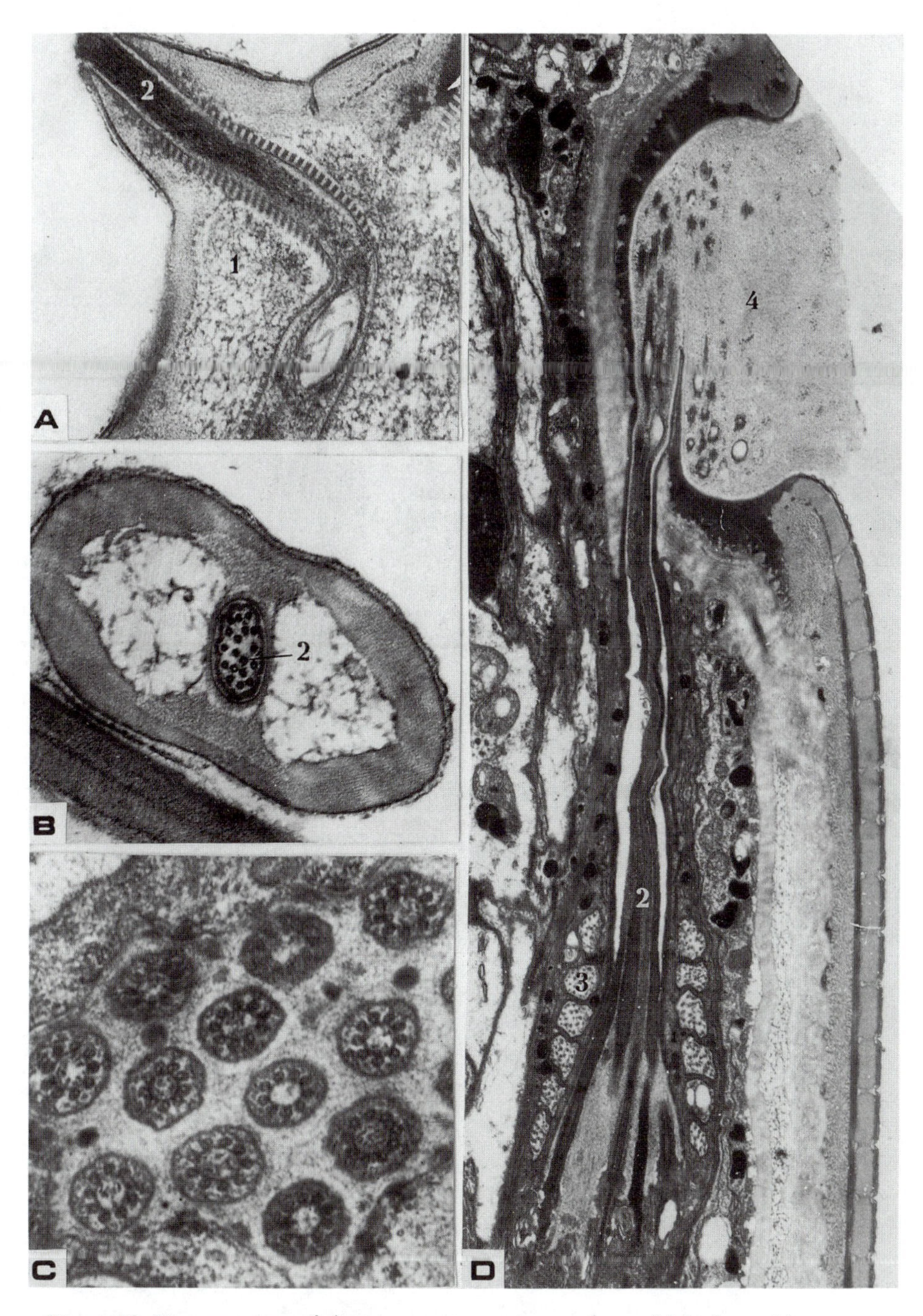

Figure 43. Fine structure of the sensory organs in members of Monhysterida: longitudinal section of labial papilla in *Sphaerolaimus balticus* (A); cross section of seta in *Daptonema setosum* (B); longitudinal section of amphidial nerve of *Sphaerolaimus balticus* (C); cross section of amphidial dendritic branches at level of basal bodies in *Daptonema setosum* (D). Magnifications: A = 33,000; B = 83,400; C = 9,800; D = 57,800. 1, cuticle; 2, dendritic branch; 3, spiral dendritic branch; 4, mucoid secretion.

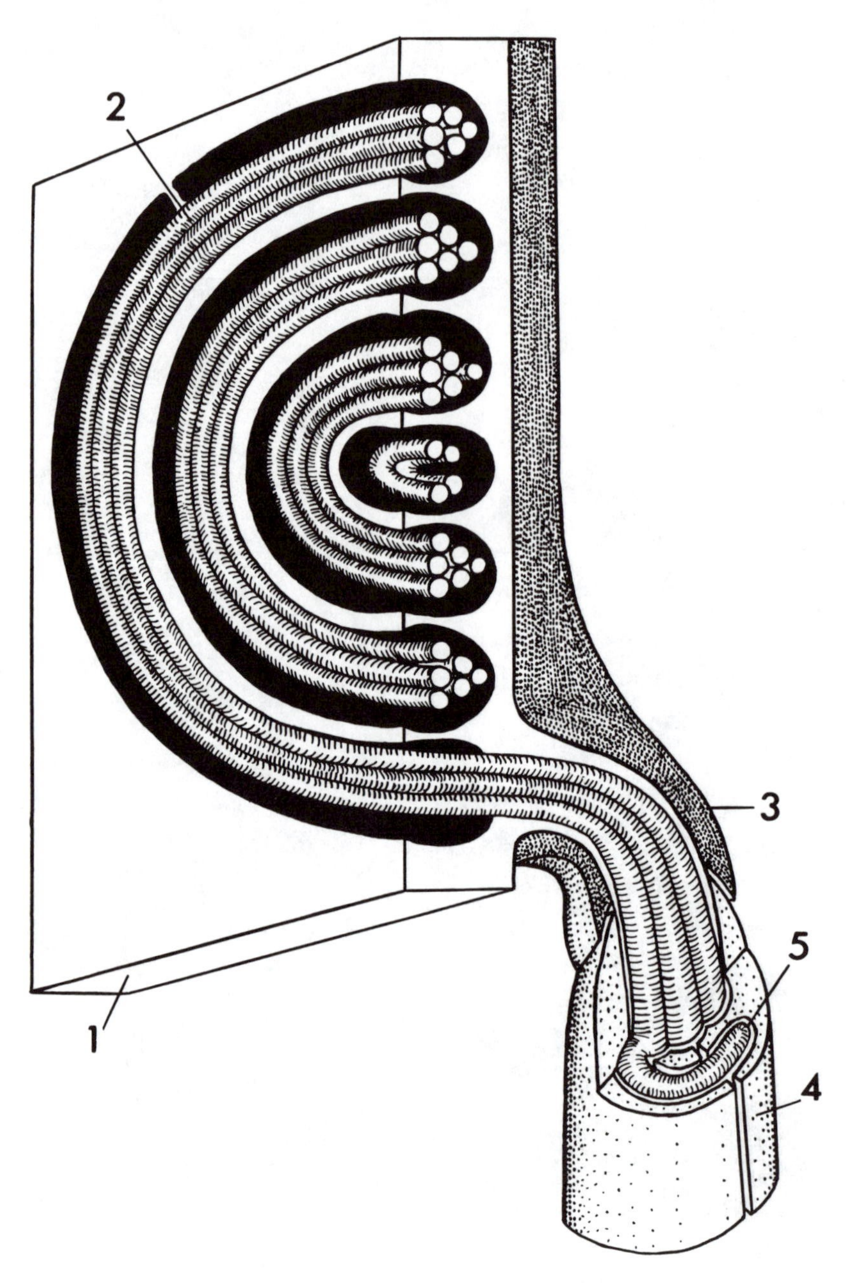

Figure 44. Diagram of the amphid of *Paracanthonchus* sp. 1, cuticle; 2, dendritic branches; 3, socket cell; 4, sheath cell; 5, dendritic branch within the sheath cell.

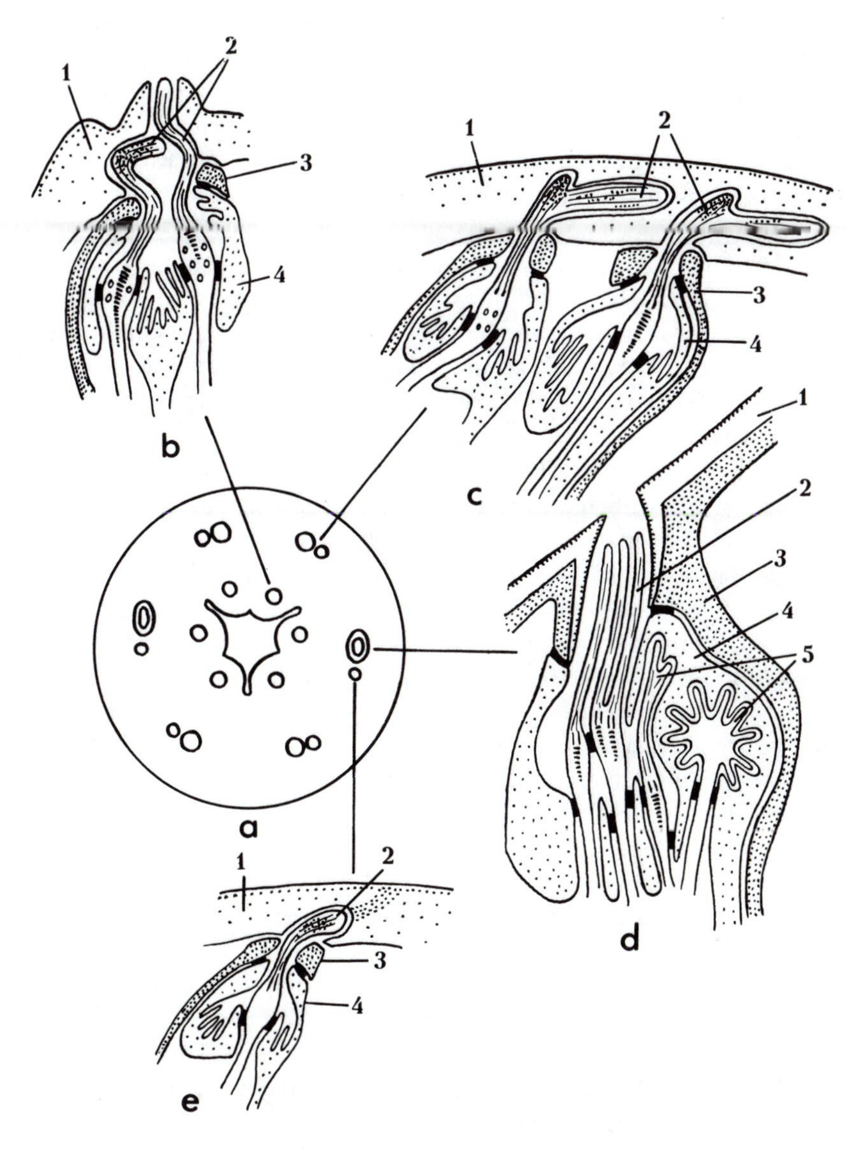

Figure 45. Arrangement and structure of sensory organs in *Caenorhabditis elegans:* arrangement of cephalic sensory organs (a); labial papilla (b); submedian cephalic papilla (c); amphid (d); lateral cephalic papilla (e) (after Wright 1980). Labels as in Figure 36 with the following exception: 5, dendritic branches within sheath cell.

The doubled submedian papillae are each innervated by two dendritic branches, which extend into the cephalic cuticle (Figure 45c). The dendritic branches contain dark material. Lateral papillae of the inner circle each contain one dendritic branch extending into the cephalic cuticle (Figure 45e). The end of the dendritic branch is connected by a strand of dark material to the outer cortical layer of cuticle. The swollen end of the dendritic branch contains an accumulation of dark material (Ward et al. 1975).

The amphid of *C. elegans* is innervated by ten dendritic branches extending from eight dendrites (two dendrites each have two branches). The amphidial nerve has three more dendrites having branches that pass into the receptor cavity of the amphid but penetrate the sheath cell and do not reach the amphidial nerve pore. In addition, there is a dendrite penetrating the sheath cell, the latter with a cavity containing about fifty microvilli (Figure 45d). Thus, the receptor apparatus of the amphid is formed by differentiated dendritic branches coming from 12 dendrites (Figure 45). The glandular sheath cell lies in the lateral ganglia at the level of the nerve ring (Ward et al. 1975; Ware et al. 1975).

Among phytoparasitic nematodes of the order Tylenchida, six labial papillae also open by way of pores lying around the oral aperture (*Ditylenchus, Tylenchus, Heterodera,* and *Meloidogyne*). In *Radopholus,* lateral labial papillae are pushed together, opening in the interior of the oral cavity. In *Rotylenchus* and *Macroposthonia,* the pores of all labial papillae are moved inside the oral cavity (Baldwin and Hirschmann 1973; Wright 1980). The labial papillae are each innervated by two dendritic branches. In *Meloidogyne* juveniles, one branch reaches the cuticular pore, and the other ends in the depth of the papilla (Endo and Wergin 1977). In adult forms both branches reach the pore (Baldwin and Hirschmann 1973).

The cephalic papillae of the inner circle in members of Tylenchida, as a rule, are not connected to the external environment and are innervated by single dendritic branches ending in the cuticle. Lateral papillae are partially or fully reduced (their dendritic branches do not reach the cuticle or are absent altogether). In *Meloidogyne,* the inner circle of cephalic papillae are entirely reduced (Endo and Wergin 1977).

The four cephalic papillae of the Tylenchida are each innervated by a single dendritic branch penetrating the cuticle and containing electron-dense material (Baldwin and Hirschmann, 1973; Endo and Wergin 1977; Wright 1980). Papillae of the outer circle are without openings.

However, papillae of the outer circle in *Aphelenchoides* open with pores, and each contains two dendritic branches—one reaches the pore and the other ends in the depth of the papilla.

Amphids among members of Tylenchida are very close in structure to the amphids of *Caenorhabditis*. The receptor apparatus of the amphid is represented by seven dendritic branches that extend to the pore of the amphid, another five branches that enter the sheath cell, and a branch with numerous (up to 200) microvilli that juts into the sheath cell.

Among nematodes of the subclass Rhabditia that are animal parasites, we will consider only two examples. In *Nippostrongylus brasiliensis*, at the head end is an arrangement of six labial papillae and four doubled submedian cephalic papillae (lateral cephalic papillae are obviously reduced). According to Wright's description (1975), labial papillae are each innervated by a single dendritic branch containing an accumulation of dense material. The branch penetrates into the cephalic cuticle, and a strand of dark material connects it to the cortical layers of the cuticle (Figure 46). Each submedian papilla, in accord with its origin from a pair of cephalic papillae, contains two dendritic branches: one of them (analogous to the papilla of the outer circle) bends and is disposed in parallel to the cuticle surface, and the other is analogous to the papilla of the inner circle. The end sections of the dendritic branches contain dense material. On the outside, the dendritic branches are surrounded by a condensation of cuticle (Figure 46). The amphids of *N. brasiliensis* are porelike. Thirteen dendritic branches extend to the pore in the cuticle. In addition, within the innervation apparatus of the amphid are two more dendrites having one branch and numerous microvilli (Figure 46). The branches and microvilli of these dendrites enter the interior of the sheath cell.

In *Heterakis gallinarum*, the oral aperture is surrounded by three secondary labia bearing six labial papillae, four doubled submedian cephalic papillae (the product of the fusion of submedian cephalic papillae of the inner and outer circles), and a pair of amphids. Lateral papillae in *H. gallinarum* are reduced and in their place are found cervical papillae that have moved anteriorly (Figure 47a). Two pairs of such papillae are found posteriorly from the secondary labia on the lateral sides of the body.

Labial papillae are pointed cuticular processes, each innervated by a dendritic branch that penetrates the interior of the cuticle at the papilla base (Figure 47b). Submedian papillae are innervated by a pair of den-

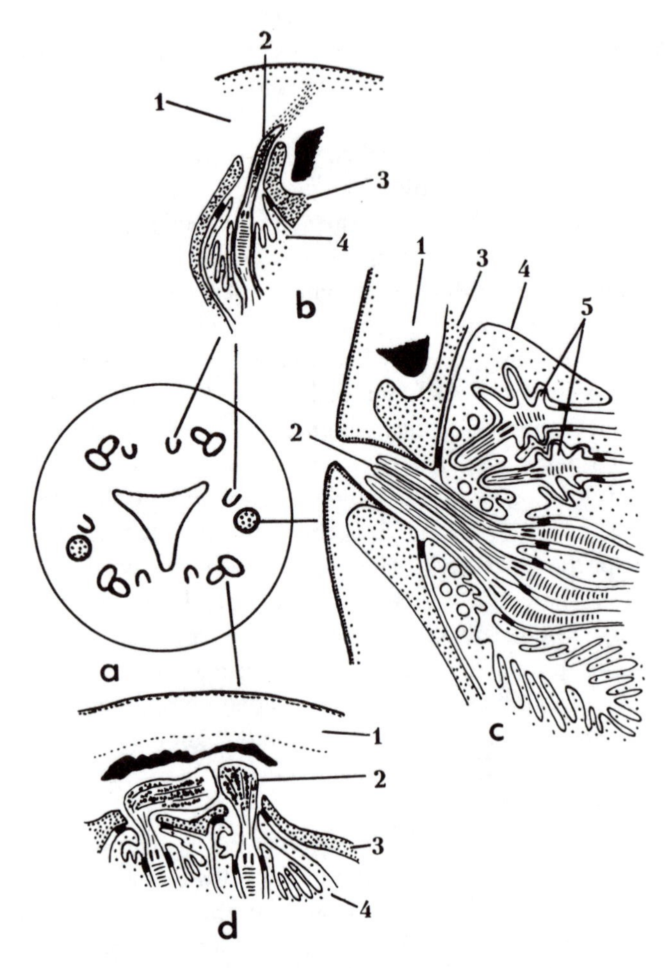

Figure 46. Arrangement and structure of sensory organs in *Nippostrongylus brasiliensis:* arrangement of cephalic sensory organs (a); labial papilla (b); amphids (c); submedian cephalic papilla (after Wright 1980). Labels as in Figure 45.

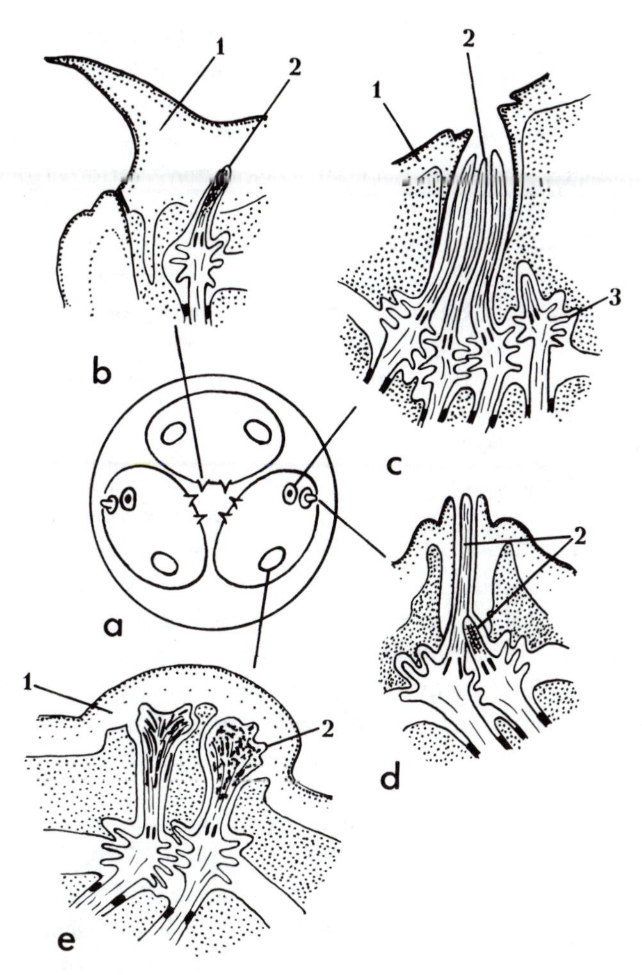

Figure 47. Arrangement and structure of sensory organs of *Heterakis gallinarum:* arrangement of cephalic sensory organs (a); labial papilla (b); amphids (c); additional lateral papilla (d); submedian cephalic papilla (e) (after Wright 1980). 1, cuticle; 2, dendritic branches; 3, dendritic branch penetrating sheath cell.

dritic branches that penetrate the interior of the cuticle (Figure 47e). The distal ends of the dendritic branches are widened and turned 90° to one another. They contain, in addition to microtubules, "quasi-geometrically" arranged strands of dense material. Additional lateral papillae, as well as cervical papillae, are represented by a cuticular process with an opening at the tip and innervated by two dendritic branches, one of which reaches the pore at the tip of the papilla and the other (containing an accumulation of dense material) ends in the interior of the papilla (Figure 47d). As in the preceding species, thirteen dendritic branches extend to the amphidial pore, and two dendritic branches penetrate the interior of an ordinary cell (Figure 47c). The structure of cephalic sensory organs in other Rhabditia, especially animal parasites, is close to that described for *N. brasiliensis* and *H. gallinarum*. All of their cephalic sensory organs, other than amphids, are devoid of openings and contain dendritic branches that penetrate the interior of the cephalic cuticle (McLaren 1976; Wright 1980).

Direct physiological experiments on nematode sensory organs have not been conducted, so judgments on the functions of cephalic sensory structures have to be made primarily on the basis of their morphology. However, the morphology is frequently ambiguous about these functions. Thus, cephalic setae and labial papillae in free-living marine nematodes are usually regarded as mechanoreceptors (zur Strassen 1904; Filip'jev 1921; Paramonov 1962). Actually, if an organ rises above the surface of the cuticle, it is easily subjected to mechanical effects. But, in Arthropoda, for example, trichoid sensillae, which have bristles similar in external appearance to the setae of nematodes, can have a chemoreceptor function or be organs of mixed reception—chemical and mechanical (Dethier 1963; Adams et al. 1965; Ivanov 1969; Slifer 1970; Schmidt and Gnatzy 1972; Ernst 1976; Seeliger 1977). These sensillae have openings at the tip for reception of a chemical agent and in the case of double reception also contain two types of dendritic branches.

Labial papillae and cephalic setae of free-living marine nematodes have openings at the tips and, in *Pontonema,* have two types of dendritic processes: threadlike, containing alveolar osmiophilic structures in the distal parts, and thickened processes containing a vacuole and an accumulation of dark material. The accumulation of dark material is characteristic of the mechanoreceptor processes of arthropods, and the thickened process in the setae and papillae of *Pontonema* can be regarded as mechanoreceptors. The threadlike processes with alveolar structures also may be mechanoreceptors, but it cannot be excluded that they have a chemoreceptor function. In this case, the papillae and setae of *Pontonema* can be organs of mixed reception. In *Sphaerolaimus*, the receptor apparatus of the papillae and setae consists of a single dendritic branch extending to the pore at the tip of the sensory organ. The presence of dense material in the distal section of the dendritic branch makes a mechanoreceptor function for papillae and setae of *Sphaerolaimus* more probable.

Evidently the papillae and setae at the head end of free-living marine nematodes are not very specialized organs that could easily change type of reception in the process of evolution. A primitive trait of the sensory apparatus of marine nematodes is that the labial papillae and cephalic setae of both circles are constructed very similarly and differ only in length.

Similarly, labial and cephalic papillae have been preserved also in phytoparasitic Dorylaimida, where these organs are represented by simple pores to which extend dendritic branches (Wright 1980). Evidently, labial and cephalic papillae in this case are chemoreceptors. Among representatives of the order Trichocephalida, such as *Capillaria hepatica*, all sensory organs are represented by simple pores at the head end to which the dendritic branches extend. Evidently all cephalic sensory organs of *C. hepatica*, excluding the lateral labial papillae, are pure chemoreceptors. Within the receptor apparatus of the lateral labial papillae are broadened dendritic branches containing an accumulation of dark material (Figure 41). Evidently the lateral labial papillae are organs of mixed reception.

In nematodes of the subclass Rhabditia, the cephalic sensory organs are more specialized. Thus, in *Caenorhabditis*, labial papillae are organs of mixed reception: chemical (accomplished by a dendritic branch extending to the pore) and mechanical (accomplished by a dendritic branch ending within a papilla). Cephalic papillae in this case are devoid of openings and are pure mechanoreceptors (Figure 45). The labial papillae of the Tylenchida also open as pores and are innervated by two dendritic branches, which supports the assumption that they are simultaneously chemo- and mechanoreceptors. Cephalic papillae of Tylenchida are pure mechanoreceptors, but, in *Aphelenchoides*, the papillae of the outer circle have openings and probably represent organs of mixed reception: chemical and mechanical. In nematodes of the subclass Rhabditia that are parasites of animals, labial and head papillae, as a rule, are without openings and serve as pure mechanoreceptors (Figures 46 and 47). Lateral papillae of *Ascaridia* and *Heterakis,* which have openings at the tip, probably fulfill the function of mixed mechano- and chemoreceptors (Figure 47d). However, these papillae, as has been noted, are not real cephalic receptors, but represent somatic receptors that have shifted to the head end.

Nematode amphids almost unanimously are regarded as organs of distant chemical reception. Despite all the variation in their external form in free-living nematodes, all amphids are composed in part of a depression in the cuticle that more or less insulates the dendritic branches from mechanical effects but does not inhibit the penetration of chemical stimuli. The sheath cells of amphids are well-developed glands, issuing a secretion that fills the amphidial fovea. Dendritic branches also are submerged in this secretion and extend into the fovea (Riemann et al. 1970; Riemann 1972). It is known that in most animals olfactory organs are connected with glands. This is because molecules of odorous substances, before reaching the dendritic branches, must be dissolved in the liquid that surrounds these branches (Vinnikov 1971).

Indirect evidence of a chemoreceptor function in amphids is also derived from greater development of amphids among males of certain free-living nematodes and of the free-living stages of certain parasites (Mermithida). Mutant forms of *Caenorhabditis* that have defective amphids are unable to detect chemical stimuli (Wright 1980).

The external form of the amphids in free-living Enoplia and Chromadoria is varied, and the significance of the form for chemical reception is still not understood. In free-living members of Enoplia, amphids have the form of a pocket and only as an exception (suborder Tripyloidina) the form of a simple spiral. The pocket-shaped amphid is not encountered in members of Chromadoria, whereas spiral, circular, slit-shaped, or loop-shaped amphids are characteristic of this subclass (Figure 48). In Rhabditia, the external form of amphid is a simplified pore.

Chemoreception, evidently, is not the only function of the amphid. In many species, in addition to dendritic branches extending to the pore of the amphidial nerve, there are branches that penetrate into the sheath cell. The structure of these branches are varied and are sometimes distinguished by great regularity as, for example, in *Sphaerolaimus* (Figure 42). The functions of these receptors, evidently, may also vary considerably. Thus, the dendrite branches extending into the sheath cell in *Oncholaimus vesicarius* fulfill a photosensory function in the eyespot,

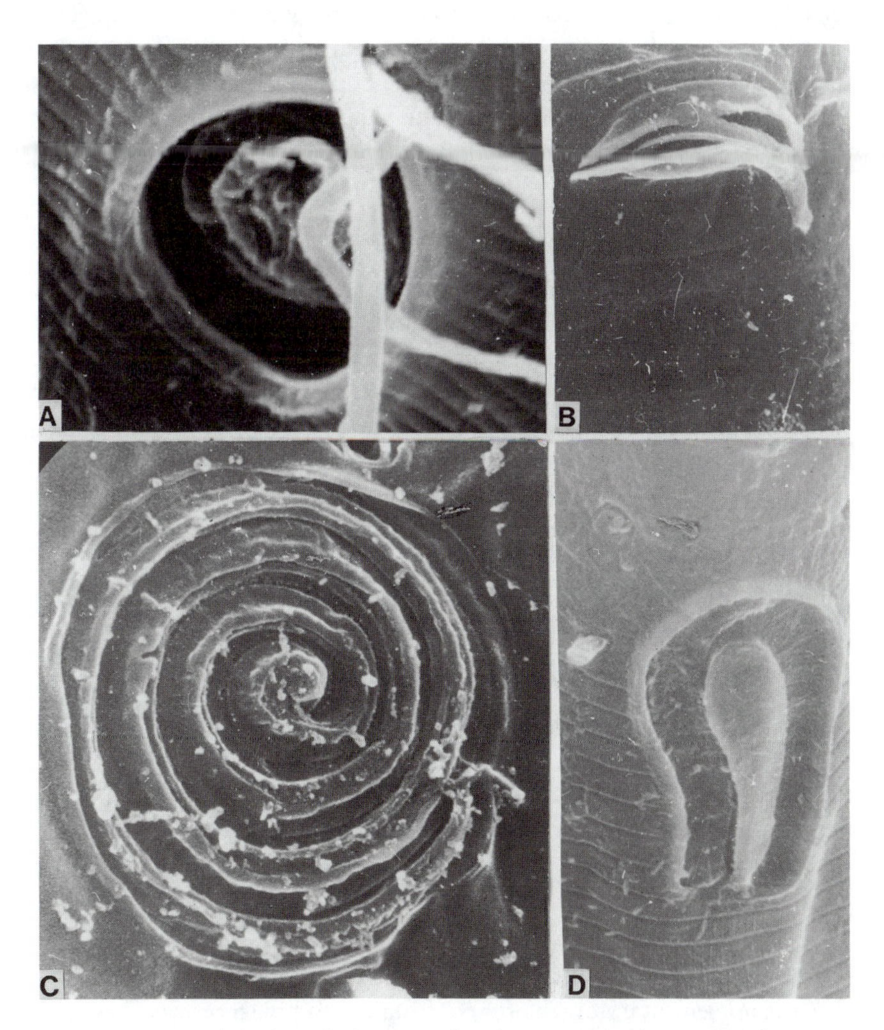

Figure 48. Examples of amphids among free-living nematodes: A, circular amphid of *Sphaerolaimus;* B, slit-shaped amphid of *Neochromadora;* C, spiral amphid of *Paracanthonchus;* D, looped amphid of *Axonolaimus.*

the pigmented granules of which are localized in the muscular cells of the pharynx (Figure 40). The eyespots of *Deontostoma californicum* are innervated from the lateral cephalic nerves and contain membrane cups characteristic of photoreceptors (Siddiqui and Viglierchio 1970). The pigmented granules of the eyes in the latter case are also localized in the marginal cells of the pharynx. Membrane-bound vesicles associated with the pigment, localized in the wall of the pharynx or outside it, are found in *Enoplus communis, Araeolaimus elegans,* and *Chromadorina* sp. (Croll et al. 1975).

At present, internal receptors of the head end have been described for only a few species of nematodes. In *Xiphinema americanum* there are two pairs of dendrites, the branches of which end in the tissues of the head end without reaching the cuticle (Wright and Carter 1980). One pair of dendritic branches ends near the lateral labial papillae. The dendritic branches of the other pair pass near the amphidial nerves, closely adjoining the sheath cell. At the level of the amphidial pocket, the distal parts of the branches bend 90°. In *Caenorhabditis* two pairs of dendritic branches of typical structure have been found that are associated with the sheath cells of lateral labial papillae (Ward et al. 1975). Two more pairs of nerve endings are arranged on the inner side of the submedian papillae of *Caenorhabditis.* Internal head receptors are also found in members of the order Tylenchida (Endo and Wergin 1977). The internal head receptors function, evidently, as proprioreceptors.

Sensory Structures of the Copulatory Organs

The male copulatory organs are represented by spicules, gubernaculum, pre- and postanal papillae, and supplementary organs. Supplementary organs are widespread among male members of free-living Enoplia and Chromodoria. They represent a number of more or less complexly organized cuticular structures arranged in a row along the center line of the ventral side in the preanal area of the body. The number of supplementary organs ranges from ten to fifteen or even as many as 20. Among Rhabditia, nonpaired, preanal supplements are encountered as an exception in individual representatives of Rhabditida and Oxyurida. Well-defined single-cell glands are related to supplementary organs. The presence of sensory dendritic branches also has been shown in supplementary organs (Hope 1974).

Paired pre- and postanal papillae are characteristic of nematodes of the subclass Rhabditia. In *Aphelenchoides,* each papilla contains two dendritic branches, one of which (chemoreceptor) extends to a cuticular pore and the other (mechanoreceptor) ends within the cuticle and is characterized by the presence of dense material (Figure 49a). In *Dipetalonema,* each pre- or postanal papilla contains one dendritic branch, widened in a mushroom shape in the distal part and containing electron-dense material (McLaren 1972). A narrow canal unites the distal part of the branch with the external environment (Figure 49b). In nematodes of the subclass Rhabditia, the males may have a complex copulatory bursa supported by ribs. At the ends of the ribs in *Pelodera* there are openings to which extend dendritic branches (Wagner and Seitz 1981). In *Nippostrongylus,* each rib is innervated by two dendritic branches (Croll and Wright 1976).

The spicules and gubernaculum are innervated by sensory neurons lying in the lateral chords of the hypodermis (Figure 31). The spicules are hollow cuticular formations opening with pores at the distal ends. Sensory dendritic branches are found in the spicules of many nematodes

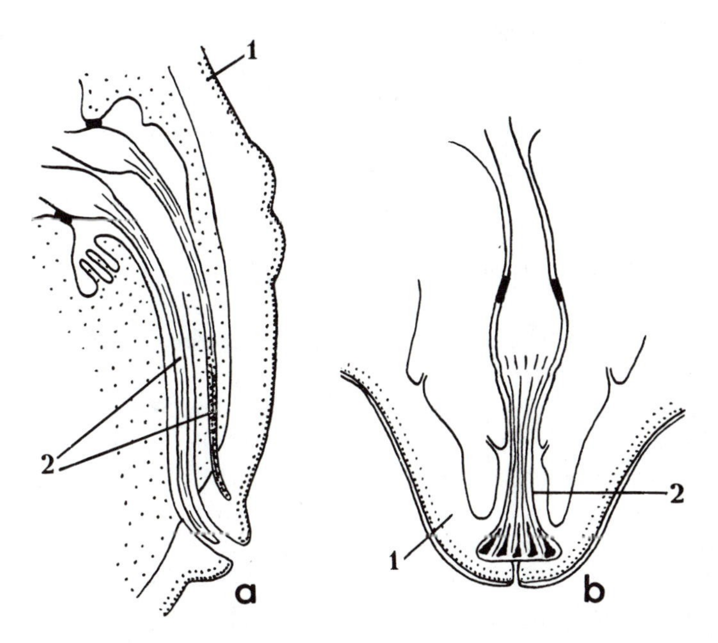

Figure 49. Structure of caudal papilla in *Aphelenchoides blastophorus* (a) and *Dipetalonema vitae* (b) (after Wright 1980). 1, cuticle; 2, dendritic branch.

(Clark et al. 1973; Dick and Wright 1974; Croll and Wright 1976; Wen and Chen 1976; Clark and Shepherd 1977; Wright 1978; Wagner and Seitz 1981). Each spicule of *Tylenchulus semipenetrans* contains two dendritic branches, one of which extends to the pore at the end of the spicule and the other ends in the proximal part of the spicule (Wright 1980). In *Heterodera* and *Pratylenchus,* single dendritic branches extend to the pore at the end of the spicule (Clark et al. 1973; Wen and Chen 1976). Parasitic nematodes of the order Trichocephalida have one long spicule with numerous pores opening near the end (Wright 1978). Each pore is connected with a small cuticular chamber to which a single dendritic branch extends.

The gubernaculum is also innervated by a pair of neurons lying in the lateral hypodermal chords (Figure 31). A pair of dendritic branches in the gubernaculum has been described in *Aphelenchoides blastophorus* (Clark and Shepherd 1977) and *Pratylenchus penetrans* (Wen and Chen 1976). The so-called "genital cone" of *Nippostrongylus brasiliensis* contains four dendritic branches (Croll and Wright 1976). In all cases studied, the dendritic branches in the gubernaculum are not connected directly with the external environment.

The supplementary organs of free-living nematodes are sensory-glandular formations and during copulation facilitate fixation of the male on the female. The dendritic branches connected to the supplementary organs evidently accomplish reception (mechanical?) at the moment the male attaches to the female.

Various pre- and postanal hollow papillae in nematodes can be chemo- and mechanoreceptors, but sometimes they are both simultaneously, as in *A. blastophorus* (Figure 49). Papillae at the ends of the bursa ribs are treated as chemoreceptors (Wagner and Seitz 1981). Dendritic branches innervating spicules are treated as chemoreceptors (Clark and Shepherd 1977; Wright 1978). At the same time, in *Pelodera,* the spicules are protruded only after the final attachment of the male to the female; therefore, in this case, it is more probable that the function is that of mechanoreceptor (Wagner and Seitz 1981). Dendritic branches in the gubernaculum not connected to the external environment more probably have a mechanoreceptor function.

Phasmids are paired lateral organs that have developed among representatives of the subclass Rhabditia (Figure 50). Externally, they appear to be paired lateral glands, very similar to those of some Plectida (Figure 50). Electron microscopic investigations have revealed the pres-

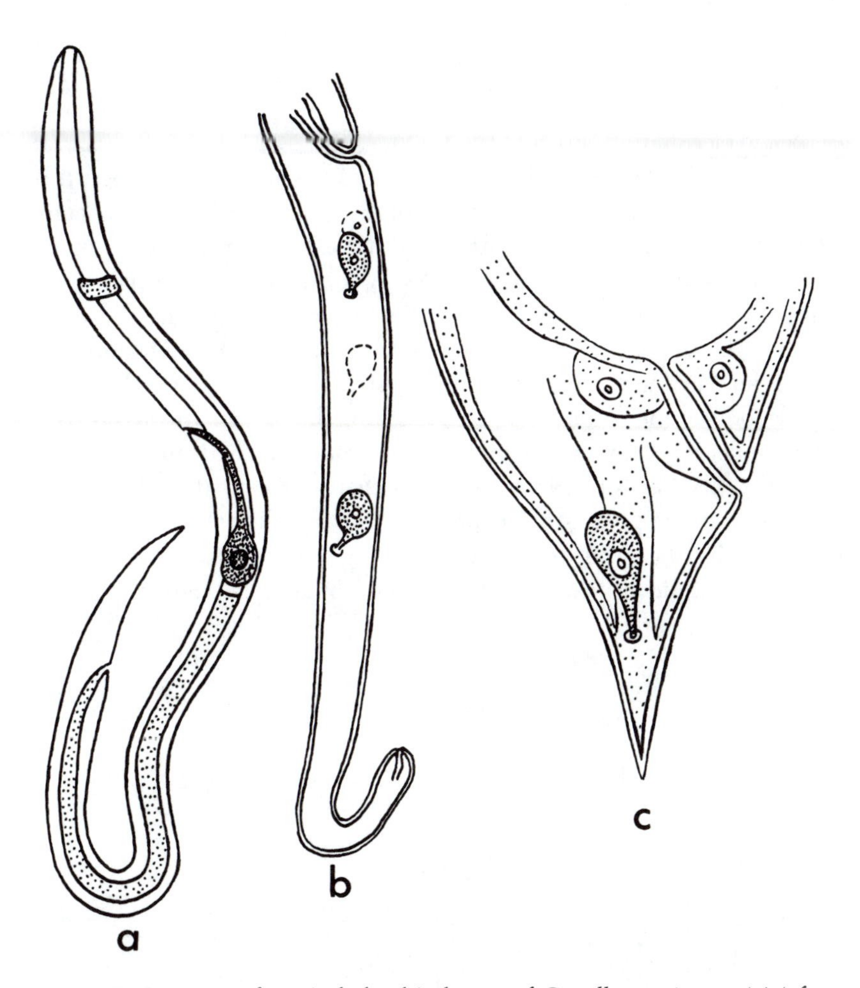

Figure 50. Structure of cervical gland in larvae of *Cucullanus cirratus* (a) (after Valovaya 1978); phasmid-shaped glands in *Bathyonchus embryonophorus* (b) (after Alekseyev and Naumova 1977); and actual phasmids in *Neoaplectana* (c) (original).

ence in phasmids of single dendritic branches connected to a pore in the cuticle (McLaren 1972, 1976). Evidently phasmids accomplish a sensory (chemoreceptor) and a glandular function. In many forms, phasmids are better developed in females than in males.

Receptors of the Body Walls

Among free-living nematodes of the order Enoplida, unusual receptors have been found in the lateral chords of the hypodermis, the metanemes (Lorenzen 1978b, 1981a, b). As shown by electron microscopy, metanemes represent a complex of a process of a primary sensory cell united with a dendrite of a secondary sensory neuron (Hope and Gardiner 1982). The process of the primary sensory cell has a depression into which juts a cilium (Figure 51). A very strong rootlet extends from the basal body of the cilium. The cytoplasm of the process of the primary sensory neuron contains electron-dense substance. The walls of the receptor cavity are formed by processes that surround the cilium. The distal part of the process of the primary sensory cell comes into synaptic contact with the dendrite of the secondary sensory neuron. The cytoplasm of the dendrite is translucent and contains numerous microtu-

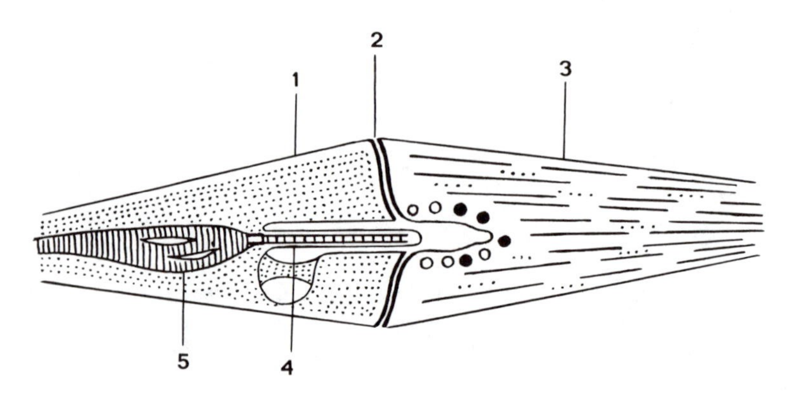

Figure 51. Structure of the metaneme in *Deontostoma californicum* (after Hope and Gardiner 1982). 1, primary sensory cell; 2, area of contact with secondary sensory cell; 3, secondary sensory cell; 4, dendritic branch; 5, ciliary rootlet.

bules. Opposite the receptor cavity of the primary sensory cell, in the dendrite of the secondary sensory neuron, there is a depression surrounded by large dark granules (Figure 51).

Metanemes evidently are a kind of proprioreceptor sensitive to the dorsoventral bending of the nematode body (Hope and Gardiner 1982). Individual dendritic branches, interpreted as proprioreceptors or as pressure receptors, are found in the hypodermis of various Rhabditia (Wright 1980).

The somatic pores and setae that are common among free-living marine Enoplia and Chromadoria should be considered the most primitive type of nematode sensory organ. These organs contain poorly specialized dendritic branches that are directly connected to the external environment. The sheath and socket cells in these organs are represented by slightly modified hypodermal cells. The somatic sensory organs, evidently, in some measure are able to accommodate both types of receptors (chemical and mechanical) and, during evolution, could have easily changed the type of reception, becoming the source for the formation of complex specialized sensory organs. It is interesting that the sheath cell in somatic organs, as a rule, has a more or less clearly defined glandular nature. A gland has significance for chemoreception but in some cases has independent significance, as is characteristic of Leptosomatidae (order Enoplida), which is devoid (perhaps primarily so) of a cervical gland (renette), and the excretory function is handled by glands connected to somatic setae.

In comparing the structure of nematode sensory organs with the sensory organs of other animals, one cannot help noting the similarity of the somatic setae of free-living marine nematodes and the so-called adhesive tubules (Haftrörchen) in Gastrotricha. Adhesive tubules, about 200 in number, extend over the whole body of Macrodasyoidea on the lateral sides. The adhesive tubules are covered by cuticle, covering a complex of three cells. One cell is sensory and has a long cilium, and from it extends an axon directed to the trunks of the nervous system. A glandular cell forms a long process (up to 10 μm), at the tip of which there is a vacuole and a pore through which a secretion is emitted to the external environment. The third cell is called a supporting cell but, in reality, it is also glandular. The secretion is emitted through an ampulla at the tip of the long process. In the cytoplasm of the glandular and supporting cells are many microtubules. The adhesive tubules are an organ of dual function. The sensory cell perceives a stimulus, and the glandular and supporting cells issue a sticky secretion that helps the animal cling to grains of sand.

The somatic sensory organs of nematodes and gastrotrichs come from the sensory-glandular organs scattered in the ciliary epithelium of their common ancestor (Figure 52). The acquisition of cuticle caused the reduction of the ciliary epithelium, the rudiments of which are preserved in nematode sensory organs in the form of dendritic branches, and the structure of which is very similar to these cilia. Nematode somatic sensory organs are not highly specialized structures, so their future evolution can occur in several directions. If a seta with an opening at the tip is capable, in principle, of reacting to both chemical and mechanical stimuli, then, to transform into a purely tangoreceptor, its opening at the tip has only to close to inhibit the entry of chemical stimuli (Figure 52d and e). On the other hand, to transform the original unspecialized organ into a pure chemoreceptor, it is necessary to isolate it from mechanical effects by submerging it into a depression in the cuticle (Figure 52c). Here, the entry of chemical agents into a cuticular pocket and from there to dendritic branches is maintained. Here also, a gland (sheath cell)

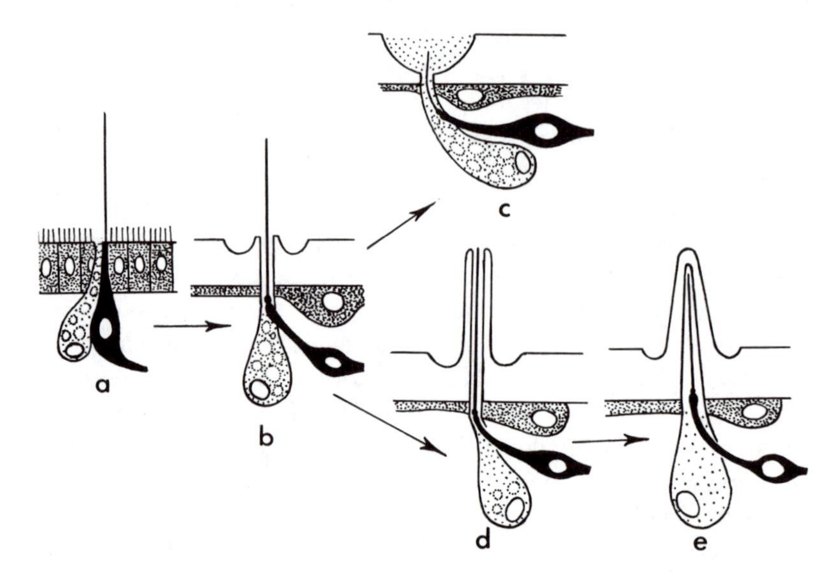

Figure 52. Hypothetical evolutionary path of nematode sensory organs: a, epithelial sensory-glandular organ; b, somatic pore; c, amphid; d, somatic setae; e, mechanoreceptive seta.

necessary for chemoreception is preserved and specialized. The described evolutionary path probably led to the formation of specialized olfactory organs in nematodes—the amphids.

The arrangement of somatic organs on the nematode body is relatively disorderly and is subject to individual variation. Nevertheless, as a whole, somatic setae are concentrated in two pairs of sublateral rows, and their arrangement is subject to the general biradial symmetry of the nematode body wall. The arrangement of cephalic sensory organs of nematodes is subject to strict regularity and makes a deep imprint on the symmetry of the whole nematode organism. The problem of nematode cephalic structure has been discussed for a long time (Schneider 1866; Filip'jev 1921; Chitwood and Wehr 1934; Chitwood and Chitwood 1950; Paramonov 1962; Belogurov and Belogurova 1975a, b; and others). However, to solve this problem, up to now, there has been insufficient factual material pertaining to the internal structure of the sensory organs and especially the development of the head structures in ontogeny.

Within the class Nematoda, there is a certain variation in cephalic structure. For Enoplida, the following arrangement of cephalic sensory receptors is most common. There is an inner circle of six labial, papilliform sensory receptors situated around the oral aperture. Posteriorly from them lies a circle of ten cephalic setae: four pairs of submedial (two pairs of subdorsal and two pairs of subventral) and one pair of lateral (Figure 53). Posteriorly from the lateral setae are situated the amphids. Such an arrangement of receptors of the head region is supported by data from scanning electron microscopy (Figures 54 and 55). The oral aperture, as a rule, lies apically. Its shape depends on the shape of the labia. In Leptosomatidae the labia are not developed, and the slightly expanded pharynx directly adjoins the anterior end of the body. The shape of the oral aperture, therefore, reflects the triangular shape of the pharyngeal lumen. In Enoplidae and a number of other forms, we find three labia, of which one is dorsal and two are subventral (Figure 54). In Oncholaimidae, the oral aperture is surrounded by six labia: two subdorsal, two subventral, and two lateral (Figure 55).

All this taken together creates a complex combination of various forms of symmetry. The oral aperture is subject to tri- or hexaradiate symmetry, the labial papillae are subject to hexaradiate symmetry, and the cephalic setal crown and amphids, to biradial symmetry. For the cutaneous-muscular sac of the preneural body section, quadri- or biradial symmetry is characteristic (see above). The least order of symmetry in the labial apparatus and labial papillae is three, and the least order of the complex of cephalic setae and amphids is two. The combination of bi- and triradial symmetry provides symmetry of the order of one, that is, bilateral symmetry. The plane of the bilateral symmetry of the nematode head end coincides with the sagittal plane of the animal. This is not a coincidence; it shows that the radial symmetry of the various elements of the head end was formed on the basis of previous, common, bilateral symmetry, characteristic of the nematode organism as a whole and inherited from bilaterally symmetrical ancestors.

However, such a head receptor arrangement is hardly primary for nematodes. Even within the order Enoplida there are forms with the cephalic setae arranged in two circles: six in the anterior one and four in the posterior (Figure 53). The two circles of setae are situated far apart in nematodes of the family Oxystominidae, or are situated at a short distance (certain Enoplidae and the family Anoplostomatidae). Thus, in all, there are three circles of receptors: (1) a circle of six labial papillae, (2) a circle of six cephalic setae, and (3) a circle of four cephalic setae. Posteriorly from where the missing lateral setae of the third circle would be found are situated the openings of the amphids. What type of

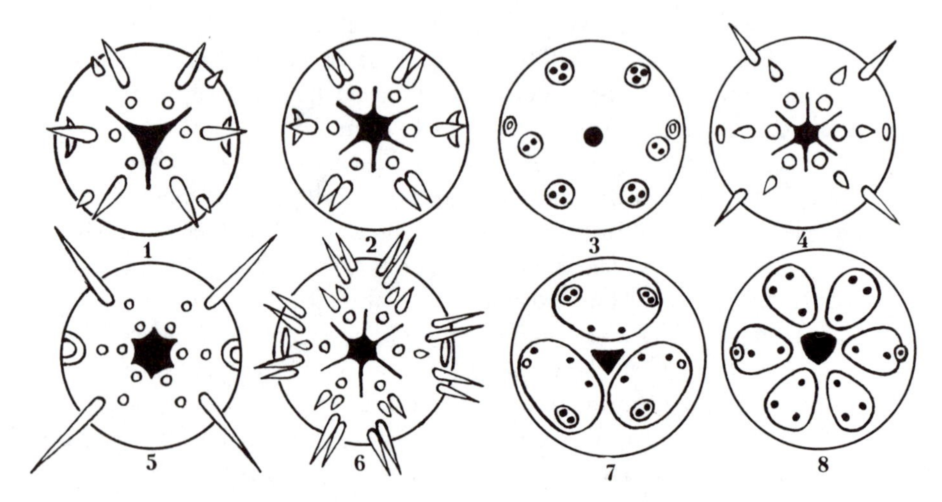

Figure 53. Examples of the arrangement of cephalic sensory structures in nematodes: 1, *Enoplus;* 2, *Oncholaimus*; 3, *Mermis;* 4, *Desmodora;* 5, *Axonolaimus;* 6, *Sphaerolaimus;* 7, *Ascaris;* 8, *Rhabditis.*

head receptor arrangement is the original one, two or three circles? What number of labia is primary for nematodes, three or six? What type of symmetry determined the establishment of the cephalic receptor complex?

To answer such questions it would be desirable to also have data on the ontogeny of cephalic structures. Interesting data can be obtained by investigating the development of *Pontonema*. In a 16-day embryo, three circles of head receptors can be observed. All three circles have the shape of tubular setae (Figure 56). The bases of the setae are situated in small depressions. The head end in the embryo is covered with a mucous hood (Figure 56). In the first circle there are six setae, in the second, there are six, and in the third, four. At the point of the missing lateral setae lie rudiments of amphids. In a 25-day embryo, the receptors of the first circle still have the appearance of setae, but they are significantly shorter than the setae of the second and third circles (Figure 56). The arrangement of cephalic receptors in three circles is retained also in the

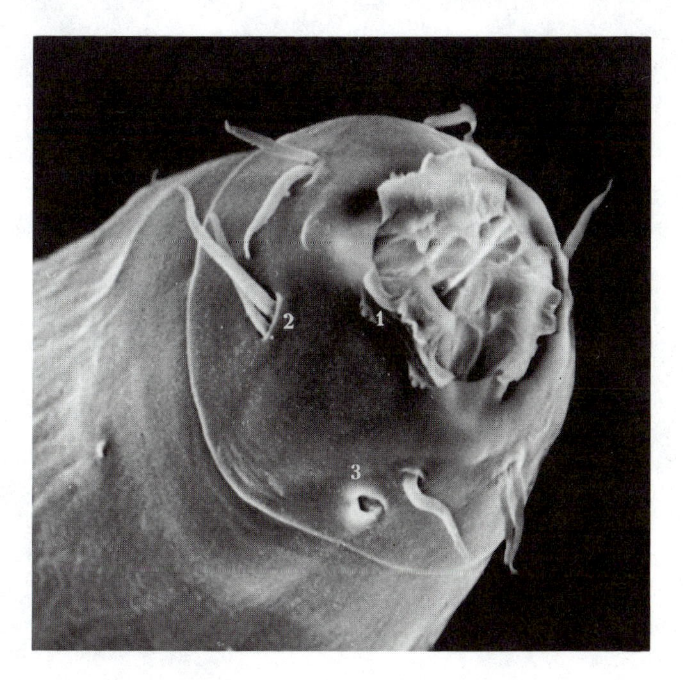

Figure 54. Arrangement of sensory organs on the head of *Enoplus*. Magnification = 1,500. 1, labial papillae; 2, cephalic setae; 3, amphids.

juveniles, but the first circle of receptors is differentiated as labial papillae (Figure 18). An important feature of the structure of the first-stage juvenile is the presence of three labia (Figure 18).

As has been noted, adult specimens of *Pontonema* have six labia and two circles of cephalic receptors. Also noted above was the great internal similarity in structure between labial papillae and cephalic setae. The fact that there are three labia in juveniles and the cephalic receptors are in the form of setae in three circles evidently has significance in

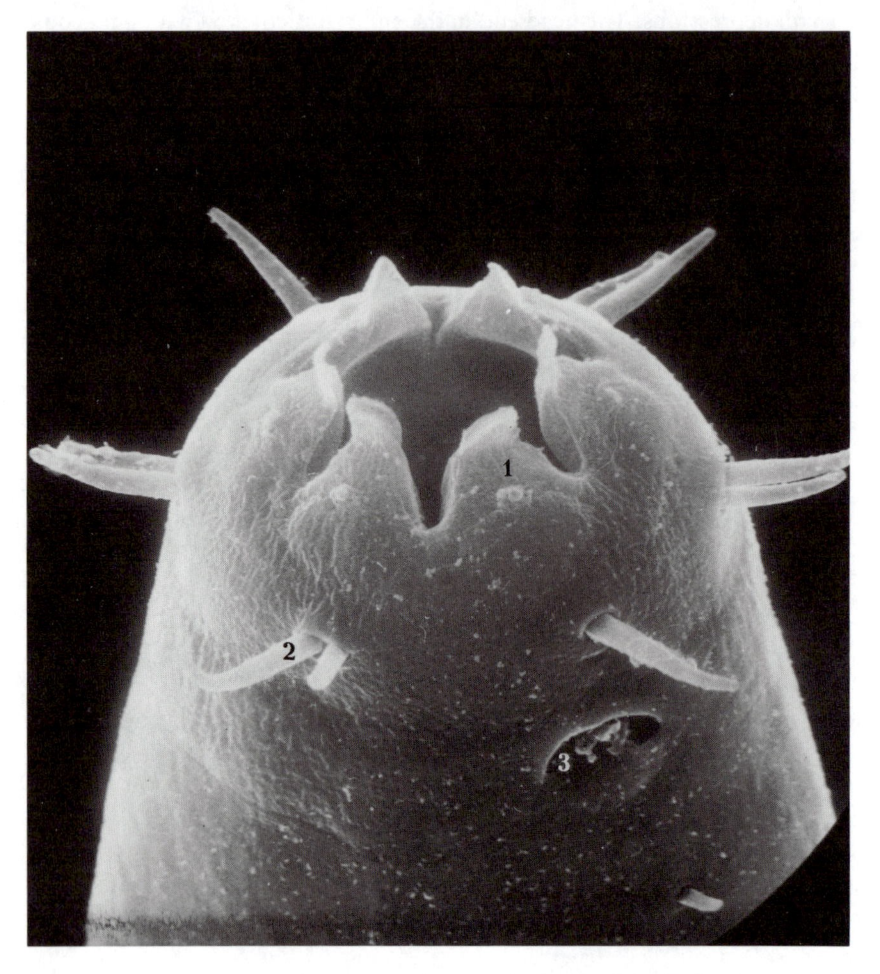

Figure 55. Arrangement of sensory organs on the head of *Oncholaimus*. Magnification = 1,750. Labels as in Figure 54.

recapitulating the original ancestral state. In this connection, it is interesting to note that among many representatives of the order Enoplida there are three labia. In general, the evolution of the labial apparatus can be interpreted as follows. The original state can be considered as one in which labia were absent and the triangular contour of the mouth simply corresponded to the lumen of the pharynx. Such a state is present in nematodes of the family Leptosomatidae. The subsequent stage is the appearance of three labia in places corresponding to the muscle sectors of the pharynx. Finally, the increase in the mobility of the labial apparatus required further elaboration of the labia, and this is how the six-labial forms appeared.

Among many representatives of the order Enoplida (*Enoploides, Enoplolaimus,* and others), the first circle of receptors has the appearance of setae. This has been found also among certain representatives of Monhysterida (certain *Theristus* and *Cobbia*), and Desmodorida (*Nudora*). Cephalic receptors in the configuration of three circles of setae evidently represents a recapitulation of the primitive state and once again shows the complete homology of setae and papillae. The evolution of cephalic receptors can be interpreted as follows:

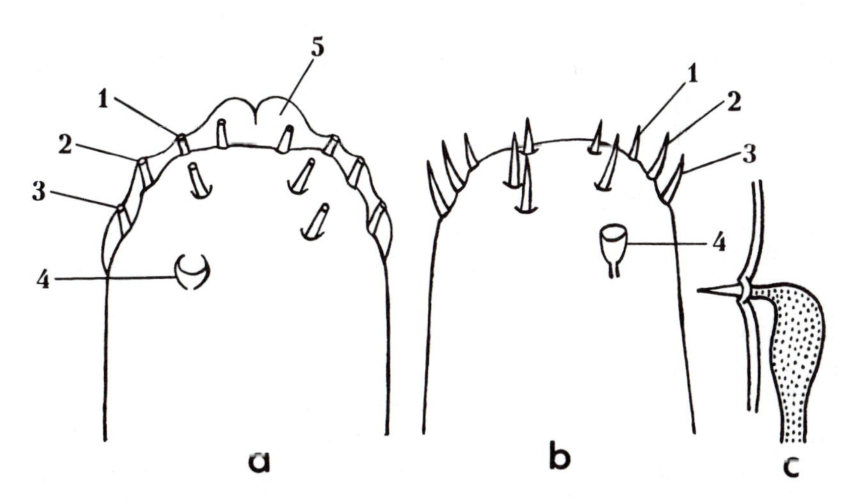

Figure 56. Development of sensory organs in ontogeny of *Pontonema vulgare:* head of 16-day embryo (a); head of 25-day embryo (b); cervical gland pore of 25-day embryo (c). 1, labial papilla; 2, anterior circle of cephalic setae; 3, posterior circle of cephalic setae; 4, amphids; 5, mucous covering.

Originally, cephalic receptors were arranged in three circles (6 + 6 + 4) and had the appearance of setae. In further evolution, there was a process of receptor concentration and differentiation. The six receptors lying at the most anterior end of the head became shortened and transformed into labial papillae. The setae of the third circle gradually shifted anteriorly and finally united with the setae of the second circle, making a total of ten setae in one circle. Within the order Enoplida, various degrees of the evolution of this process can be seen in the families Oxystominidae, Anoplostomatidae, and Enoplidae. Usually, even in cases where the setae are arranged almost in one circle, their origin is evident in their different lengths. Among representatives of various groups of Enoplida (Enoplidae, Anoplostomatidae, Enchelidiidae, and some Oncholaimidae) six setae of the second circle are longer than the four setae of the third circle (Figure 53).

Among nematodes of other orders, the arrangement of cephalic receptors in three circles is retained. Among a majority of Chromadorida we encounter a circle of six labial papillae and the second circle of receptors is also represented by six papillae and the third circle by four setae (Figure 53). In Desmodorida is found a primitive state where all three circles of receptors are represented by setae (*Chromaspirina*), but usually the first two circles are represented by papillae and in the third are four more or less long setae.

Four long setae in the third circle is characteristic for members of Desmoscolecida and many Araeolaimida (Figure 53).

In Monhysterida are encountered all of the above variants, the arrangement of sensory organs (in two or three circles), and the ratio of setae to papillae having little taxonomic weight, which, perhaps, indicates the shortcomings of the systematics of the order. Here, an interesting phenomenon is observed, where cervical setae are united with cephalic setae, forming six to eight groups of a few each (Figure 57). In some cases, the circle of cephalic setae is reduced and is mixed with groups of cervical setae (*Anticyathus, Steineria*).

The organization of the head end among representatives of various groups of free-living nematodes, of course, deserves significantly more detailed examination; however, this would require too much space. We will note only that within the confines of almost every order there are both primitive and advanced forms. The evolution in cephalic structure among various groups of free-living nematodes proceeded in parallel, preserving certain common tendencies. The chief evolutionary trend was the concentration and specialization of receptors. Thus, characteristic of Enoplia are the differentiation of the first circle of receptors as labial papillae and the merger of the second and third circles into one crown of ten cephalic setae. Chromadorida, Desmodorida, Desmoscolecida, Araeolaimida, and some Monhysterida are characterized by the transformation of the first and second circle receptors into papillae and by the preservation of four setae of the third circle (although in a number of families there is also a crown of ten cephalic setae). The arrangement of receptors in two circles of the 6 + 10 type is also preserved among parasitic representatives of the Enoplida (orders Mermithida, Trichocephalida, and Dioctophymida), although the receptors themselves have the appearance of papillae (Figure 53).

It is characteristic of Rhabditia to have a reduction of the primary

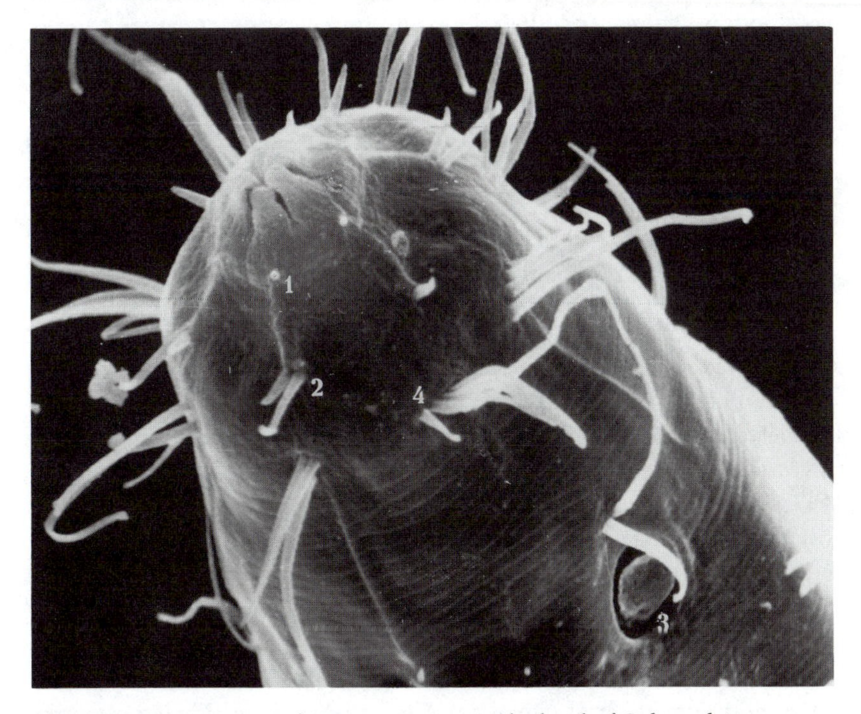

Figure 57. Arrangement of sensory organs on the head of *Sphaerolaimus*. Magnification = 1,500. Labels as in Figure 54 except as follows: 4, groups of cervical setae.

labia, a transformation of all receptors into papillae, and a further concentration of receptors at the head end. The so-called "labia" of Secernentea represent a product of growth of head tissues and bear not only the receptors homologous to labial papillae, but also the receptors homologous to both circles of cephalic setae. Thus, for example, three secondary labia are characteristic of *Ascaris*. The real labial papillae are reduced (retaining only the corresponding nerve endings); on the dorsal labium lies a pair of double papillae; on the subventral labia, one double and one single (lateral); and on the subventral labia are situated the amphidial openings (Figure 53). Thus, the scheme of arrangement of *Ascaris* sensory organs completely fits that typical of nematodes (i.e., 6 + 10) including labial papillae.

This work does not permit examination of the evolution of cephalic structures among members of Rhabditia. Let us note only that, as in other groups of nematodes, the evolution of cephalic structures in this subclass proceeds in parallel in various orders. The evolution of cephalic receptors is characterized here by their concentration at the most anterior end of the body, their specialization, and especially in the parasitic forms the gradual reduction of the receptors, beginning from the anterior circles.

The cephalic structure that we examined above as the most primitive for nematodes itself represents a product of long evolution. In the nematode cephalic structure it is as though there are two tiers of structure with different types of symmetry: the first and second circles of receptors together with the labial apparatus are subject to triradial symmetry and the setae of the third circle and amphid to biradial symmetry.

From the time of Schneider (1866), who formulated the principle of triradial symmetry in the nematode anterior end, various views have actively developed as to the nature of the general type of symmetry for all structures of the nematode anterior end. According to Chitwood and Chitwood (1933, 1950) and Chitwood and Wehr (1934), the original nematode head structure exhibited hexaradiate symmetry: the oral aperture was surrounded by six labia and there was an inner circle of six papillae (one on each labium) and an outer circle of 12 papillae (two to each labium), plus a pair of amphids. However, as the authors of this concept themselves noted, there is no nematode in which the rudiments of an additional pair of lateral papillae have survived. Therefore, in the opinion of de Coninck (1965) and Crofton (1966) the most ancestral form probably consisted of six labia, six papillae in an inner circle, six outer papillae, four cephalic papillae, plus a pair of amphids; that is, a symmetry of mixed character.

The reviewed points of view, however, did not take into account yet another possibility. Perfect hexaradiate symmetry in the arrangement of cephalic sensory organs is lost by the absence of an additional pair of lateral setae. But just at the place for this missing pair lie the amphids! This thought evidently occurred many times to various nematologists because it was so obvious, but only Andrassy (1976) ventured to defend it. Of course, at first glance, it is difficult to accept—the amphids and papillae are too different. However, as we have seen, amphids came from the same common prototype as papillae and setae (Figure 52). Judging only by morphology, the primitive pocketlike amphid is a seta or papilla implanted in a depression of the cuticle. All the varieties of amphids among the most evolutionarily advanced nematodes is a result of their further specialization as chemoreceptors.

Remote nematode ancestors had unspecialized sensory-glandular organs (like the tubiform somatic setae of Leptosomatidae) scattered without any special order over the whole anterior end (Figure 58). The oral aperture repeated the triangular form of the lumen of the pharynx, attached directly to the anterior end of the body.

The next stage was the ordering of the arrangement of the sensory-glandular setae lying close to the anterior end, forming three tiers of receptors and triradial symmetry in the organs of the pharyngeal system, causing the same order of symmetry in the circles of receptors (six receptors in each circle). The further evolution of the receptor apparatus of the head end is related to the differentiation of the receptors: the lateral receptors of the third circle were submerged into depressions of the cuticle and took on the function of remote chemoreception. The remaining receptors fulfilled the function of tangoreceptors and contact chemoreception. In the next stage, which we can observe in a number of presently living forms, there is a concentration of receptors at the head end and a merger of the second and third circles into one circle of ten setae (Figure 58). This process can go even further, resulting in the transfer of cervical setae to the head end (order Monhysterida). Among evolutionarily advanced nematodes, further specialization of cephalic sensory organs is observed, reaching the highest expression among certain parasitic representatives of the subclass Rhabditia, where all cephalic sensory organs are transformed entirely into tangoreceptors, and only the amphids remain chemoreceptors.

The combination in one organ of both sensory and glandular elements also makes possible an evolution in the direction of the intensification of the glandular function. Thus, the increased development of glands simultaneously with the structural transformation of the somatic setae themselves led to the formation of the special attachment setae characteristic of nematodes of the order Desmoscolecida and the family Draconematidae (Schepotieff 1908a, b; Shepot'yev [Schepotieff] 1908; Steiner 1916; Filip'jev 1921; Stauffer 1924). At the base of these setae lies a rather strongly developed gland, the sticky secretion of which is emitted through a pore at the tip of the setae. Sometimes at the end of the setae there is a broadening in the shape of a plunger or goblet. The

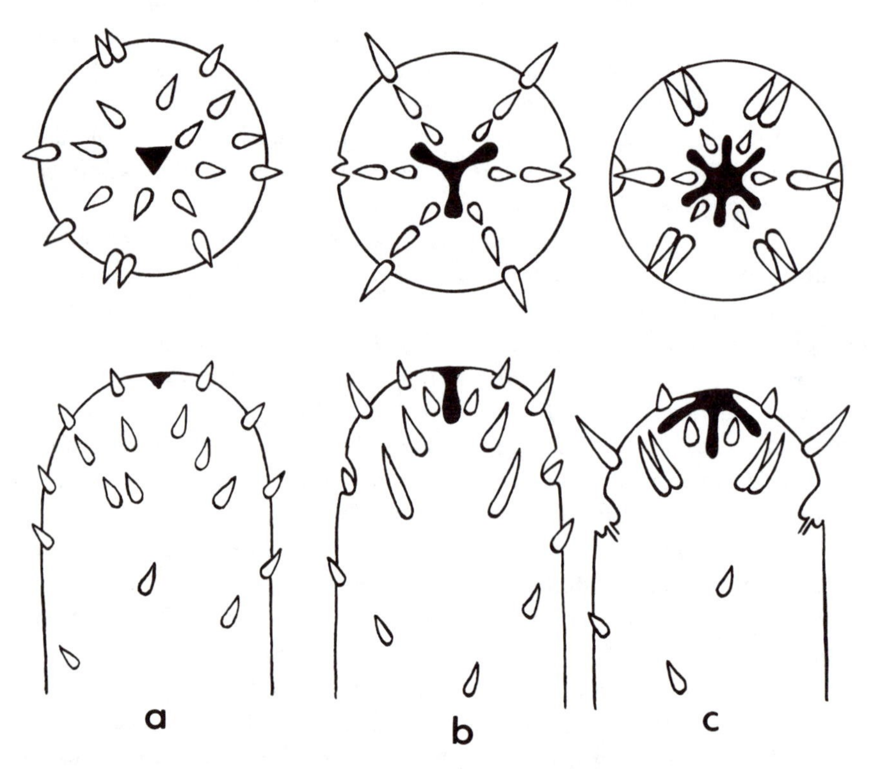

Figure 58. Hypothetical evolutionary path of sensory organ arrangement at the head end of nematodes: a, unordered arrangement; b, arrangement of sensory organs in three circles and the beginning of differentiation in lateral receptors of the third circle as chemoreceptors; c, 6 + 10 arrangement of sensory organs typical of marine nematodes.

structure of such setae is understood from the type of locomotion of these nematodes. Thus, nematodes of the family Draconematidae, while attaching themselves to substrata alternately with anterior and posterior groups of setae, move like a caterpillar (Stauffer 1924). The intensification of the glandular function also takes place in the male supplementary organ, which simultaneously retains its sensory function. The gland secretion evidently has a sticky property and provides attachment of the male to the female during copulation.

Of course, not all glandular structures in nematodes are necessarily derivatives of sensory-glandular organs. An abundance of glandular structures is characteristic of the integuments of primitive Bilateria. In Turbellaria we find various epidermal glands. In Gastrotricha, along with sensory-glandular attachment glands, there is also a system of cutaneous glands. However, the combination of sensory and glandular formations is so characteristic of nematodes that, even in such purely glandular nematode organs as the cervical gland or renette, it can be suspected that it is a transformed somatic seta. Among Enoplida, the cervical gland is a massive cell from which a duct extends to the excretory pore situated on the ventral side anterior or posterior to the nerve ring. A large nucleus contains one or two nucleoli, and a strong Golgi apparatus participates in forming secretion granules (Narang 1970). The duct is connected to the pore by a cuticularized tube. In the vicinity of the pore the cuticle forms a depression devoid of the striated layer (Narang 1970). The pore is surrounded by a special hypodermal cell as is seen in the study of the hypodermal mosaic. These features also belong to somatic setae. The homology of the cervical glands associated with somatic setae is also underscored by the fact that, in the ontogeny of *Pontonema,* at the location of the pore of the cervical gland is situated a seta connected with the cervical gland canal (Figure 56c).

Thus there is a basis for saying that the cervical gland is evolutionarily derived from a single-cell gland associated with a somatic seta. The further evolution of the cervical gland is related to the intensification of its secretory function. This at once reflects on the morphology of this gland. It increases in size and has a large "nucleus" with a massive nucleolus. In a number of forms, there is a multiplication of glandular cells that constitute the renette. Thus, for example, in *Sabatieria,* the renette has three cells: one ventral and two subventral (Riemann 1977a). Paired glandular cells are described in the renettes in *Anticyclus* of Monhysterida, *Prionchulus* of Mononchida, and even in the juveniles of Mermithida (Kaiser 1977). The list of such examples could be continued.

Intensification of function of the cervical gland leads to change in its form. For a majority of Enoplia and Chromadoria, a simple saclike cervical gland is characteristic, but in some Enoplida and Monhysterida it becomes paddle-shaped (Figure 59).

The transition to habitation in soil and in fresh water causes a functional shift of the cervical gland and leads to the morphological complication of the cervical gland—the formation of a canal system. The canal of the renette in Rhabditia lies in the lateral chords of the hypodermis. The cervical gland in Rhabditia is a complex formation containing several nuclei. In some cases (Figure 59g and i) the renette has two secretory cells, and the whole system consists of two symmetrical halves. In other cases, the strictly secretory gland is singular, but there are additional nuclei corresponding to the reservoir and secretory duct (Figure 59l).

Chitwood and Chitwood (1950) considered the branched cervical gland of Rhabditia original to nematodes, but Steiner (1921) even compared it to the protonephridial excretory system of flatworms and rotifers. This comparison, of course, is incorrect. The branched cervical gland of Rhabditia undoubtedly comes from the usual hypodermal gland, just as does the saclike cervical gland of Enoplia and Chromadoria. In the ontogeny of some Rhabditia, the simple saclike cervical gland precedes the complex branched excretory system. Thus, the preparasitic juveniles of *Cucullanus* have a simple saclike cervical gland, whereas the adult forms have a complex H-shaped cervical gland (Figures 50a and 59d). The formation of the branched renette evidently demonstrates one of the means for solving the problem of osmotic regulation in the transition from life in the sea, where a majority of Chromadoria live, to habitation in fresh water and soil moisture. It is interesting that the tendency to form a branched cervical gland has been traced in certain freshwater and soil Plectida, from which, most probably, come all the basic ducts found among members of Rhabditia. Besides the cervical secretory system in certain Rhabditia, an additional excretory system has developed in the posterior half of the body (Figure 60).

Just as in the case of the cervical excretory system, the caudal excretory system is represented by canals lying in the hypodermal chords; and evolutionarily, it comes from one of the hypodermal glands of the

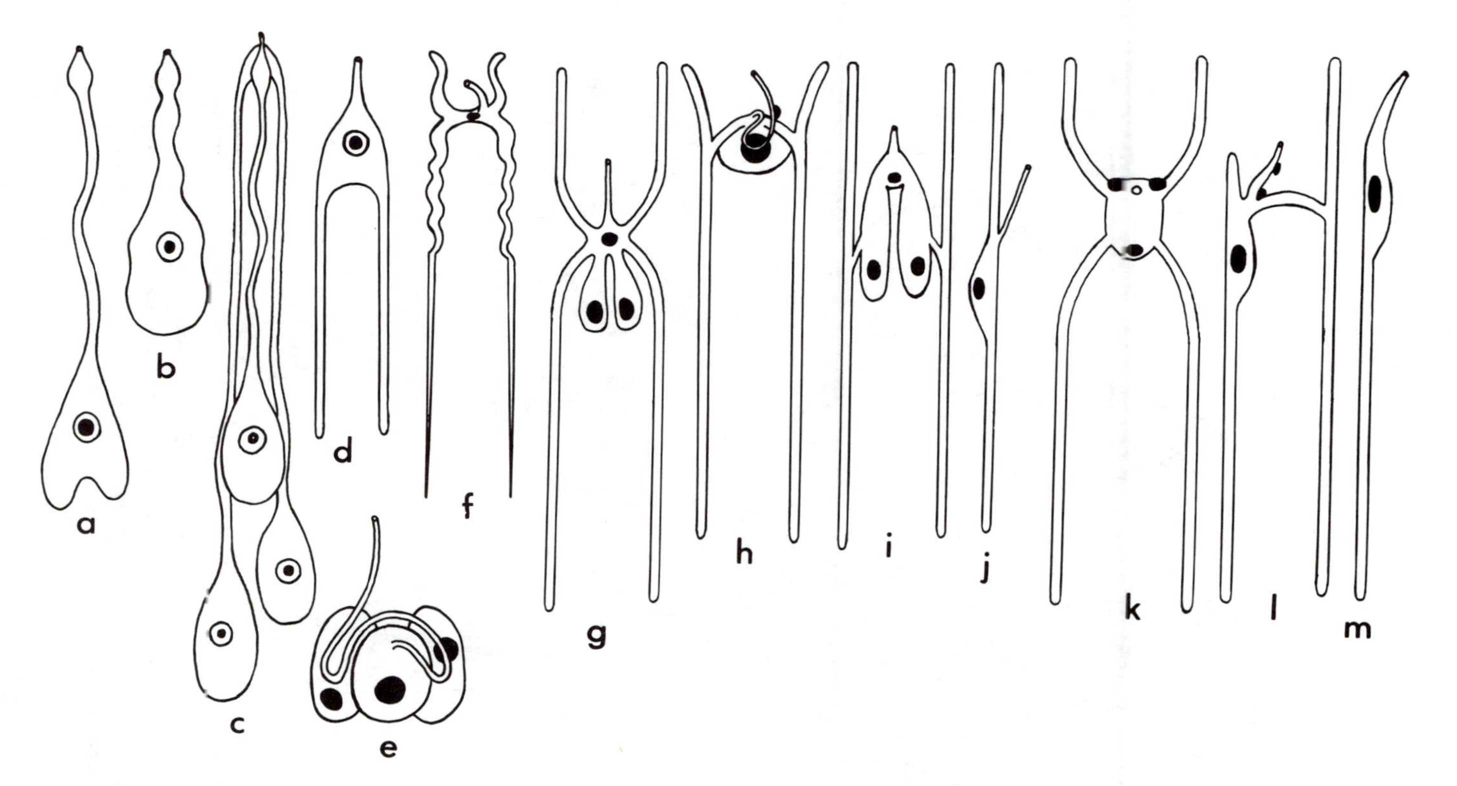

Figure 59. Types of excretory systems in nematodes (after Riemann 1977a [c]; original [k]; remaining figures after Chitwood and Chitwood 1950): a, *Phanodermopsis longisetae;* b, *Chromadora quadrilinea;* c, *Sabatieria celtica;* d, *Anonchus mirabilis;* e, *Anaplectus granulosus;* f, *Camallanus lacustrus;* g, *Rhabditis strongyloides;* h, *Rhabditis dolichura;* i, *Oesophagostomum dentatum;* j, *Tylenchida* sp.; k, *Hammerschmidtiella* sp.; l, *Ascaris* sp.; m, *Anisakis* sp.

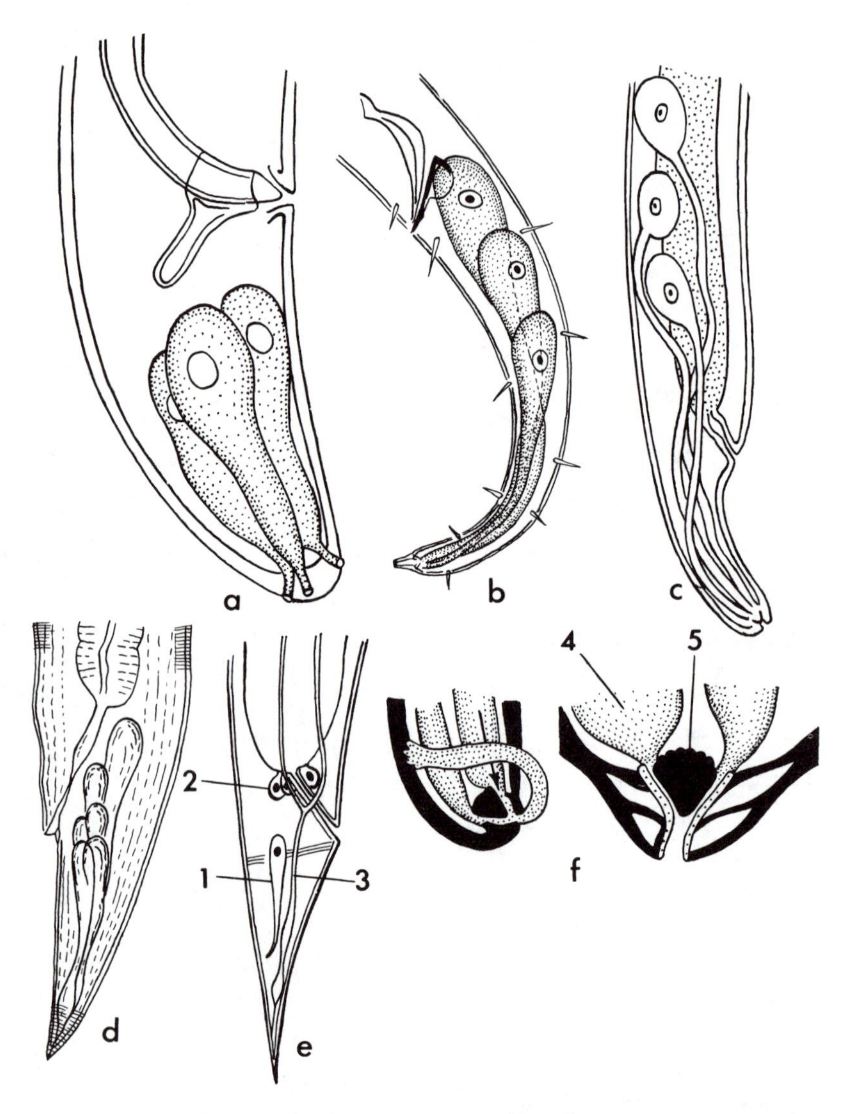

Figure 60. Caudal gland of nematodes (after de Coninck 1965 [b and d]; after Belogurov and Listova 1977 [c and f]; original [a and e]): a, *Diplopeltula;* b, *Chromadora;* c, *Oncholaimus;* d, *Anisakoides;* e, *Rhabditis;* f, operation of spinneret in specimen of Enoplida. 1, phasmids; 2, rectal glands; 3, canals of the accessory excretory system; 4, ducts of caudal glands; 5, caudal cone.

posterior half of the body. Characteristic of free-living nematodes of the subclasses Enoplia and Chromadoria are caudal glands that form a complex organ, the spinneret. The function of the spinneret is the excretion of a sticky substance that helps the nematode cling to substrata. Recently it has been shown that the secretion of the caudal gland is used by free-living nematodes to stick to particles of detritus, which the nematodes then eat (Riemann and Schrage 1978).

The cytology and ultrastructure of the caudal glands are close to those of the cervical gland. In the cytoplasm of the caudal glands, endoplasmic reticulum, a Golgi complex, is strongly developed, and the terminal sections of the cell have microvilli (Lippens 1974a). Opinions on the question of the organization of the terminal apparatus of the spinneret contradicted until recently (Golovin 1901; Filip'jev 1921; Chitwood and Chitwood 1950). Belogurov and Listova (1977) partially clarified this matter. The ducts of the canal glands open into the cuticular canal (ampulla), which can close tightly from the inside like a cork. It is a special cuticular structure of the caudal cone (Figure 60f). Muscles are attached to the caudal cone that can extend it, and thus the glandular secretion pours into the lumen of the ampulla. When relaxing the muscle, the caudal canal under internal pressure plugs the lumen of the ampulla from inside (Figure 60f). According to Belogurov and Listova (1977), the apparatus of the spinneret was formed by the concentration of hypodermal glands. Actually, in certain nematodes of the order Araeolaimida, each of the three caudal glands opens independently, and the complicated terminal apparatus of the spinneret is absent (Figure 60a). With the concentration of caudal glands, a section of cuticle lying between the pores of the glands is submerged inside and turned into the cuticular tail cone.

The body of the caudal glands can lie within the tail section (Figure 60b) or extend anteriorly into the region of the gut and gonads (Figure 60c). In parasitic Enoplia and all Rhabditia, the caudal glands are reduced. Nevertheless, glandlike structures in the tail of certain Anisakoidea can be regarded as the rudiment of a spinneret (Figure 60d).

Body Cavity and Phagocytic Cells

The presence of a wide body cavity is considered one of the more characteristic indications of the class Nematoda. The nematode body cavity is called a primary cavity, a schizocoel, or pseudocoel, to distinguish it from the secondary body cavity or coelom of higher Bilateria. The name schizocoel is the most preferable, as it better reflects the origin of this cavity and is commonly accepted in current zoological literature.

The schizocoel does not have its own walls of cellular material, and it represents an interspace between organs filled with liquid. The schizocoel is ascribed an important role in the creation of internal pressure, the circulation of fluids with their dissolved substance, and the maintenance of the organism's internal environment.

However, it should be recalled that the idea of the schizocoel as a widened body cavity filled with liquid arose from research on large parasitic nematodes. As early as the beginning of this century, in research on free-living marine nematodes of the order Enoplida, it was found that such a body was absent in these nematodes (Linstow 1900; Jägerskiöld 1901; Stewart 1906). Free-living nematodes are small worms; therefore, their body organs lie close to each other. The internal spaces remaining between the organs are filled with a dark-colored substance of an unknown nature (Linstow 1900; Jägerskiöld 1901; Stewart 1906; Korotkova and Agafonova 1975). The problem of schizocoel structure in primitive nematodes requires solution by contemporary methods.

With the aid of a light microscope, the schizocoel studied in *Pontonema* has a homogenous appearance with small (0.8–1.6 μm) inclusions. With an electron microscope, the schizocoel content looks like a noncellular substance that fills all the interspaces between organs and individual cells (Figure 61). In it there sometimes occurs electron-translucent spheres in which osmophilic materials may occur. Here and there are noted drops of osmophilic substance and sections of bundles of innervating processes of muscle cells.

The use of the high magnification of an electron microscope shows that the basic substance of the schizocoel consists of a thick network of irregularly arranged, thin, interwoven fibers. Histochemical research shows that the basic substance of the schizocoel contains total and basic proteins and neutral mucopolysaccharides (Malakhov and Agafonova 1978). The presence of proteins and mucopolysaccharides represents a rather typical combination that is characteristic of the basic substance of connective tissue in animals, but the schizocoel lacks any structured fibrils (collagenous or elastic) that are also typical of connective tissue. Evidently a similar structure is also present in the homogenous content of the schizocoel revealed at the optical level in other marine nematodes

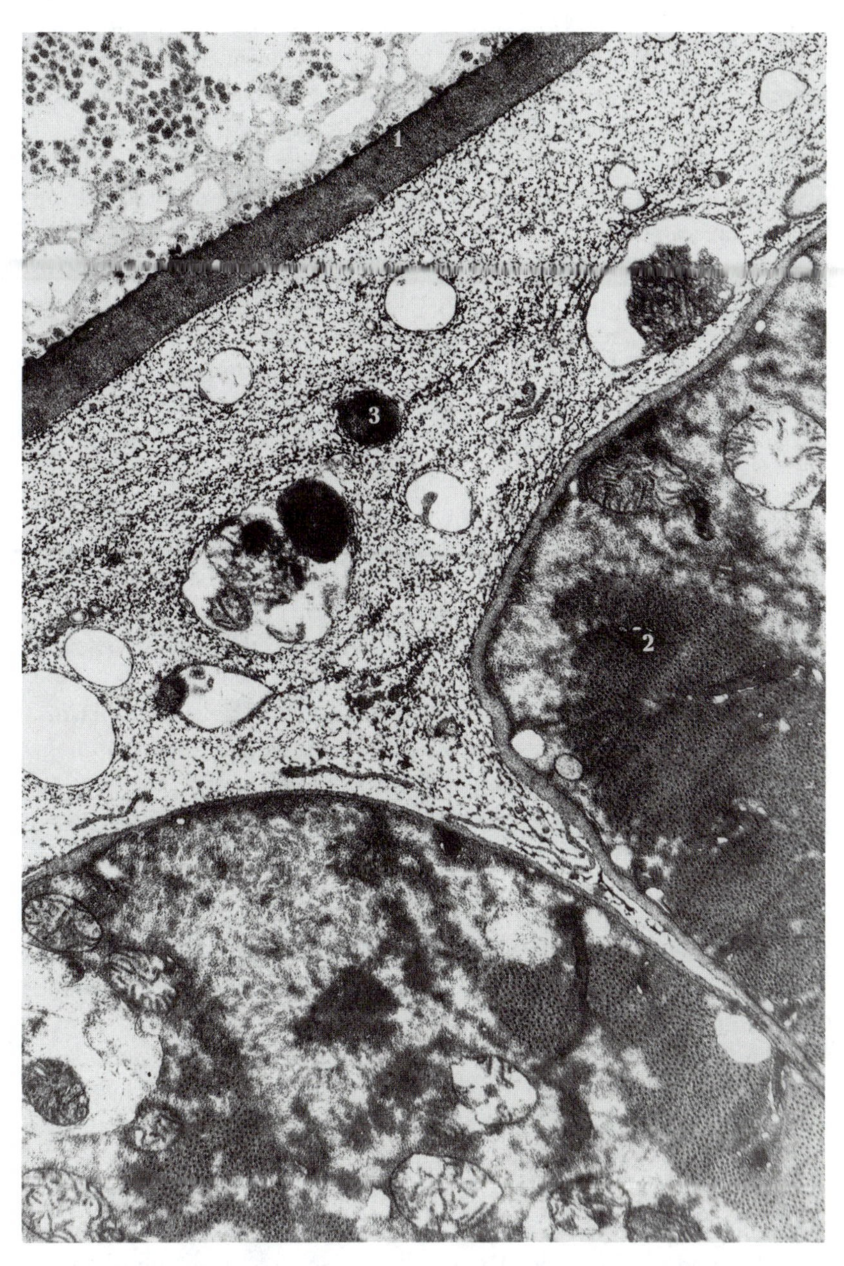

Figure 61. Structure of the basic substance in the schizocoel of *Pontonema vulgare*. Magnification = 11,000. 1, basal membrane of the intestine; 2, muscle cells; 3, basic substance.

(Linstow 1900; Jägerskiöld 1901; Stewart 1906; Korotkova and Agafonova 1975). Analogous content has been found in ultrastructure research of the body cavities of other marine nematodes of the order Enoplida—*Deontostoma* (Siddiqui and Viglierchio 1977).

Among most nematodes, the schizocoel does not contain significant accumulations of dense substance. Small free-living and parasitic nematodes have a schizocoel in the form of narrow slit-shaped spaces between organs. In large parasitic nematodes, the schizocoel is a widened cavity filled with liquid. However, these nematodes also have been described as having strands uniting the internal organs to the body wall (Chitwood and Chitwood 1950). With electron microscopic study, these strands reveal a great similarity to the elastic fibers of the connective tissue of vertebrates (Wright et al. 1972).

Also a component of the body cavity are cellular elements: the so-called coelomocytes and phagocytic cells. The number of these cells varies from a few, among free-living marine nematodes of the order Enoplida, to several among parasitic forms. Among free-living forms, the best-studied body-cavity cells are within representatives of the order Enoplida. With light microscopy, these cells look like round structures largely associated with the hypodermal chords.

The ultrastructure of these cells is distinctive. The cell cytoplasm is almost wholly filled with round vacuoles with electron-dense homogenous contents (Figure 62). The nucleus is pushed to the periphery and the chromatin is concentrated in lumps. Mitochondria and other organelles are virtually absent.

The cells termed coelomocytes or phagocytic cells easily absorb vital dyes, which accumulate in their cytoplasm in the form of chromophilic spheres (Shimkevich 1898; Golovin 1901). The reaction of these cells is acidic. Also, the absorption of dyes by these cells has provided grounds for calling them phagocytic.

The number of phagocytic cells is sharply reduced in nematodes of the subclass Rhabditia (Chitwood and Chitwood 1950). Reduction in the number of phagocytic cells leads to their increased complexity, an increase in size, and intensification of function, especially in the case of parasitic nematodes. In *Ascaris* we find a total of four large cells shaped somewhat like a star situated in the anterior part of the body on the lateral hypodermal chords. These cells are able to absorb dyes and to phagocytize bacteria introduced into the body cavity (Nassonov 1900 and Metalnikoff 1923, as cited by Chitwood and Chitwood 1950). The

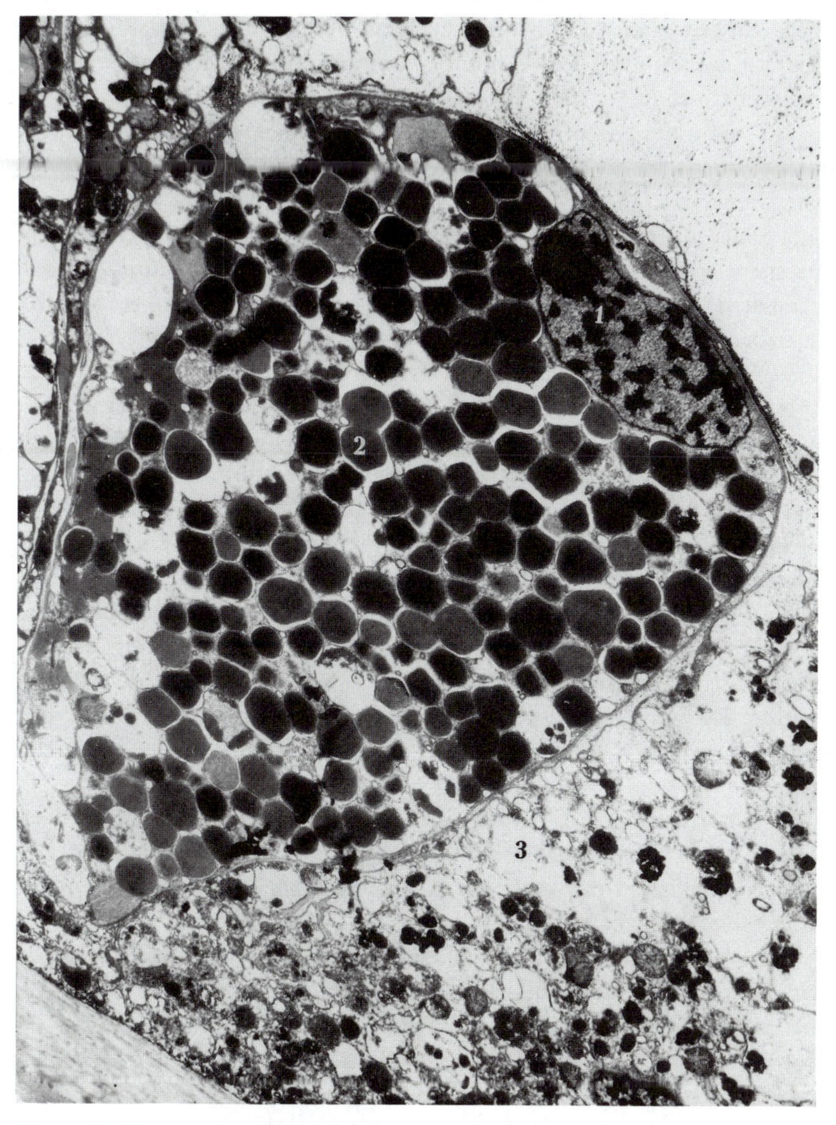

Figure 62. Phagocytic cells in *Pontonema vulgare*. Magnification = 7,500. 1, nucleus; 2, granules; 3, hypodermis.

function of the phagocytic cells in nematodes evidently consists of the absorption of certain metabolic products and foreign particles.

In addition, special branched "fat cells" around the intestine have been described in certain marine forms of the order Enoplida (Türk 1903; Rauther 1907, 1909), but the accuracy of these observations needs reaffirmation.

Thus, among lower nematodes, the schizocoel is still not developed, and this structure is evidently undergoing evolution within members of the Nematoda. There is also the question of high internal pressure usually believed to be created by the schizocoel. In reality, the high internal pressure is not necessarily related to the presence of a more spacious schizocoel. Parasitic nematodes have high pressure in the liquid of the schizocoel, but such pressure, naturally, is also present in all tissues of the animal. Therefore, all tissues and organs of the worm create the pressure that is transferred to the cuticle. Nematodes have, as described, a special type of movement based on antagonistic interaction between the stretched cuticle and the longitudinal musculature, there being practically no changes in volume and, consequently, no significant transfer of liquid within the organism. The nature of nematode movement stands in sharp contrast with the peristaltic movement characteristic, for example, of coelomic worms. In the peristaltic means of movement there are significant redistributions of pressure and volume among various parts of the body and, consequently, substantial transfers of liquid within the body. Therefore, the presence of the inner cavity filled with liquid is a necessary condition for the peristaltic means of movement (Clark 1964).

The situation in nematodes is different. Here, the cavity is not necessary, but high internal pressure—the pressure of the liquid in the cavity or turgor pressure of tissues—is created all the same by either or both. It doesn't really matter which.

Many marine nematodes do not have a real body cavity, and the stretched state of the cuticle is maintained by the high pressure of internal tissues. The presence of such pressure is easy to demonstrate: by cutting across the nematode the contents of the intestine and sexual tubes are expelled into the external environment with great force. The organs themselves remain closely adhered to each other by a basic substance that fills all spaces between organs of the body.

In other small nematodes the basic substance is already reduced, the number of phagocytic cells is reduced, and narrow slit-shaped spaces of the schizocoel appear between organs. The tissue turgor and the pressure of the liquid in the narrow schizocoel cause the stretched state of the cuticle.

The enlarged, fluid-filled schizocoel occurring in the large parasitic nematodes should be considered a secondary phenomenon. Nematodes were originally small animals. Evidently the original small size of nematodes relates to the reduction of connective tissue elements: there simply wasn't space for the tissue in the narrow nematode body between the compressed organs. The secondary increase in body size in parasitic forms naturally could not recover the layer of parenchymatous connective tissue that was lost in evolution. That is why the wide space between the organs is filled with liquid.

Of course, the enlargement of the schizocoel had definite advantages that reinforced it in evolution. The wide schizocoel filled with liquid aids the even distribution of internal pressure to the cuticle. But in addition, a spacious schizocoel becomes a kind of liquid internal medium for the organism. The originally small nematodes have no special distributive apparatuses. With increase in body size in parasitic nematodes, the accomplishment of the distributive function becomes extremely complex. The voluminous schizocoel of large parasitic nematodes creates a kind of internal liquid medium for the organism and provides the transportation of food substances and metabolic products. Sometimes, as noted in Mermithida, the schizocoel becomes a place for storing protein granules (Poinar et al. 1970). Thus, the schizocoel of parasitic nematodes is becoming a polyfunctional structure, permitting them to reach unusually large sizes.

Digestive System

Nematodes have a complete digestive system. In their evolution the development of a complete gut created the conditions for development of cavity-type digestion. In contrast, the saclike intestine characteristic of lower Bilateria prohibits wholly cavity-type digestion. Only with a complete intestine in which food masses move in one direction is the possibility created for dividing the intestine into zones with various phases of and complete cavity-type digestion. Therefore, the acquisition of a complete gut was the most important morphological precondition for improving the physiology of digestion and the most important aromorphosis in the evolutionary development of invertebrates.

The nematode digestive tract is formed in three main sections: the ectodermal pharynx, the endodermal middle intestine, and the ectodermal posterior intestine.

At the present time, we have available theoretical and experimental investigations pertaining to the mechanism of the working of the pharynx (Mapes 1965a, b, 1966; Roggen 1970, 1973). The pharynx represents a pump that operates according to a special principle. Two concentric tubes have radially oriented muscular fibers between them that contract to open the lumen of the pharynx, counteracting the hydrostatic pressure within the nematode body. After relaxation of the muscles, and under the effects of the hydrostatic pressure within the body and within the pharyngeal walls, the lumen closes. The process of the contraction of the pharyngeal muscles goes from anterior to posterior, thus moving the food through the pharynx into the middle intestine. A reverse flow of food from the middle intestine to the pharynx is prevented by a muscular cardia valve.

A common type of pharyngeal structure and function remains constant within the boundaries of the whole class Nematoda. Independent of the type of nutrition (herbivores, predators, detritus feeders, and parasites) and arrangement of the oral cavity, the nematode pharynx is constructed according to a single principle, and the structure of the pharynx displays even greater constancy than the means of locomotion. The triradiate symmetry of the pharynx determines the symmetry of the whole anterior end including the structures of the oral cavity (mandibles, onchia, etc.), the labial apparatus, and the arrangement of sensory organs. The well-known principle of "triple symmetry" of the anterior end of nematodes, which was established by Schneider, has resulted from the triradiate form of the pharyngeal lumen, which is the only suitable arrangement for a muscular pump from a biomechanical point of view. The triradiate symmetry of the pharynx evidently is the oldest acquisition of nematodes, predetermining the type of symmetry for the whole head end. At the same time, the triradiate symmetry of the pharynx contrasts with the biradial symmetry of the cutaneous-muscular sac, and the combination of these two types of symmetry gives a symmetry of the first order, that is, bilateral symmetry. The plane of the bilateral symmetry is the sagittal plane of the worm, and this circumstance shows that the radial symmetry of both the pharynx and the cutaneous-muscular sac was secondarily superimposed on the basic bilateral condition of the ancestors of nematodes: as a consequence of the biomechanics of, first, the muscular pump, and, second, undulatory locomotion.

Theoretically, one can imagine several types of pharynges that would operate according to this principle. The pharyngeal lumen can have a 2, 3, 4 . . . n-radial form. However, calculation shows that in all cases when $n > 3$, the functioning process of the pharynx appears unstable; that is, after muscle relaxation, full closure of the lumen is not necessarily achieved. With $n = 2$, the length of the muscle fibers changes too much, roughly by a factor of seven compared with the change when $n = 3$. Therefore, theoretically a pharynx with a triradiate lumen appears to be the most advantageous (Roggen 1973). In practice, this pharynx type is the one found in nematodes. The pharyngeal operational cycle is shown in Figure 63b.

High hydrostatic pressure is necessary for successful operation of the pharynx. In small nematodes, the biomechanics of the undulating movement requires a reduction of the intracavity pressure, and this unfavorably affects the operation of the pharynx. Special calculations done by Roggen (1970) show that in small forms we should observe either relative increase in pharynx length as compensation or a transition to another means of locomotion. Actually, in small nematodes and juveniles, the relative length of the pharynx is significantly greater than in large forms; this is expressed in a decrease in the second index of de Man (b = body length/pharynx length). In small juveniles the length of the pharynx constitutes about half of the total length of the digestive tract. In certain groups where the nematode body is especially small, a transition is observed toward an atypical means of locomotion (Desmoscolecida, Criconematidae).

A similar pharyngeal structure occurs in various groups of animals. In all cases, the pharynx, operating as a type of muscular pump, has a triradiate lumen, and all the considerations in the case of the nematode pharynx are equally applicable. In most cases this pharyngeal structure has developed completely independently of the nematode pharynx. We encounter it in Turbellaria (Dalyelloidea, Otoplanidae), Lophophorates (Ectoprocta), Gastrotricha, Kinorhyncha, Tardigrada, and others (Figure 63c). It was originally proposed that the nematode pharynx wall has a syncytial structure (Goldschmidt 1905, 1906; Hsu 1929), but with the electron microscope the presence of a cell structure in the pharynx became obvious (Yuen 1967; Lee 1968; Wright 1972, 1974a; and others).

The triradiate symmetry of the pharynx was clearly expressed in the arrangement of the cells that make up the pharyngeal wall. The pharyngeal wall is composed of muscular, marginal, nerve, and glandular cells. At the anterior end of the pharynx, in the area of its juncture with the stoma, epithelial cells have been described in certain species (Albertson and Thomson 1976). The basic structure of the pharyngeal wall is composed of muscular and marginal cells. In all, a cross section reveals six such cells: three muscular and three marginal (Figure 63a). The muscular cells lie in the dorsal and subventral sectors of the pharynx, while the marginal cells lie opposite the radii of the pharyngeal lumen; that is,

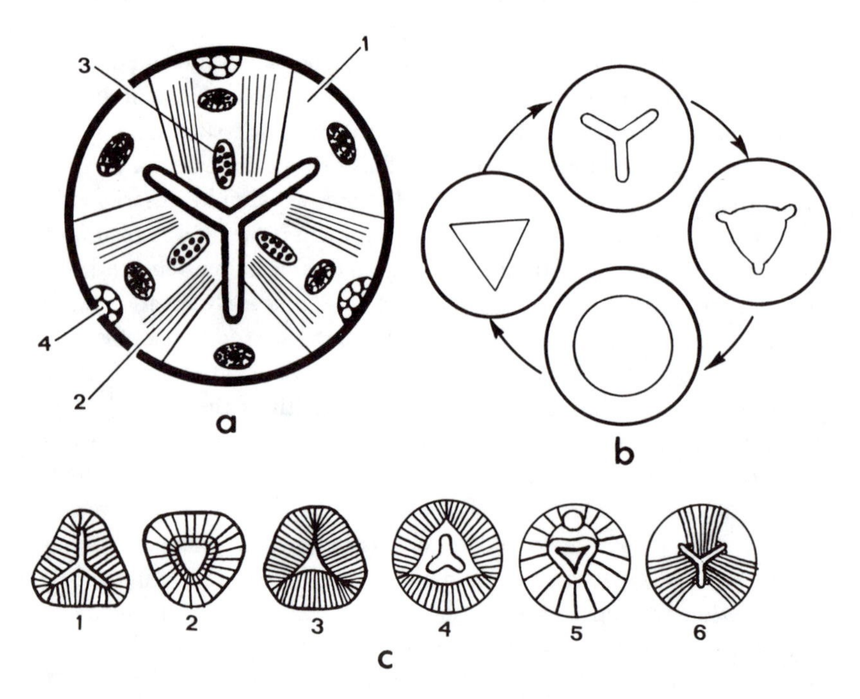

Figure 63. The structure and function of the pharynges with trihedral lumina in various animals (after Roggen 1973 [c]; original [a and b]): a, diagram of the nematode pharynx in cross section: 1, marginal cells; 2, muscular cells; 3, pharyngeal gland ducts; 4, nerves; b, operational cycle of a pharynx; c, pharynges of the "nematode" type in various animals: 1, Dalyelloidea, Turbellaria; 2, Otoplanidae, Turbellaria; 3, Macrodasyoidea, Gastrotricha; 4, Kinorhyncha; 5, Cyclostomata, Ectoprocta; 6, Tardigrada.

ventrally and subdorsally. The ducts from three digestive glands pass between the muscle cells.

The dorsal gland of the pharynx always consists of one cell, the nucleus of which lies in the posterior part of the pharynx. There are two subventral glands, and often both cells are united by a common duct. The pharynx is innervated by three basic nerves, occupying the dorsal and subventral positions, that is, adjoining the muscular cells (Figure 63a). All the components of the pharynx wall are surrounded on the outside by a basal lamina.

The muscular cells contain radially oriented myofilaments that are attached to the inner cuticle of the pharynx and to the basal lamina. The regular arrangement of actin and myosin fibrils provides for the regular alternation of A, I, and H zones, so that as a whole the structure of the pharyngeal muscle bundles corresponds to the structure of cross-striated musculature (Roggen 1973). Pharyngeal muscles of nematodes are unique in that they are only one sarcomere in length (Roggen 1973; Albertson and Thomson 1976).

The marginal cells do not contain contractile filaments. In their cytoplasm are strong bundles of supporting fibers attached by means of a hemidesmosome to the inner cuticle and outer basal lamina (Bird 1971; Albertson and Thomson 1976; Siddiqui and Viglierchio 1977).

The nematode pharynx is a formation of few cells. Even in free-living marine forms of the order Enoplida, it consists, in cross section, of six cells. In the Rhabditia, the number of cells in the pharynx is strictly constant (Voltzenlogel 1902; Goldschmidt 1909; Martini 1916; Pai 1928; Schönberg 1944; Chitwood and Chitwood 1950; Wessing 1953). In *Caenorhabditis elegans,* the pharynx contains thirty-four muscle, nine marginal, nine epithelial, five glandular, and twenty nerve cells (Albertson and Thomson 1976). In large parasitic nematodes, the cell size grows greatly, but their number does not increase. The functional cell hypertrophy caused by increased load greatly affects the glands. Thus, in *Ascaris,* the ploidy of the nucleus of the pharyngeal glands grows by a factor of tens of thousands.

The pharyngeal nervous system is composed of three longitudinal nerves connected to the ring and sector commissures (Goldschmidt 1910; Martini 1916; Hsu 1929; Frenzen 1954; Bogoyavlenskii et al. 1974). The pharyngeal nervous system includes motor and intermediate neurons of several types, and also neurosectory cells (Albertson and Thomson 1976). Nerve fibers are surrounded on all sides by muscle cells. Along the path of a nerve fiber are synaptic areas in which stimulation is transferred to adjoining muscle tissue.

The most primitive type of pharyngeal structure is one in which the pharynx is represented by a muscular tube of regular thickness with the anterior end attached to the cuticle of the head end of the body and the posterior end joined to the middle intestine. Such a type of pharynx is widespread among nematodes of the order Enoplida. The pharynx is not subdivided into any sections, and the nuclei of muscular and marginal cells are distributed evenly over the whole pharynx (Chitwood and Chitwood 1950). The radii of the pharyngeal lumen gradually narrows without forming a tube. Members of Enoplida have a simple pharyngo-intestinal valve with the trihedral cross section of the lumen.

Among some marine Enoplida with a narrow and elongated trophic-sensory section to increase the suctorial force of the pharynx, the basal part of the latter widens and sometimes acquires a wavy outline (*Phanodermopsis, Polygastrophora*). However, a real bulb is almost never present. In Mononchida, the pharynx is more or less regular in thickness, but in Dorylaimida, where the development of a stylet requires a sharp increase in the suctorial force of the pharynx, the latter is distinctly divided into a narrow anterior half and a broad posterior half.

A simple cylindrical pharynx is characteristic of a number of groups of Monhysterida (Monhysteridae, Sphaerolaimidae), and in Siphonolaimidae and Linhomoeidae a muscular swelling is formed in the basal part of the pharynx, which, in members of Siphonolaimidae, has developed into a real bulb. In Monhysterida, in contrast to Enoplida, the nuclei of the pharyngeal tissue are grouped. The radii of the pharyngeal lumen are narrow, and the pharyngo-intestinal valve is flattened in the dorsoventral direction.

In the pharynx of a majority of the members of Chromadorida and Desmodorida, there is a strong basal bulb, the nuclei are segregated, the radii of the lumen are narrow, and the pharyngo-intestinal valve is trihedral in section.

In Araeolaimida, the basal area of the pharynx is almost always swollen, the radii of the lumen end with tubes or without them, and the valve is trihedral. Among members of the order Plectida can be found forms with an almost cylindrical pharynx (*Aphanolaimus*), but most forms have a pharynx with complex regional differentiations. Among members of the Plectida, the pharynx has a somewhat wider anterior part (corpus), a narrow middle part (isthmus), and a massive bulb. The radii of the lumen end with tubes, and the pharyngo-intestinal valve is flattened in the dorsoventral direction.

The most complex type of pharyngeal differentiation is also encountered in Plectida, where the pharynx is divided into a procorpus, metacorpus, isthmus, and bulb. The rhabditid pharynx is very close to the plectid pharynx and can be regarded as the result of increased complexity of the latter. In reality, the similar structure of the pharynx in Plectida and Rhabditia evidently represents the result of parallel evolution. The plectid bulb operates as a bellows, forcing the contents into the middle intestine, whereas in Rhabditia, muscular paddles are developed that push the food through (Chitwood and Chitwood 1950; Maggenti 1963). An important distinction of the rhabditid pharynx, which does not permit its immediate separation from that of the Plectida, consists of the fact that Rhabdita have a trihedral pharyngo-intestinal valve and the Plectida have one flattened in the dorsoventral direction. The rhabditid type of pharynx is characteristic of members of Tylenchida, common among members of Oxyurida, and it is characteristic of the juveniles of the order Strongylida. In the latter case, however, the characteristic regional differentiation is lost in the adult stage.

In nematodes of the spiroascarid branch, the pharynx is a simple cylinder. In Cucullanidae, its basal part is widened, and in Camallanidae, Spiruridae, and Filariidae, the pharynx is divided into a relatively narrower anterior part and a widened posterior part, containing, in representatives of the latter two groups, hypertrophied multinuclear glands. The pharyngo-intestinal valve is flattened in the dorsoventral direction (Cucullanidae, Spiruridae) or trihedral (Camallanidae, Ascaridida), and sometimes may be reduced (some Filariidae).

The arrangement of pharyngeal glands has an important bearing on the separation of nematodes into major taxa, but, unfortunately, it has been insufficiently studied. The most complete information on this subject is contained in the well-known monograph by Chitwood and Chitwood (1950). The original number of pharyngeal glands in nematodes evidently is five and they are arranged according to a bilaterally symmetrical plan: one gland lies in the dorsal sector and two in each subventral sector. The five pharyngeal glands are characteristic for representatives of the order Enoplida (Figure 64); in Enoplidae and Oncholaimidae the ducts of the subventral glands of each sector merge and open with a common opening. In all, in Enoplidae and Oncholaimidae, there are three pharyngeal gland openings (one corresponds to the

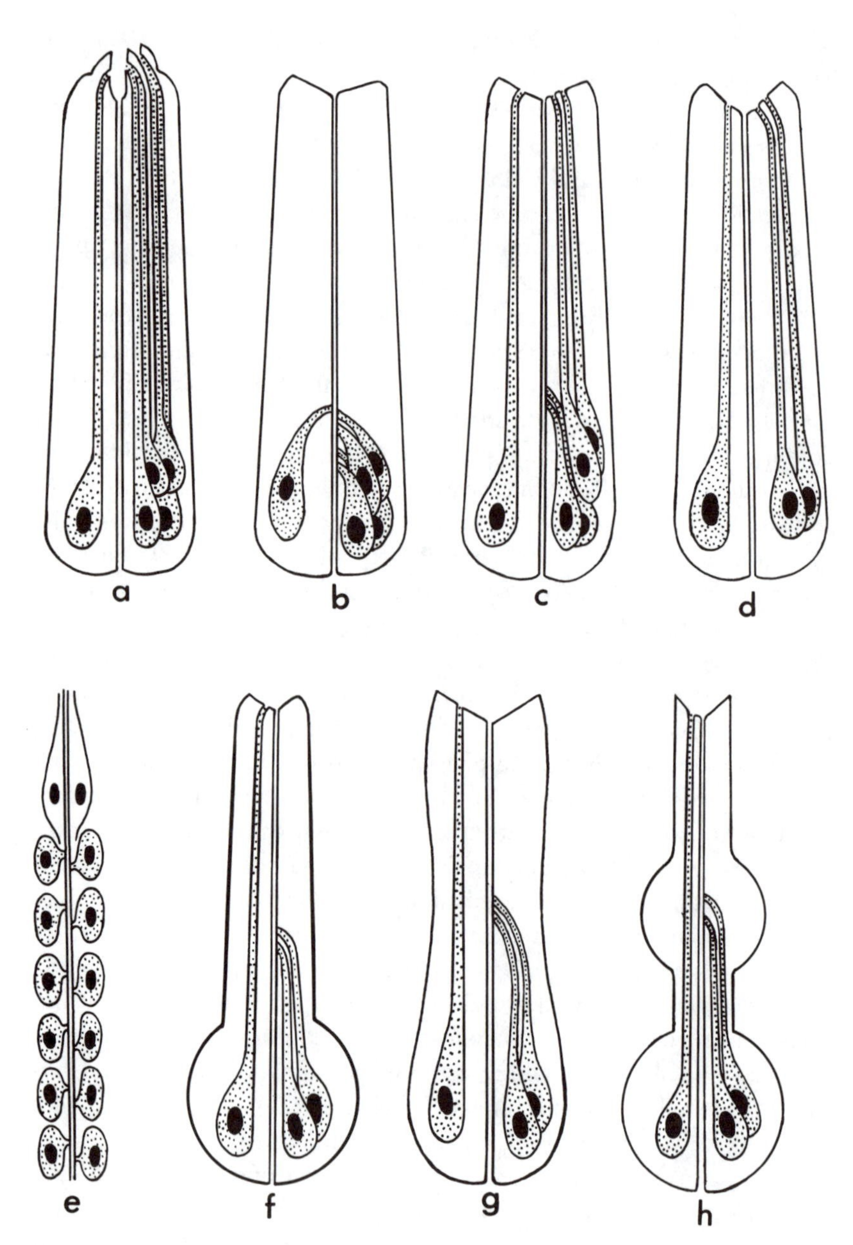

Figure 64. Basic arrangements of the glands in pharynges of nematodes: a, *Oncholaimus* type; b, *Mononchus* type; c, *Tripyla-Eleutherolaimus* type; d, *Eurystomina-Anticyclus* type; e, *Mermis* type; f, *Chromadora* type; g, *Cucullanus* type; h, *Rhabditis* type.

dorsal gland and two to two pairs of subventral glands), which are situated in the anterior part of the pharynx or directly in the stoma. The subventral glands are not equivalent structurally and physiologically. Often, one of the subventral glands in each sector is larger than the other. In each pair of subventral glands in Oncholaimidae, one contains an acidophilic secretion and the other a basophilic secretion. This provides a basis for proposing that one of the glands produces proenzyme and the other is an activator (Jenning and Colam 1971). Each of the five pharyngeal glands in members of Tripylina opens independently: a dorsal and a pair of subventral in the anterior part of the pharynx and another pair of subventral in the posterior part of the pharynx (Figure 64).

Among nematodes of the orders Mononchida and Dorylaimida, all five glands open into the posterior part of the pharynx (Figure 64b). A surprising elaboration of the subventral pharyngeal glands is observed in Mermithida and Trichocephalida, the pharynx of which is a stichosome transformed into a glandular organ (Figure 64e). However, in the anterior part of the pharynx of the Trichocephalida are preserved three independent glands that open behind the nerve ring. Preparasitic juveniles of Mermithida have three independent pharyngeal glands. Characteristic of nematodes of the order Dioctophymida are three large, dichotomously branching glands that open into the anterior part of the pharynx. Among some forms of the order Monhysterida, which occupies the most primitive place in the classification of the subclass Chromadoria, in *Eleutherolaimus* for example, there are five pharyngeal glands opening as in Tripylina. In contrast, there are three glands opening into the anterior part of the pharynx in *Anticyclus* and *Paralinhomoeus* (Riemann 1977b). For the remaining forms of the subclass Chromadoria, as for all Rhabditia, a common type of pharyngeal gland arrangement is characteristic: the single dorsal gland opens into the anterior part of the pharynx, while the subventral glands open into the posterior part of the pharynx (Figure 64f, g, and h).

Rather wide variation is also observed relative to the attachment of the pharynx to the anterior end of the body. The most primitive state is with the pharynx attached directly to the cuticle of the anterior edge of the nematode head end. The oral cavity and labia are absent, and the shape of the oral aperture corresponds to the shape of the pharyngeal lumen. The primary absence of labia is common to lower Enoplida (Leptosomatidae, Oxystomonidae, some Tripylina, and probably some Araeolaimida (*Diplopeltis*). The development of the oral cavity leads to the appearance of primary labia—purely cuticular extensions covering the entrance of the digestive system. In the primarily nonlabial forms, the labial papillae are arranged simply along the edge of the trihedral mouth. After the appearance of the primary labia, the labial papillae remain at their base and never become a part of the labial folds themselves. This is one way in which primary labia differ from head tubercles (secondary labia) that are common to many free-living and parasitic nematodes. The original number of labia corresponds to the three sectors. This number is widespread among Enoplida, particularly in the family of the same name. However, more often a variant is encountered—the splitting of each labium into two and the appearance of six primary labia. It is interesting that in the ontogeny of the six-labial forms, there are originally three labia (see Chapter 5).

Primary labia in many groups of nematodes are reduced in some degree or another and replaced by secondary labia, or cephalic tubercles, which represent growths of cephalic tissue. Secondary labia (cephalic tubercles) have cephalic receptors. Cephalic tubercles have developed independently in three groups of nematodes. Six cephalic tubercles (along with primary labia) have developed in Mononchida and have achieved especially marked development in Dorylaimida. The primary labia in Rhabditia have been reduced or have grown together with the six secondary labia. Three secondary labia are a rather frequent occurrence among the parasitic nematodes of the orders Ascaridida, Oxyurida, and Strongylida. Two secondary lateral labia have developed independently in representatives of Spirurida and Strongylida. A rarer occurrence is the four secondary labia among representatives of Spirurida.

The oral cavity is absent among lower Enoplida (Leptosomatidae, Oxystominidae, Tripylidae, and Anticomidae) and here this absence is primary. The oral cavity is absent or hardly noticeable among *Diplopeltis* and *Araeolaimoides* of Araeolaimida and *Camacolaimus* of Plectida. Morphologically, the oral cavity represents a widening of the lumen of the anterior part of the pharynx. In part of Leptosomatidae, the anterior region of the pharyngeal lumen is slightly widened and represents a rudiment of an oral cavity. A similar rudimentary oral cavity, surrounded on all sides by pharyngeal muscular tissue, is found in Anticomidae, many Tripylina, and *Monhystera* of Monhysterida, and in a number of other forms. The muscular tissue surrounding the walls of the oral cavity provides them with mobility. In Leptosomatidae and Tripylidae, cuticular protuberances—onchia—have developed on the wall of the oral cavity. The oral cavity of Enoplida, as surrounded on all sides by pharyngeal muscle tissue, contains three strong mandibles formed by the cuticularization of the pharyngeal cuticular covering.

A detached oral cavity comes about when the muscular tissue recedes from the anterior part of the pharyngeal covering. This process proceeds in parallel among various groups (Figure 65) but nonetheless leads to the formation of three basic regions of the oral cavity: the basal region of the stoma, enveloped by pharyngeal muscle tissue; the middle part, free from the muscle tissue and formed by more or less cuticularized plates of the stomorhabdion; and the apical part, directly adjoining the labia. The basal part is mobile as a consequence of the presence of pharyngeal musculature, and here is where onchia or teeth may develop that take part in the processing of food. Therefore this is also known as the onchial region of the stoma. The middle part, known as the buccal region of the stoma, may have immobile onchia. The apical part of the stoma, the vestibular region, can have cuticular bulges—odontia. Any attempt at a complete survey of nematode stoma structure would take up too much space, so this discussion will be limited to characteristic examples.

Onchia in the basal part of the stoma are the only armament of the oral cavity in *Rhabdodemania, Prismatolaimus,* and other lower enoplids. In Oncholaimidae, with their large oral cavity, only immobile on-

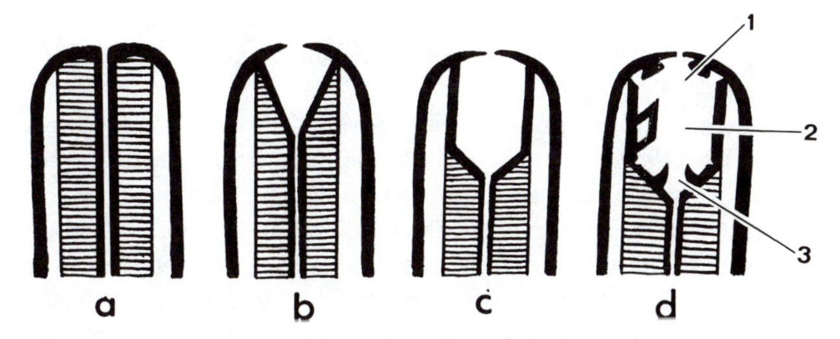

Figure 65. Basic stages of stoma formation in nematodes: a, complete absence of stoma; b, widened anterior part of pharyngeal lumen and labial formation; c, absence of pharyngeal muscular tissue and division of stoma into three sections; d, development of armament in stoma sections. 1, vestibular part; 2, buccal part; 3, basal part.

chia in the middle (buccal) part of the stoma have developed. A huge dorsal onchium, together with a set of additional onchia, adorn the middle part of the stoma in Mononchida. In Dorylaimida, an odontostyle has developed in the stoma that is homologous to one of the onchia of Mononchida (Chitwood and Chitwood 1950). Stylets, which are homologous to those of Dorylaimida, are characteristic of the juveniles of Mermithida (Christie 1936; Chitwood and Chitwood 1950; Richter 1971; Poinar and Hess 1974), and also of juveniles of Trichocephalida (Fülleborn 1923; Chitwood and Chitwood 1950) and Dioctophymida (Lukasiak 1930).

The oral cavity of Chromadorida and Desmodorida consists of two basic parts: a basal part, enveloped by the pharyngeal musculature and equipped with mobile teeth, and a wide vestibulum, usually equipped with twelve odontia.

Complex compartmentation has been achieved for the oral cavity of Monhysterida, although lower representatives of this order still have a completely undeveloped stoma (*Monhystera*), and, in the stoma of Sphaerolaimidae, the basal, middle, and apical parts are easily distinguishable. The vestibulary (apical) part of the stoma is equipped with cuticularized thickening. In Linhomoeidae, on the other hand, cuticularized rings, rows of teeth, and so forth, develop in the basal section of the stoma.

Among Araeolaimida, the oral cavity reaches a rather complex structure in Axonolaimidae. In Plectida, the most complexly structured cylindrical stoma is encountered in Plectidae. The plectid stoma is rightly regarded as a transitional stage to one of the more complex types of nematode oral cavity structures—that of Rhabditidae (Chitwood and Chitwood 1950).

The rhabditid stoma, as is well known (Chitwood and Chitwood 1950; Osche 1952; Paramonov 1962), is constructed of five elements: cheilostom, protostom, mesostom, metastom, and telostom. The cheilostom corresponds to the apical part of the stoma (vestibular) and can be equipped with odontia; the proto- and mesostoma together correspond to the buccal section of the stoma; the metastom and telostom are enveloped by pharyngeal musculature, and they often have small onchia or large movable teeth. Elements of the rhabditid stoma are easy to trace in the structure of the stylet of Tylenchida formed by all parts of the stoma except the cheilostom.

The rhabditid type of stoma is characteristic of juveniles of Strongylida and a number of Oxyurida (Thelastomidae, for example). Yet there is no basis for looking upon the complex stoma of Rhabditida as the original one for nematodes in general as do the Chitwoods (Chitwood and Chitwood 1950). The rhabditid stoma is not even characteristic for the whole of subclass Rhabditia. Thus, nematodes of the spiroascarid branch never have had the characteristic compartmentation of the stoma into five regions. In many Ascaridina and Filariina, the stoma has a rudimentary character, and this may not be a consequence of secondary reduction.

Digestion in nematodes begins as early as in the oral cavity under the action of enzymes from the pharyngeal glands. Digestion in the oral cavity was the basis for development of a special external digestion among phytoparasitic nematodes (Paramonov 1962). However, the main phase of digestion and food absorption in nematodes takes place in the middle intestine. The middle intestine in nematodes is a simple tube of one layer of cells. It is relatively short, does not have significant diverticuli (if a small growth in Ascaridata is excluded), and is devoid of additional glands. The secretion of enzymes and all stages of digestion and absorption of food are accomplished by the cells of the middle intestine wall. On the one hand, this indicates the relatively primitive organization of the endodermal section of the nematode digestive tract and, on the other hand, suggests a high degree of physiological activity on the part of the cells.

The apical surface of the cells of the middle intestine is equipped with microvilli that form a complete brushlike border. It is notable that, in some nematodes of the order Dorylaimida, along with microvilli in the intestine, there are real synchronously moving cilia (Zmoray and Guteková 1972). Structurally, the cells of the middle intestine differ little from one another, although some of them evidently are absorptive and others secretory. All types of cells are equipped with microvilli and, evidently, at first function as absorbing cells. At a definite stage of the cell cycle, they begin to secrete by merocrine or holocrine means, then die and are sloughed off into the lumen of the intestine (Jenning and Colam 1971). Digestion occurs in the lumen of the intestine and by contact with the surface of the microvilli, where maximal enzymatic activity is concentrated. Among many nematodes, lysosomes are found in the cells of the middle intestine (Colam 1971a, b; Bogolepova 1976), and in some forms there are digestive vacuoles, indicating the presence of elements of intracellular digestion (Colam 1971b; Riley 1973; Bogolepova 1976).

As a rule, a constant cellular composition in the middle intestine is absent. This is related to the necessity for making up for the death of cells resulting from secretion or damage to the intestinal epithelium. Characteristic of free-living nematodes of the order Enoplida are a high level of physiological regeneration of the intestinal epithelium (Jenning and Colam 1971) and ease of regeneration of whole sections of the middle intestine after their experimental destruction (Korotkova and Agafonova 1975). The renewal of intestinal epithelium is also characteristic of some large parasites of animals (Anisimov and Tokmakova 1973; Anisimov and Usheva 1973).

Free-living nematodes of the enoplid branch and also most representatives of Chromodorida and Desmodorida have a polycytose intestine containing hundreds and thousands of cells. In many large parasites of animals the number of cells in the middle intestine grows to many millions (miriocytose intestine). An oligocytose intestine is encountered in many members of Monhysterida, Axonolaimidae, and Plectida. Nematodes of the order Rhabditida and other groups of nematodes related phylogenetically with Rhabditida (Tylenchida, Strongylida) have an oligocytose intestine with strict constancy in the cell composition. Thus, for example, in *Turbatrix aceti*, the middle intestine has a total of eighteen cells (Pai 1928). In secondary increase in size among some Strongylida, the number of cells in the middle intestine remains constant, but the sizes grow to gigantic proportions: 4 mm in length and 0.5 mm in width. Such gigantic cells, as a rule, are multinuclear.

Reserve nutrients are deposited in the cells of the middle intestine: glycogens, fats, and protein granules. The middle intestine also functions as a secretory organ. In the cells of the middle intestine are encountered numerous granules, sphero-crystals, and crystals, containing a number of inorganic components ($CaSO_4$, Fe^{+++} and Fe^{++}, and others). Evidently these granules can be expelled from the organism together with the dying cells of the intestinal epithelium or be buried in cells.

The posterior intestine is a short tube of ectodermal origin, lined with cuticle, and having a trihedral or uneven lumen. A ringed muscular valve separates it from the middle intestine. Special rectal glands can be associated with the posterior intestine.

Thus the evolutionary transformation of the intestine has proceeded in parallel in various groups. As the basis for various ceca, we encounter forms with a simple cylindrical pharynx, undeveloped stoma, and polycytose middle intestine. In nematodes of the enoplid branch, the pharynx preserves a simple structure and a real bulb does not appear; there are three or five pharyngeal glands, although in most forms three reach predominant development and open near the stoma; and the middle intestine is polycytose. Characteristic of nematodes of the chromadorid-monhysterid branch is the development of a bulb, and three pharyngeal glands open at different levels: the dorsal one near the stoma, the subventral ones in the posterior half of the pharynx, and the intestine is poly- or oligocytose. The most complex regional specialization is reached by the pharynx in nematodes of the rhabditid-strongylid branch of the subclass Rhabditia; the pharynx is divided into four parts, the glands open as in the previous case, and the intestine is oligocytose. Nematodes of the spiroascarid branch of the subclass Rhabditia preserve a relatively simple pharyngeal structure, the arrangement of glands is typical for Rhabditia, and the intestine is poly- or myriocytose.

Reproductive System

Nematodes are dioecious. They have internal fertilization, which in nematodes has clearly ecological bases. External fertilization, as is known, requires a large concentration of ova and spermatozoa in the water and is attended by large losses of sexual products, which, in turn, requires a high degree of fertility. Free-living nematodes, which are the original and most numerous ecological group within the class, are microscopic in size. Among small forms, the relative sizes of ova (in comparison with the body sizes of the animals) grow because of the existence of a natural limit in the ovum size, and the number of ova produced, consequently, falls. Internal fertilization represents a necessary adaptation directed at the economy of scarce sexual products in small forms. Let us note that small body sizes caused the transition to complex copulatory apparatuses and sexual ducts and internal fertilization among a large number of groups of animals occupying a primitive position in the classification of animals. Internal fertilization and an unusually complex sexual system are characteristic of the most primitive Bilateria—Turbellaria—and many groups of small organisms: Gastrotricha, Rotifera, Kinorhyncha, and others.

Internal fertilization, as a rule, leads to the development of a rather complicated sexual apparatus, comprising sexual ducts, glands, and complex copulatory structures. Nevertheless, the nematode sexual apparatus is constructed relatively simply. Basically, both sexes have two sexual tubes.

The sexual tubes of the female are differentiated longitudinally: oogonia are produced near the distal, blind end of the gonoducts, and growing oocytes move along the gonoduct to the seminal receptacle. After fertilization, the oocytes arrive in the uterus, which opens to the external environment through the vagina. The seminal receptacle represents a more or less differentiated region of the gonoduct and, more rarely, a process from its wall. In free-living forms, one of the gonoducts is directed anteriorly and the other, posteriorly.

The female gonoducts can be straight or curved. In the second instance, the ovaries may be antidromous or homodromous (Lorenzen's terms, 1978a). Antidromous are the most common type of curved ovaries in free-living nematodes. In this case, in the region of the curve, a blind process is formed in which the largest oocytes are accumulated. The oocytes move gradually in the direction of the blind process, depending on growth, because of the pressure on the part of the sexual tubes and from the more distal accumulation of growing cells. From the blind end of the ovary the oocytes reverse direction so that the opposite end of the oocyte is now the forward end. They then force themselves between the wall of the gonoduct and the series of younger oocytes, and move into the uterus. This segment is dominated by oocytes, owing to a unique amoeboid activity that strongly changes the shape of the oocytes (Jägerskiöld 1901; Türk 1903; Stewart 1906).

Antidromous gonoducts are common to almost all Enoplida, all Mononchida and Dorylaimida, Chromadorida and Desmodorida (besides Comesomatidae, Aponchiidae, and some Microlaimidae), certain Monhysterida, and parts of Araeolaimidae and Plectida. Among Rhabditia, this type of gonoduct is characteristic of Diplogasteridae. The curved homodromous ducts are characteristic of many Rhabditia. In this case the blind process is absent, and oocytes always move with the same end forward (Figure 66I). Straight ovaries are encountered sporadically in various groups of nematodes and are characteristic of all Desmoscolecida, various groups of Araeolaimida, and individual representatives of other orders.

In many groups of nematodes, one of the gonoducts is independently reduced. Lorenzen (1978a) calculated that within forty-two families of nematodes there are forms among which are preserved only the

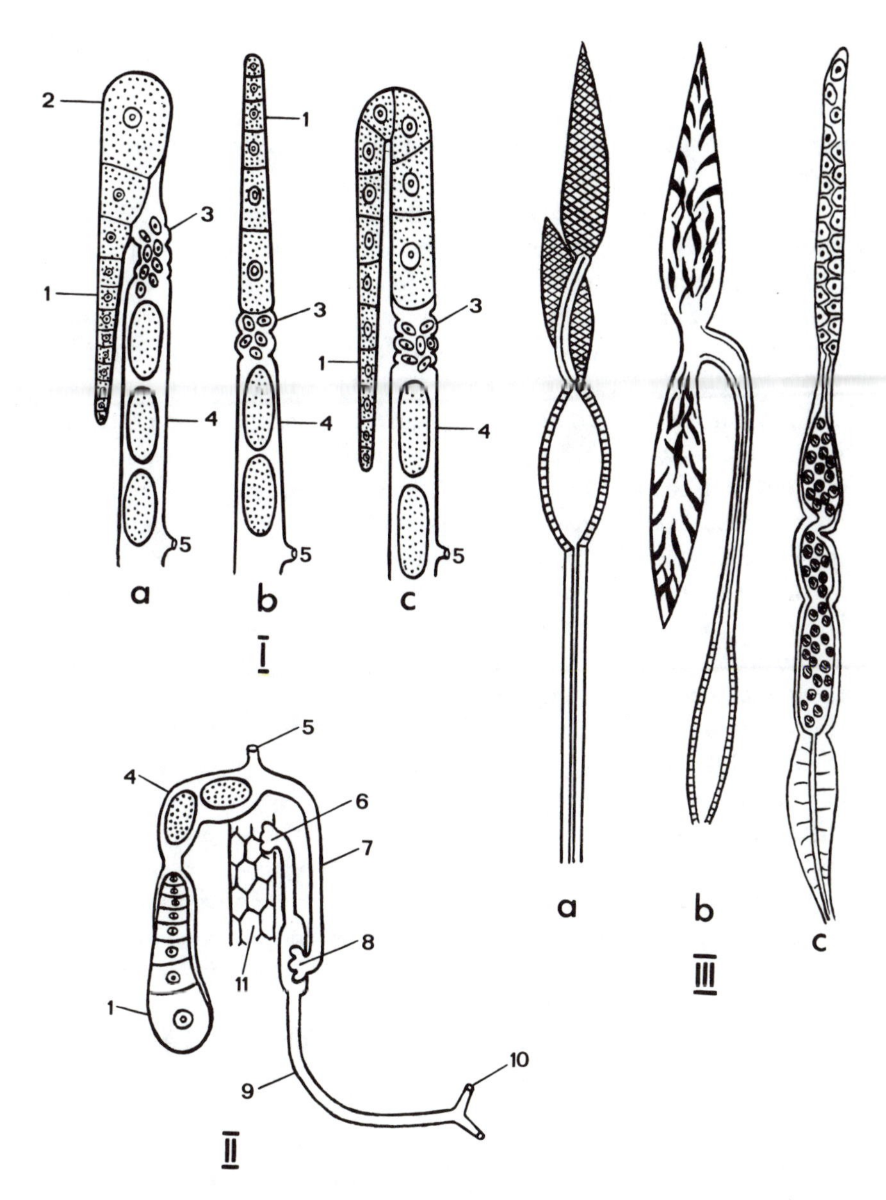

Figure 66. Structure of the nematode reproductive system (original [I]; after Belogurov and Belogurova 1977 [II]; after Chitwood and Chitwood 1950 [IIIa and b]; after de Coninck 1965 [IIIc]: I, types of structure of female reproductive systems: a, curved antidromal; b, straight; c, curved homodromal; II, structure of the female sexual system of *Oncholaimium* with one gonoduct; III, examples of male gonads and gonoducts: a, *Anticoma;* b, *Tobrilus;* c, *Rotylenchus*. 1, ovary; 2, blind process; 3, seminal receptacle; 4, uterus; 5, vulva; 6, osmosium; 7, uterine canal; 8, uvette; 9, main tube; 10, pores; 11, intestine.

anterior gonoduct, and within eighteen other families there are nematodes with only a single posterior gonoduct.

The female reproductive system has contact with the external environment only by means of the vulva, which is simultaneously both a copulative and a parturitive opening. In only one group of marine Enoplida—the family Oncholaimidae—the female sexual system has a complex tubular organ, the so-called "de Manian system" (de Man 1886; Filip'jev 1921; Cobb 1930; Kreis 1934; Rachor 1969; Belogurov and Belogurova 1977). This tubular organ represents a system of canals uniting the ovary with the external environment and through an osmosium with the intestine. The de Manian system strikingly resembles the complex sexual system of Turbellaria (for example, Monocelidae of the Proseriata), among which the female reproductive system is united with the external environment by two openings, one of which is copulative and the other parturient. The reproductive system of such forms is also in contact with the intestine through a special genito-intestinal canal, the structures resembling an osmosium developing at the point of contact between this canal and the intestine. The functions and origin of the de Manian system are utterly enigmatic. It is speculative to see in it a direct remnant of the complex reproductive system of the Turbellaria—which are like antecedents of nematodes. However, there has been a description of the copulation of *Oncholaimus oxyuris,* in the process of which the male introduced its spicula into one of the openings of the de Manian system and injected sperm, which subsequently moved into the main canal of the de Manian system. From the main canal, the sperm go into the uterus, where they fertilize the oocytes (Maertens and Coomans 1979). Thus, the terminal pores of the de Manian system serve as a seminal receptacle, and the female vulva serves only as a parturient opening (Figure 66II).

The reproductive system of the male is also constructed of paired gonoducts—testes. As a rule, one of the testes is directed anteriorly and the other posteriorly, although infrequently testes may extend in the same direction (*Anticoma*). The reduction of one testis is a rather frequent occurrence among free-living nematodes of both the Enoplida and Chromadorida and is encountered among representatives of forty families (Lorenzen 1978a). All Rhabditia have only one testis (Figure 66III).

The proximal sections of the testes are united in the common sperm duct, which opens together with the posterior intestine into the common cloaca. In nematodes of the Enoplida, the sperm duct is equipped with ejaculatory musculature, which is absent in nematodes of the Chromadorida-Monhysterida lineage and Secernentea.

The male copulatory organs—spicula and gubernacula—represent paired, strongly cuticularized structures equipped with sensory nerve endings and are brought into movement by special musculature.

The structure of the reproductive system varies widely within the class. Paired gonoducts are undoubtedly a primary condition, widespread among primitive groups of nematodes. The reduction of one gonoduct is a phenomenon arising independently among various groups of nematodes and always represents a secondary condition. The sexual apparatus of nematodes with one gonoduct always develops from an originally paired rudiment, which is an argument in favor of original pairs in the sexual system (Chitwood and Chitwood 1950). Completely unfounded is the postulation of Andrassy (1976) in which one tube is regarded as the original sexual apparatus.

Much more complicated is the question as to which type of ovary to consider the original: curved antidromous, curved homodromous, or straight. Curved antidromous ovaries are widely distributed among primitive groups of free-living nematodes. According to their structure, antidromous ovaries are adapted to producing mature oocytes one by one and are characteristic of free-living nematodes with low fecundity. Low fecundity is an original characteristic of nematodes related to the small size of free-living representatives of this group. The functioning of antidromous ovaries is related to the amoeboid activity of the oocytes—a very archaic trait, characteristic of the sexual cells of primitive Metazoa. Straight and curved homodromous ovaries can possibly be regarded as derivatives of curved antidromous ovaries. Curved homodromous ovaries are adapted in the best way to produce a large quantity of ova and are widely distributed among parasitic nematodes of the subclass Rhabditia and are never encountered among truly free-living nematodes.

2

Development

Embryonic Development

The developing nematode ovum is one of the classical subjects of embryology. The first explanation of the characteristics of fertilization and meiosis in animals was accomplished through research on the nematode ovum (Beneden 1883; Beneden and Neyt 1887). The phenomenon of chromatin diminution was discovered and studied in detail in nematodes (Boveri 1888, 1899; Herla 1894; Meyer 1895; Zoja 1896; Bonnevie 1901; Walton 1918; Fogg 1930; King and Beams 1938; Lin 1954; Moritz 1967; and others). Ideas about mosaic development and the early determination of blastomeres also resulted from research on nematodes (Boveri 1887, 1888, 1899, 1910; zur Strassen 1896, 1898, 1904; Martini 1903; Müller 1903; Stevens 1909; Hogue 1910; Bonfig 1925; Pai 1928; Dunschen 1929; Seck 1937; von Strassen 1959).

At the same time, as P. P. Ivanov wrote (1937, p. 170), "The development of round worms provides a great deal for the understanding of general embryological problems, but provides almost nothing for the understanding of the morphology and phylogeny of round worms themselves." This situation evidently came about because the subjects of embryonic studies were specialized parasitic and saprobiotic forms. Practically all species of nematodes whose embryogeny has been studied belong to the subclass Rhabditia, most representatives of which are parasites of animals and plants or are saprobionts. Embryonic development in free-living nematodes began to attract attention only in the 1970s (Malakhov and Cherdantsev 1975; Malakhov and Akimushkina 1976; Malakhov 1981).

Male gametes in nematodes have an unusual organization. Spermatogenesis begins in the distal sections of the testes. In nematode parasites of animals, the spermatogonia may be tied by cytoplasmic bridges to the central non-nuclear zones of the cytoplasm—the rachis (McLaren 1973). In free-living forms, the rachis is not developed in the testes. While advancing through the testes, so-called membrane organelles are formed in the spermatogonia and then in the spermatocytes. They arise by the transformation of Golgi complexes and include crystalline granules, bundles of microfibrils, and tubular and lamellar membrane structures (Foor 1970; Beams and Sekhon 1972; McLaren 1973). Spermatocytes undergo first and second divisions of maturation. After the second division of maturation, the spermatocytes turn into spermatids, while a reduction of cytoplasm occurs. Each spermatid forms a cytophore in which appears a significant part of the ribosome, elements of endoplasmic reticulum, and Golgi complexes. The cytophore opens and degenerates. According to some observations, it may be phagocytized by cells of the wall of the testis (Favard 1961). Membrane organelles lose their fibrils, which transfer to the cytoplasm of the spermatid. Membrane organelles are situated near the surface, often at one end of the forming sperm. With a light microscope, they look like granules strongly refracting light.

After completion of meiosis, the chromosomes merge into a general chromatin mass. A nuclear membrane is found in enoplid spermatozoa only (Figures 67A, 68b). Spermatozoa of other nematodes have no nuclear membrane. Sperm nuclear material is concentrated into a dense granule (Figure 68c). Sperm nuclear material in *Aspiculuris* and *Nippostrongylus* forms a thin envelope around the bundles of microtubules in the false tail of spermatozoa (Figure 68e). In some species the nuclear material has the appearance of a complicated network in the central part of the spermatozoan (Figure 68d). In some instances, as for example in *Dipetalonema,* the chromosomes do not merge in a general nuclear mass, but preserve individuality up to the moment of fertilization (McLaren 1973).

The formation of egg cells begins at the distal region of the ovaries.

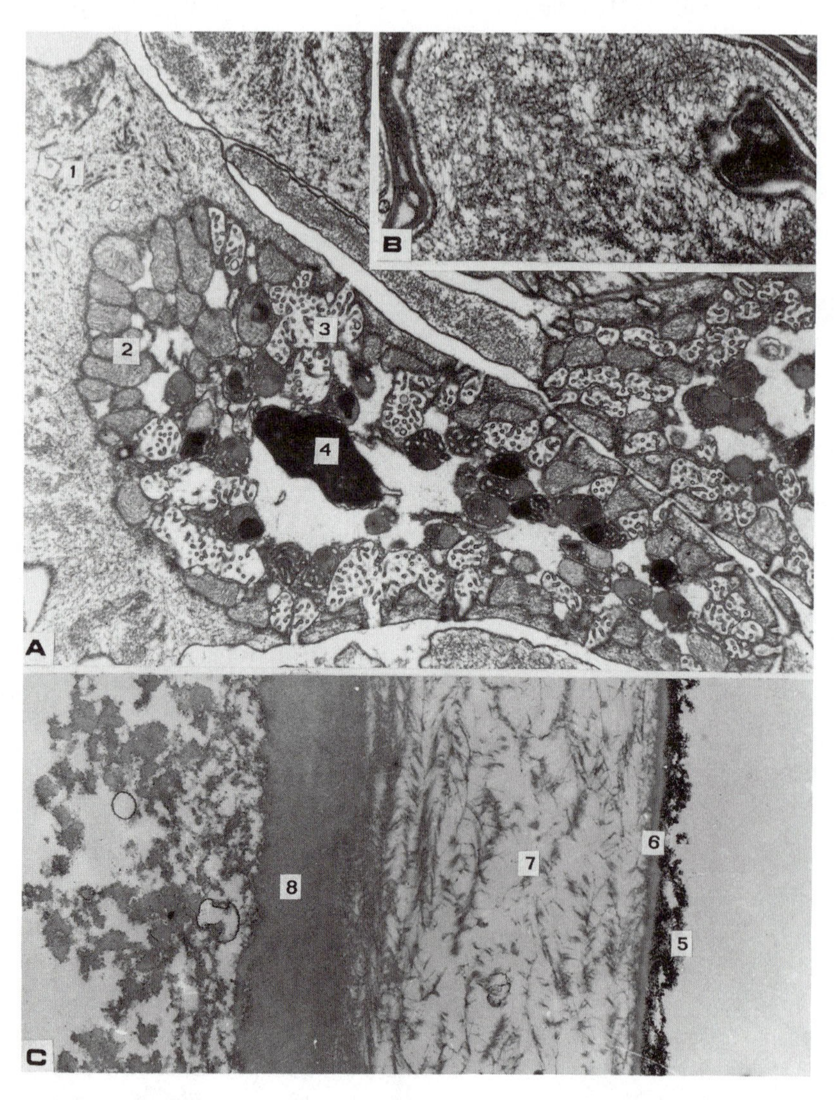

Figure 67. Ultrastructure of the spermatozoon of *Enoplus demani*. Magnifications: a = 10,000; b = 25,000; c = 16,800. A, spermatozoon in the female *receptaculum semini;* B, part of a pseudopodium; C, section through the egg shell. 1, pseudopodium; 2, mitochondria; 3, membrane organelles; 4, nucleus; 5, uterine layer; 6, vitelline layer; 7, chitinous layer; 8, lipid layer.

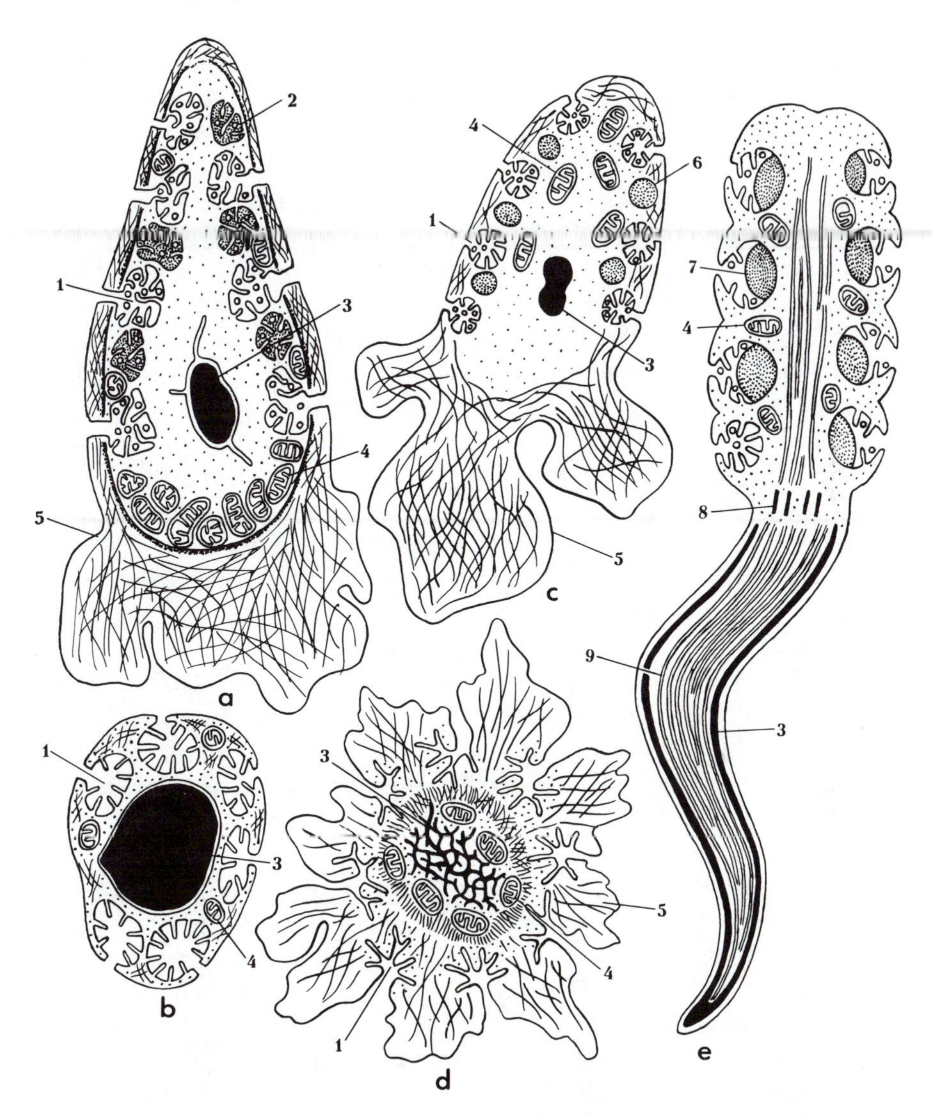

Figure 68. Diagrams of the ultrastructure of nematode spermatozoa (Figure a is original; b, d, and e after Baccetti et al. 1983; and c after Foor 1970): a, *Enoplus;* b, *Mesacanthion;* c, *Gnathostoma;* d, *Xiphinema;* e, *Nippostrongylus.* 1, membrane organelles open to external environment; 2, internal membranous organelles; 3, nucleus; 4, mitochondria; 5, pseudopodia; 6, granules; 7, crystalloid bodies; 8, centrioles; 9, microtubules.

The distal end of each ovary is occupied by oogonia originating in the cells of the ovary wall. Among nematodes of the subclass Rhabditia, the oogonia are tied by cytoplasmic bridges to the central strand of non-nuclear cytoplasm—the rachis. As the female sexual cells grow, they move through the ovary. In them accumulates two sorts of inclusions. The first are inclusions containing substances that serve as a reserve for the development of the embryo: glycogen, proteins, and lipoprotein and lipid granules. The second group includes so-called refractive and shell granules. The refractive granules contain lipids and the shell granules, proteins and polyphenols (Wharton 1979).

In the ovary, the oocytes are enveloped only in cytoplasmic membrane, on the surface of which may be a layer of granular material. In the ovary, oocytes undergo prophase of the first meiotic division. Fertilization occurs during the passage of oocytes through more or less modified sections of the gonoduct, the seminal receptacle, in which sperm are preserved. The morphology of sperm preserved in the female seminal receptacle usually to some degree or other differs from that in the terminal sections of the male sexual system. Sperm in the seminal receptacle have a rounder form and their membranous organelles are usually connected to the cytoplasmic membrane.

According to a description by McLaren (1973), on contact with sperm the cell membrane of the oocyte invaginates and captures the sperm. At the point of contact, the layer of granular material on the surface of the oolemma disappears. Possibly this is caused by a discharge of the contents of the membranous organelles of the sperm. The function of the membrane organelles, thus, is somewhat analogous to functions of the acrosome in sperm of other multicellular animals (Figure 69).

After fertilization, the plasma membrane of the oocyte exfoliates, and under it is formed a new cytoplasmic membrane, now covering the fertilized nematode egg. The fertilized eggs arrive in the uterus, where the egg envelope is formed. According to Wharton (1979), this happens as follows. The exfoliating plasma membrane of the oocyte is a vitelline membrane, on the surface of which outer and inner uterine layers are formed by secretions of the uterine epithelium. The uterine layers have a complex structure and define the external relief of the nematode egg envelope. In the space between the vitelline membrane and the cytoplasmic membrane of the egg cell, there is a secretion of envelope granules (Figure 70) from which are formed the nonchitinous components of the chitinous layer of the egg envelope. The chitin evidently is synthe-

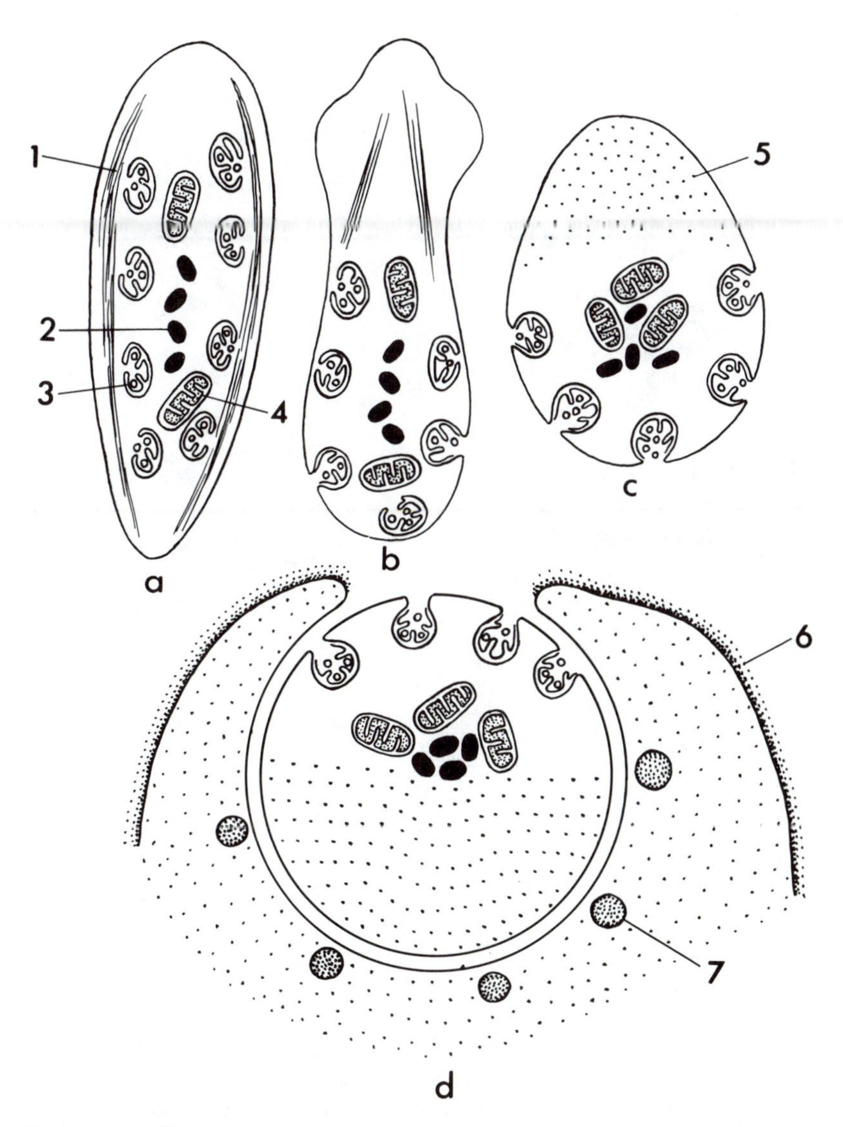

Figure 69. Differentiation of sperm and the fertilization process in *Dipetalonema vitae* (after McLaren 1973): a–c, spermatid differentiation; d, penetration of sperm into ovum. 1, fibrils; 2, chromosomes; 3, membrane organelles; 4, mitochondria; 5, granular cytoplasm; 6, grainy material on the surface of the ovum; 7, granules within the ovum.

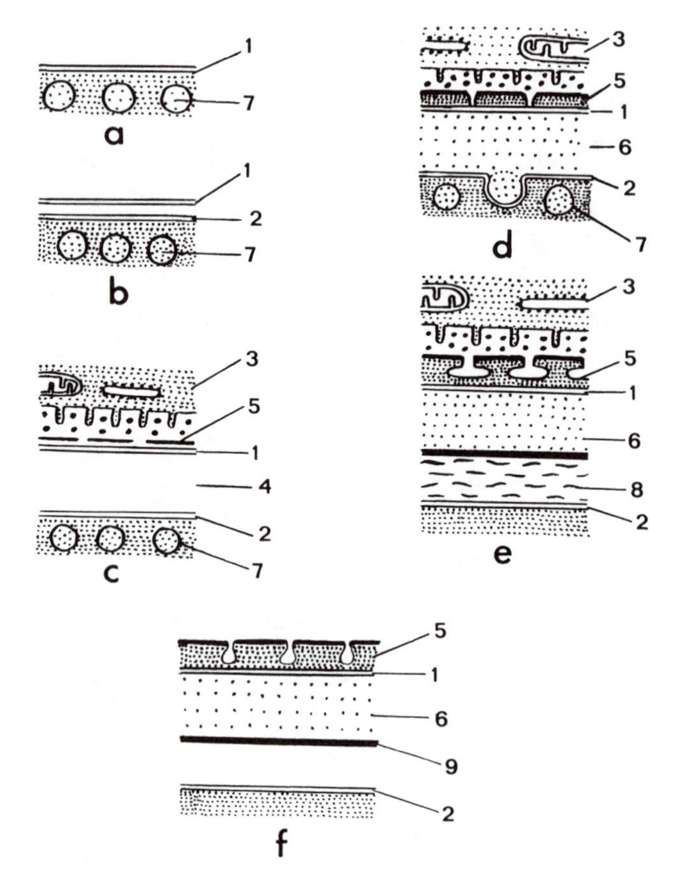

Figure 70. Diagram of the formation of the egg envelope in nematodes of the order Oxyurida (after Wharton 1979): a, oolema of unfertilized oocyte; b, exfoliation of the old oolemma after fertilization and the formation of a new one; c, beginning of formation of uterine layers; d, secretion of granules containing material of the chitinous layer; e, formation of lipid layer; f, completely formed envelope. 1, old oolemma (yolk layer); 2, new oolemma; 3, epithelium of uterus; 4, perivitelline space; 5, uterine layers; 6, chitinous layer; 7, envelope granules; 8, lipid components; 9, lipid layers.

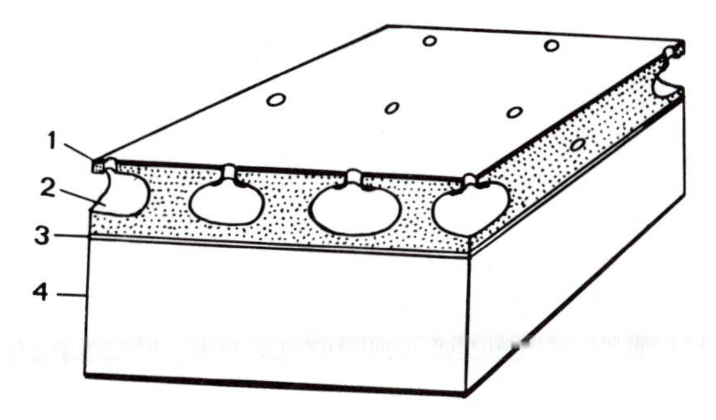

Figure 71. Diagram of the egg envelope in *Hammerschmidtiella diesingi* (after Wharton 1979). Scale = 1 μm. 1, outer uterine layer; 2, internal uterine layer; 3, vitelline layer; 4, chitinous layer.

sized from carbohydrates that are formed from the breakup of glycogen granules. After the completion of the formation of the chitinous layer, the ovum cytoplasm withdraws from the internal surface of the envelope. In the space formed, lipids are secreted that settle on the internal surface of the envelope and form an internal lipid layer (Figure 70e). Thus, in a nematode egg envelope, there can be secretions of external and internal uterine layers, the vitelline layer coming from the plasma membrane of the oocyte, a chitinous layer, and a lipid layer (Figures 70 and 71). This structure of the envelope is characteristic of parasitic nematodes, the ova of which are able to withstand the effects of various chemical agents without harm to the embryo. Egg shells of free-living marine nematodes consist of the same layers, although they have simpler structure than those of parasitic nematodes (Figure 71).

After sperm penetration, the female nucleus moves to one of the poles of the egg cell and undergoes meiotic division. In oviparous free-living nematodes, the fertilized ova accumulate in the uterus, the meiosis being retarded in the telophase of the first division. In ovoviviparous species, meiosis, the merging of pronuclei, and embryonic development take place in the female uterus.

Small free-living nematodes deposit only one ovum at a time. Oviposition by larger forms may involve up to fifty ova. Deposits by free-living nematodes are represented by chains or clusters of ova stuck together (Figure 66). Some free-living marine nematodes show concern for progeny and carry the ova on the body stuck to the cuticle (Ott 1976). The fecundity of parasitic forms reaches many thousands of ova.

Nematode ova are devoid of any clearly defined signs of animal-vegetal polarity. The yolk, as a rule, is distributed evenly through the entire content of the ovum. The point of emergence of the polar body varies strongly in different forms. They may emerge at the anterior end, as in *Oswaldocruzia* and *Strongylus* (Spemann 1895) or close to the anterior end, as in *Rhabditis pelio* (Schönberg 1944) and *Diplogaster* (Ziegler 1895). In *Parascaris* and *Rhabdias* the polar bodies emerge on the dorsal side (von Strassen 1959). In *Heterodera,* their emergence takes place on the ventral side (Strubell 1888).

In *Ascaris lumbricoides* the first polar body emerges at the anterior end of the ovum, and the second, on the dorsal surface (Boveri 1887). There is a wide range of individual variability relative to the point of polar body emergence. In *Diplogaster,* the polar bodies may emerge not at the anterior, but at the posterior end of the ovum (Ziegler 1895). In 10–20% of the ova of *Rhabdias,* the polar bodies emerge at the posterior end of the ovum (von Strassen 1959).

The even distribution of the yolk and the absence of a definite position of the polar bodies often have led to errors in determining the poles of a nematode ovum. Such errors at times have been made by Goette (1882) and zur Strassen (1892). We too failed to avoid such an error in our first investigations of the embryogenesis of marine nematodes (Malakhov and Cherdantsev 1975; Malakhov and Akimushkina 1976).

Embryonic Development of the Subclass Enoplia

The most primitive type of development among nematodes has been found among free-living marine species of the order Enoplida. The ova of these nematodes have an ellipsoid form and are covered by dense chorion, and the polar bodies appear near one end of the ovum (Figure 72a). The furrow of the first cleavage in ova of Enoplida occurs perpendicular to the longitudinal axis of the ovum and divides it into two equal blastomeres: the anterior *AB* and the posterior *CD* (Figure 72b). After division is complete, the interphase blastomeres display rapid cytoplasmic activity apparent by the varied processes and lobes that distort the shape of the blastomeres as they appear and disappear (Figure 72c). Interphase blastomeres also occur in all successive stages of cleavage.

The second cleavage proceeds according to any one of several variations, each one of which may occur among ova produced by different

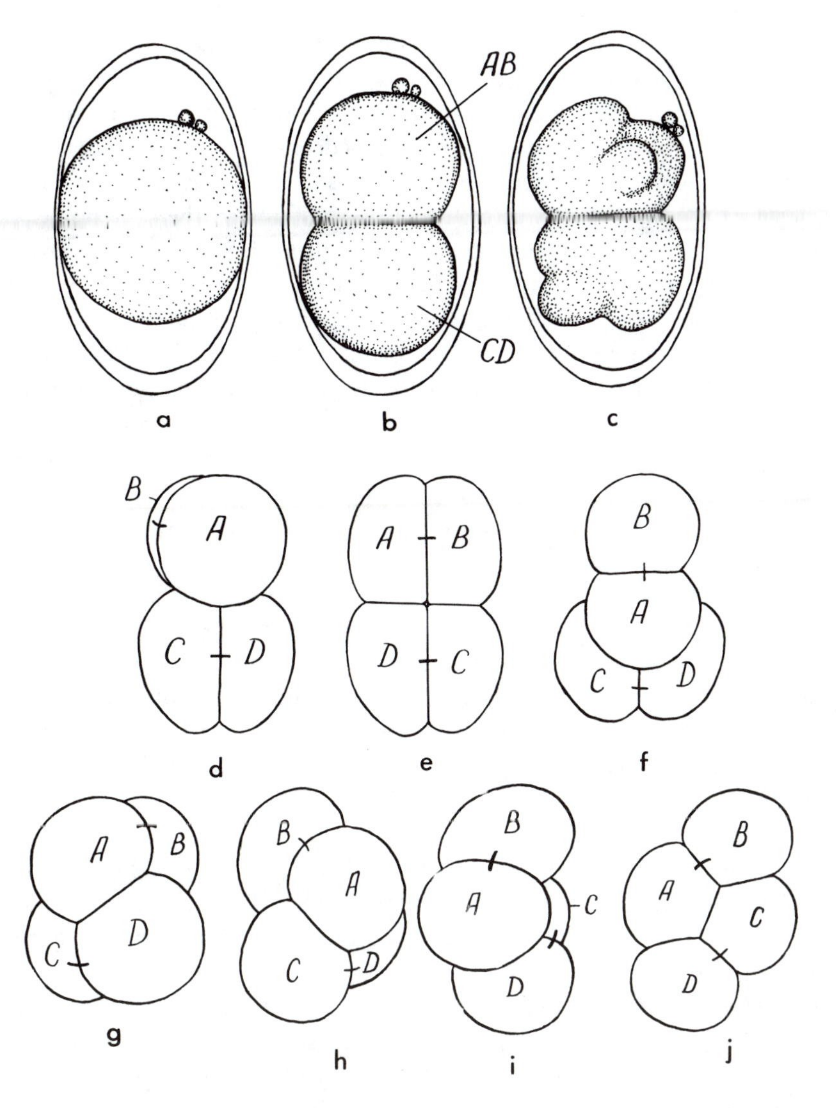

Figure 72. Early cleavage in *Pontonema vulgare:* a, ovum before the first division; b, ovum just after the first division; c, interphase activity of blastomeres; d–f, different variants of blastomere arrangement after the second division; g, h, different views of the tetrahedral figure; i, j, different views of the rhombic figure.

females of the same species and even among ova deposited by one female. After the second cleavage, there are tetrahedral, flat, or, as an exception, solid T-shaped configurations of the blastomeres (Figure 72d, e, and f). After completion of the second division, the blastomeres migrate: the tetrahedral figures become rhombic or asymmetrical tetrahedral figures; the flat figures become tetrahedrons or rhombuses; and the T figures become rhombuses. Thus, as a result of morphogenetic movements of the blastomeres after the second division in the development of Enoplida, two basic configurations are formed: rhombic and tetrahedral. The arrangement of the blastomeres in the rhombus is by no means always flat, and at the same time, the angle between the sister pairs of blastomeres in the "tetrahedron," as a rule, is less than 90° (Figure 72g–j). How can these two configurations be distinguished? In tetrahedral figures, each of the blastomeres of the embryo is in contact with three others. In the rhombic configuration, two of the four blastomeres are each in contact with two other blastomeres. The balance between these two configurations depends on the temperature in which development takes place.

The furrows of the third cleavage in the rhombic figure go more or less parallel to one another and in parallel with the plane of the rhombus (Figure 73a and b). In the tetrahedral figure, the arrangement of the cleavage furrows is more varied. One of the most common variants of the third cleavage division is that in which the cleavages in the blastomeres of each sister pair go parallel to one another but perpendicular to the cleavages of the other sister pair (Figure 73c and d).

In other instances, in one sister pair, the cleavages occur parallel to one another and, in the other, at an angle to one another (Figure 73e and f). Finally, there are also instances where in both sister pairs, the furrows of the third cleavage occur at an angle to one another (Figure 73g and h).

After division, the blastomeres are mixed relative to one another so that in a majority of cases the same eight-blastomere figure is formed in which the descendants of the four blastomeres of the previous stage are variously arranged. The directions of the mixed blastomeres after division are shown by arrows in Figure 73.

During transitions to the fourth cleavage, one of the blastomeres of the embryo lags in the division and the rest of the blastomeres grow around it (Figure 74). This blastomere provides the beginning for the entire endoderm of the embryo. Subsequently, the sinking of this blas-

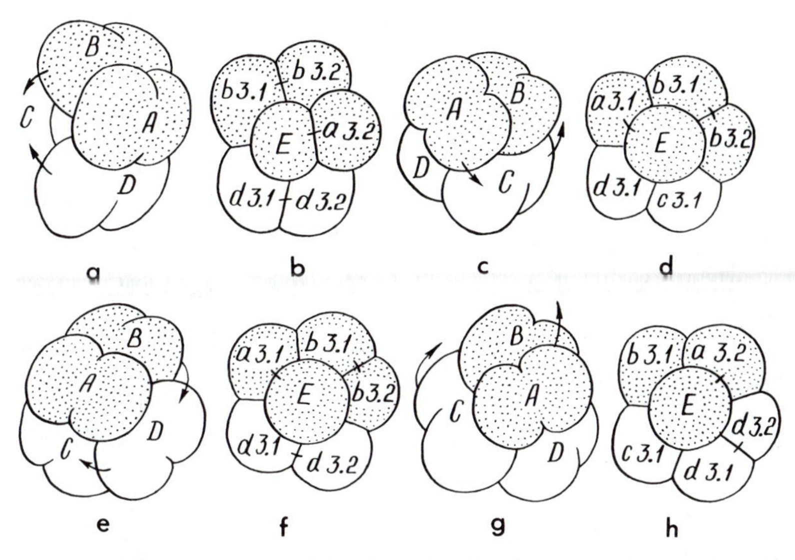

Figure 73. Different variants of the third cleavage division in the embryo of *Pontonema vulgare*. Explanations are in the text. Descendants of blastomere *AB* are dotted.

tomere at the fifteen-cell stage can be regarded as the beginning of gastrulation. The further multiplication of cells leads to a more definite formation of a blastopore, which occupies the center of the future ventral surface of the embryo.

The position of the descendants of the first four embryo blastomeres relative to the endodermal blastomere at the fifteen-cell stage varies. In some cases, the descendants of all four blastomeres have contact with the endodermal blastomeres, and in other cases only three have contact (Figure 74). These differences are also preserved in the later stages of development and are expressed in the different position of the descendants of the first four blastomeres of the embryo around the blastopore. It is possible to note only certain general regularities in the arrangement of blastomeres at the later stages of development. The descendants of blastomeres *A* and *B* are arranged in front and at the sides on the ventral side of the embryo, and the descendants of blastomeres *C* and *D* are arranged on the ventral side behind the blastopore and take up a large part of the aboral gastrula corresponding to the dorsal side of the future embryo.

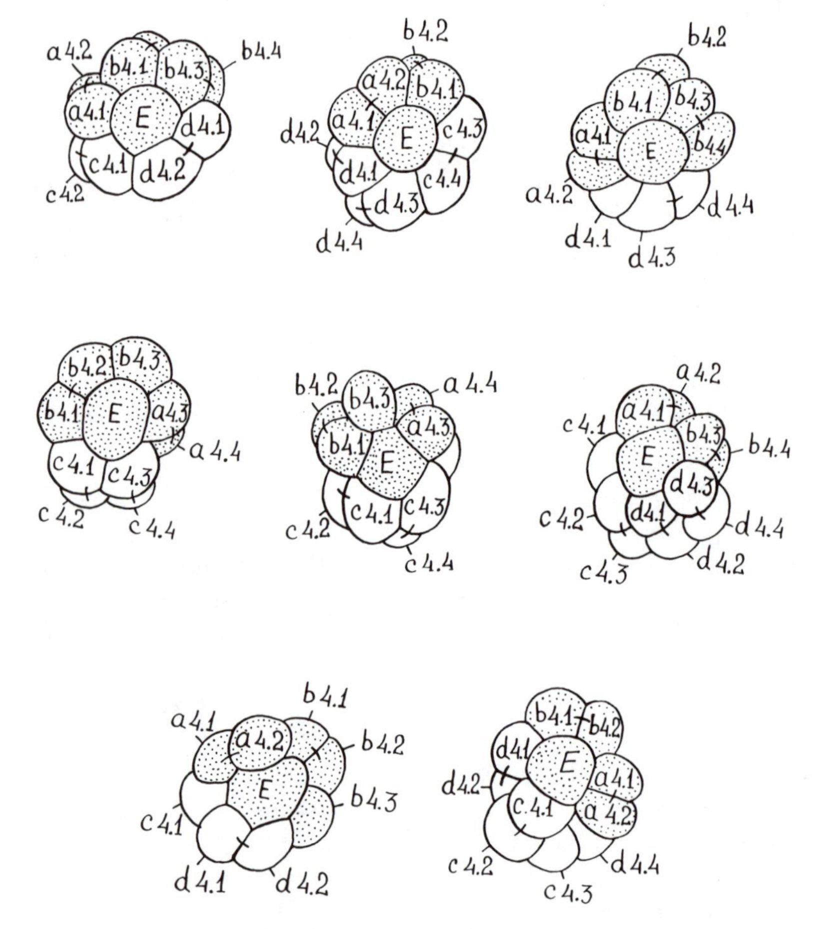

Figure 74. Different variants in the arrangement of blastomeres at the fifteen-cell stage in *Pontonema vulgare*. Explanations are in the text. Descendants of blastomere *AB* are dotted.

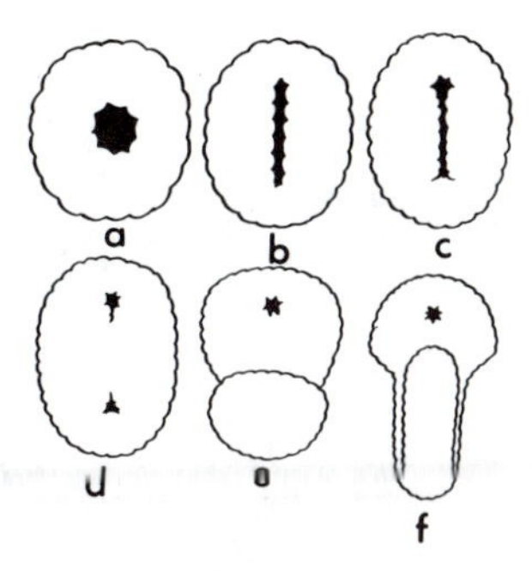

Figure 75. Formation of a slit-shaped blastopore (a and b), its closure (c and d), and flexure of the embryo (e and f) in *Pontonema vulgare*.

The embryo preserves the organization of the gastrula up to the stage of about 400 cells. The blastopore at first has a round shape, but then stretches out and becomes slit-shaped. At this stage, the descendant cells of the blastomeres bordering the blastopore move inward from the edges of the blastopore. The cells migrating from the lateral edges of the slit-shaped blastopore provide the beginning of the future mesoderm, and the cells submerged at the anterior end probably represent the precursor of the stomodeum.

In further development, the lateral edges of the blastopore approach and connect with the middle. Two openings, one at the anterior and the other at the posterior end of the embryo, are persistent remnants of the blastopore (Figure 75). The anterior opening provides the beginning of the definitive mouth, and the posterior one, the definitive anus. After the closing of the blastopore, the embryo begins to flex. Rudiments of a stomodeum, middle intestine, and posterior intestine can be noted at later stages of the embryo. In the posterior third of the embryo body, two large cells appear lying to the right and left of the rudiment of the middle intestine (Figure 76). Probably these two cells are precursors of the reproductive system, but it is not clear from which blastomere of the embryo they come. In still later embryos can be found the rudiment of the nervous system, represented by groups of cells around the stomodeum and one row of narrow columnar cells along the abdominal side of the embryo (Figure 76g).

Dioctophymida is an order of parasitic nematodes including very specialized forms, which in the adult stage are parasites of vertebrate

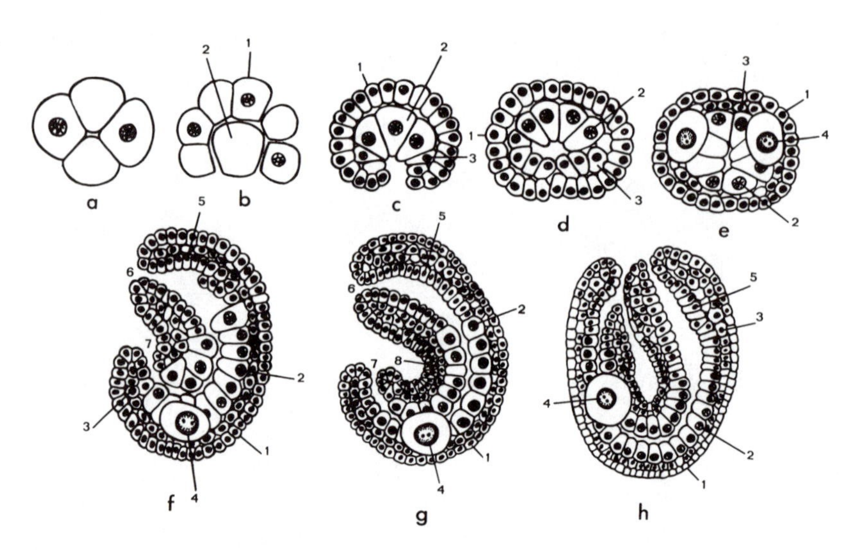

Figure 76. Histological structure of the embryo of *Enoplus demani:* a, cross section at eight-blastomere stage; b, cross section at fifteen-blastomere stage; c, cross section of embryo with slit-shaped blastopore; d, cross section of embryo at the blastopore closure stage; e, cross section through late embryo; f, g, and h, sagittal sections through late embryos at successive stages of development. 1, ectoderm; 2, endoderm; 3, mesoderm; 4, primordial sex cells; 5, rudiment of stomodaeum; 6, mouth; 7, anus; 8, primordial abdominal nerve trunk.

animals. The embryonic development of Dioctophymida has been studied only in the case of *Eustrongylides excisus*, a parasite of cormorants (Malakhov and Spiridonov 1983). The first cleavage is almost uneven. The spindles of the second division are oriented mutually perpendicular to form a tetrahedral figure (Figure 77). Part of the embryos undergo transformation to a rhomboid figure, and others enter the third cleavage directly from the tetrahedral figure.

The plane of the rhombus coincides with the future sagittal plane of the embryo. Blastomere *FESt* occupies the anterior end of the embryo, blastomere *BD* the posterior, blastomere *C* the abdominal side, and blastomere *A* the dorsal side. In relation to the figure of the tetrahedron, the plane of the future bilateral symmetry passes through blastomeres *A*, *FESt, bcd*, and βΓδ (Figure 77).

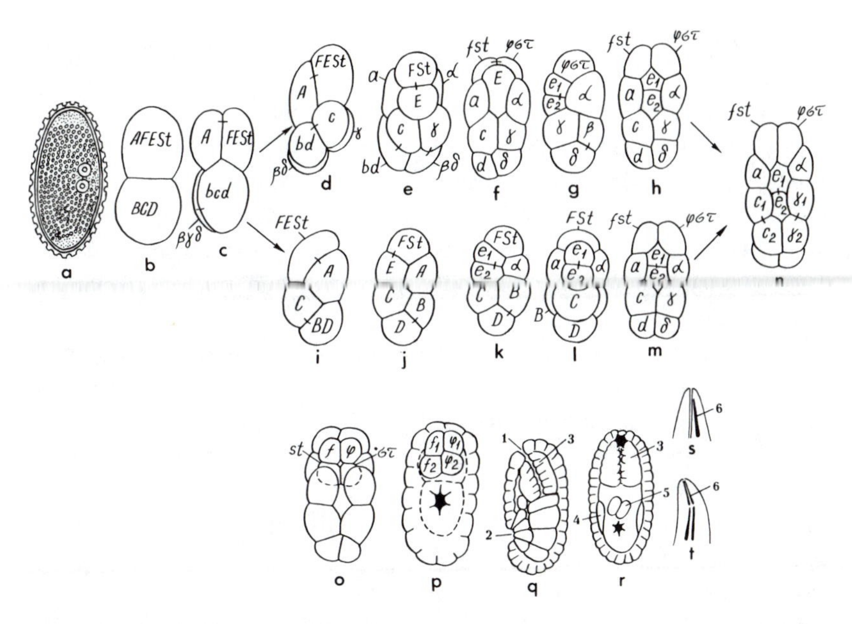

Figure 77. Embryonic development of *Eustrongylides excisus*. Illustrations c, d, g, i–k, q, s, and t are lateral views. The remainder are ventral. a, fertilized egg; b, two-blastomere stage; c, four-blastomere stage after division; d–h, division of the tetrahedral configuration; i–m, division of the rhombic configuration; n–p, gastrulation; q and r, late embryo; s and t, formation of the stylet. 1, mouth; 2, anus; 3, precursor of stomodaeum; 4, precursor of mesoderm; 5, precursors of the reproductive system; 6, stylet.

Embryos with the rhombic configuration divide in the following manner. Blastomere *BD* is divided by a slanting furrow into a more dorsal cell and a more ventral one. Further, blastomeres *B*, *C*, *D*, and also, a little later, *BM*, are divided by furrows passing through the sagittal plane of the embryo. Blastomere *FESt* divides into anterior cell *FSt* and posterior *E* (Figure 77j). Then, blastomere *E* divides into anterior blastomere e_1 and posterior e_2, and blastomere *FSt* divides into right *fst* and left ϕδτ cells (Figure 77m).

In the tetrahedral figure, blastomeres *bcd* and βΓδ divide into dorsal cells bd and βδ and ventral cells *c* and δ. Cell *A* divides into right cell *a* and left α. Blastomere *FESt* divides into anterior blastomere *FSt* and posterior *E*. Cells *bd* and βδ divide again into cells *d* and δ, lying more ventrally, and cells *b* and β, lying more dorsally. At this time, the configurations resulting from both the rhombus and tetrahedron are identical.

Blastomere *E1* and *E2* represent a rudiment of endoderm. These cells gradually sink inside the embryo, while cells descending from blastomeres *A* and *C* grow over them (Figure 77n and o). Blastomeres *fst* and $\phi\sigma\tau$ divide into anterior cells *f* and ϕ, providing material for the frontal ectoderm, and posterior cells *st* and $\sigma\tau$, providing material for the stomodeum. The latter gradually sink under and are overgrown laterally by the ectodermal cells.

The ectoderm creeping in from the sides gradually joins at the mid-ventral line. Only two openings remain in the blastomere: one at the anterior end of the embryo, which becomes the mouth; and the other near the posterior end, which is surrounded by cells descending from blastomeres *A* and *C*. Through the posterior opening of the blastomere there is a migration inside the embryo of cells providing the beginning of the embryonic mesoderm and progenitor cells of the reproductive system, but it is not clear from which blastomeres of the embryo they come. In the late embryo, the primary sex cells are situated to the right and left of the intestine as in Enoplida. In juveniles of Dioctophymida there is a stylet (Figure 77s and t).

Representatives of the order Trichocephalida have elongated ova with plugs at the poles. The first cleavage is uneven and divides the ovum into a small anterior blastomere *AESt* and large posterior blastomere P_1 (Figure 78). Then blastomere P_1 divides into a small posterior blastomere P_2 and large anterior blastomere *BM*. At this stage, all three blastomeres of the embryo are arranged in one line (Figure 78c). At the next stage, the spindles of divisions in blastomeres *AESt* and P_2 are oriented in the sagittal plane of the embryo, and in blastomere *BM* in the frontal plane. The unusual six-cell stage is very characteristic for nematodes of this order (Figure 78d). Then blastomeres *bm* and $\beta\mu$ continue to divide, forming an ectomesodermal rudiment. Descendants of blastomere *ESt* provide the beginning for the stomodeum and endoderm, and the descendants of blastomeres *C* and *D*, the caudal ectoderm. Gastrulation is accomplished by the creeping of descendants of blastomeres b_1 and β_1 over endodermal blastomere *E* (Figures 78g and h).

The origin of mesoderm in these nematodes is not fully clear, but it probably comes from descendant cells of blastomeres *bm* and $\beta\mu$. The oral opening is placed among descendants *a* and α near the anterior end of the embryo. The anal opening lies on the ventral side of the embryo.

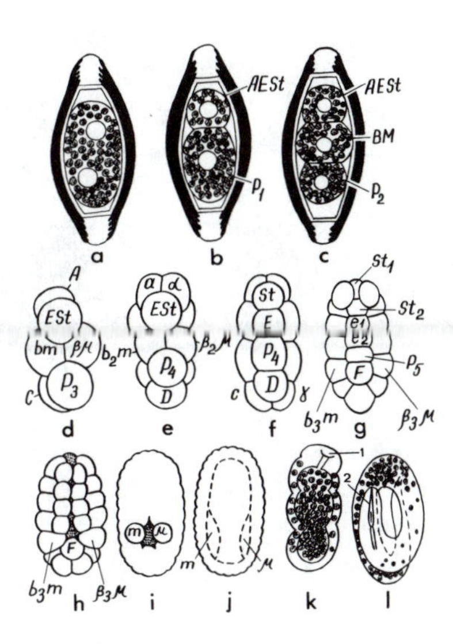

Figure 78. Embryonic development of *Trichocephalus trichurus:* a, fertilized ovum; b, two-blastomere stage; c, three-blastomere stage; d, six-blastomere stage; e, ten-blastomere stage; f, twelve-blastomere stage; g, twenty-one-blastomere stage; h, gastrulation; i, blastopore closure; j, late embryo; k, mobile embryo; l, formation of larva with stylet. 1, mouth; 2, rudiment of stylet.

The presence of a stylet is characteristic of juveniles of the order Trichocephalida (Figure 78).

Among representatives of the order Mononchida, cleavage is orderly and is characterized by early formation of bilateral symmetry. The first division leads to the formation of two equal blastomeres, the anterior of which contains material for the ectoderm, stomodeum, and endoderm and is designated *AFESt*; and the posterior contains material for the ectoderm, mesoderm, and reproductive system, and it is designated BMP_2 (Figure 79). In the second division, there is formed a T-shaped figure that transforms into a rhomboid figure (Figure 79b and c). At this stage the sagittal plane of the future embryo is determined, coinciding with the plane of the rhombus. Further division is strictly bilateral. Subsequently, a chain of blastomeres is formed on the ventral side of the embryo: *F* (containing material for the frontal ectoderm), *ESt* (containing material for the endoderm and stomodeum), P_3 (sexual rudiment and part of the posterior ectoderm), and C (containing material for the posterior ectoderm). Four cells are formed on the dorsal side: *a* and α (ec-

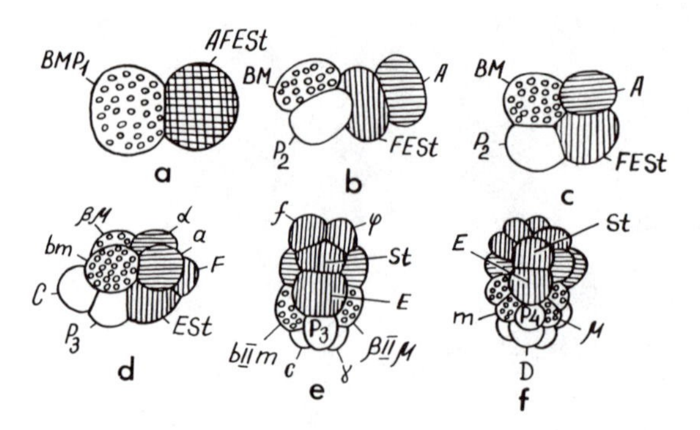

Figure 79. Cleavage of blastomeres in *Prionchulus* sp. (after Drozdovskiy 1969): a, two-blastomere stage; b, T-shaped stage; c, rhombic stage; d, eight-blastomere stage; e and f, further cleavage.

toderm) and *bm* and βμ (ectoderm and mesoderm). In distinction from all other nematodes, the isolation of endoderm material in the development of Mononchida takes place not after the third, but only after the fourth division (Drozdovskiy 1968). The mesoderm of the embryo comes from two cells descending from dorsal blastomere *BM*.

The development of nematodes of the order Dorylaimida is very similar to that of members of Mononchida (Drozdovskiy 1975). The distribution of the potentials between the first blastomeres of the embryo here are the same as in Mononchida: the anterior blastomere contains material for the ectoderm, stomodeum, and endoderm; and the posterior, material for the ectoderm, mesoderm, and sexual rudiment. The figure of the rhombus appears after the second division, the plane of which corresponds to the sagittal plane of the embryo. Further cleavage is completely identical with that of Mononchida.

Among representatives of the order Mermithida, the development of *Gastromermis* has been studied (Malakhov and Spiridonov 1981). The first cleavage in *Gastromermis* is uneven: the anterior blastomere *AESt*, containing material for the ectoderm, stomodeum, and endoderm, is larger than the posterior blastomere BMP_2, containing material for the ectoderm, mesoderm, and sexual rudiment (Figure 80). After the second cleavage, a rhombus figure is formed, the plane of which coincides with the sagittal plane of the future embryo (Figure 80b and c). Further along, on the ventral side of the embryo, a chain of cells is formed, *St, E,* P_3, and *C*, and on the dorsal side are arranged the four cells *a*, α, *bm*, and

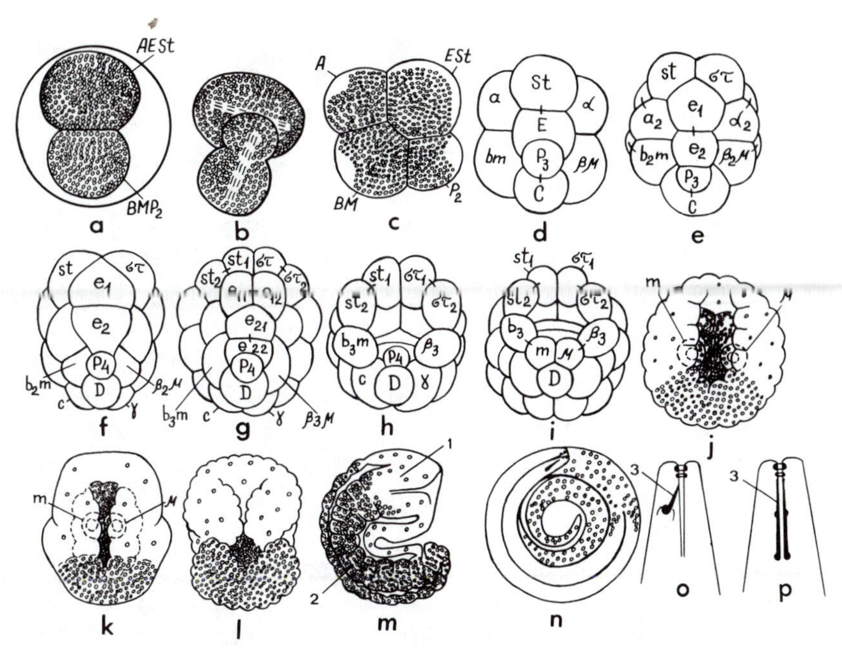

Figure 80. Embryonic development in *Gastromermis hibernalis:* a, two-blastomere stage; b and c, second division; d, eight-blastomere stage; e, fourteen-blastomere stage; f and g, further cleavage; h and i, gastrulation; j–l, blastopore closure; m, tadpole stage; n, nearly formed larva; o and p, stylet formation. 1, stomodeal primordium; 2, rudiment of midgut; 3, stylet.

$\beta\mu$ (Figure 80). Blastomere *St* subsequently provides material for the stomodeum, and blastomere *E* contains material for the endoderm. Material for the caudal ectoderm is localized in blastomere *C*. Blastomere P_3 subsequently divides into a small cell P_4 (sexual rudiment) and a larger cell *D*, which contains material for the ectoderm of the posterior part of the body (Figure 80e). Gastrulation proceeds by epibolic submersion of cells descending from blastomere *E* (Figure 80h and i). The two cells lying to the right and left of the blastopore—descendants of blastomere *BM*—divide in a transverse direction, and mesodermal cells *m* and μ result. At this stage, a reduction of the yolk begins in the majority of blastomeres. The yolk remains only in cells descendant from blastomeres *E, C, D*, and b_3m and $\beta_3\mu$ (Figure 80j and k). Subsequently, there begins a rapid multiplication of ectodermal cells and cells devoid of yolk that are descendants of blastomeres *A* and *BM*. The ectoderm grows from the lateral sides to the ventral side of the body. Two cells rich in yolk, *m* and μ, submerge into the blastopore. The blastopore adopts a slit shape (Figure 80j and k). Gradually, the ectoderm closes in on the mid-ventral line. Two openings remain in the blastopore, one at the anterior and the other at the posterior end of the embryo (Figure 80).

Subsequently, cells rich in yolk from the posterior end of the embryo multiply intensely and spread over the abdominal and lateral sides of the embryo up to the level of the rudimentary stomodeum, displacing the primary ectoderm coming from cells *A* and *BM*. Only along the midventral line there is preserved a string of cells free of yolk—probable descendants of blastomeres *A* and *BM* (Figure 80m).

An interesting feature in the development of members of Mermithida is the formation of a ventral, cuticularized, juvenile stylet (Figure 80m). Later in development, the stylet moves to the lumen of the stomodeum.

The embryonic development among members of Mermithida is very similar to that found within Mononchida and Dorylaimida. The only difference is that in Mermithida, as in all nematodes, the endodermal blastomere is isolated after the third division, whereas in Mononchida and Dorylaimida it is after the fourth. This may turn out to be only an apparent difference. According to the origin and arrangement, the blastomeres designated as frontal ectoderm in the development of Mononchida and Dorylaimida correspond exactly to those cells in the development of Mermithida that we interpret as stomodeal. An important feature in the development of Mermithida is the presence of a stylet similar to that of Dorylaimida.

The embryonic development of nematodes of the order Marimermithida has remained unstudied until now.

Embryonic Development of the Subclass Chromadoria

Strong cytoplasmic activity in the fertilized ovum is characteristic of the representatives of the order Chromadorida that have been investigated (Figure 81a). During the first cleavage division, near one of the poles of the mitotic spindle, there is formed a protuberance of transparent cytoplasm devoid of yolk granules (Figure 81b). With the subdivision of the ovum into two blastomeres, all of the transparent cytoplasm resides in an embryonic blastomere that we designate as *AB*. The first cleavage is even: blastomere *AB*, containing material for ectoderm, is equal in size

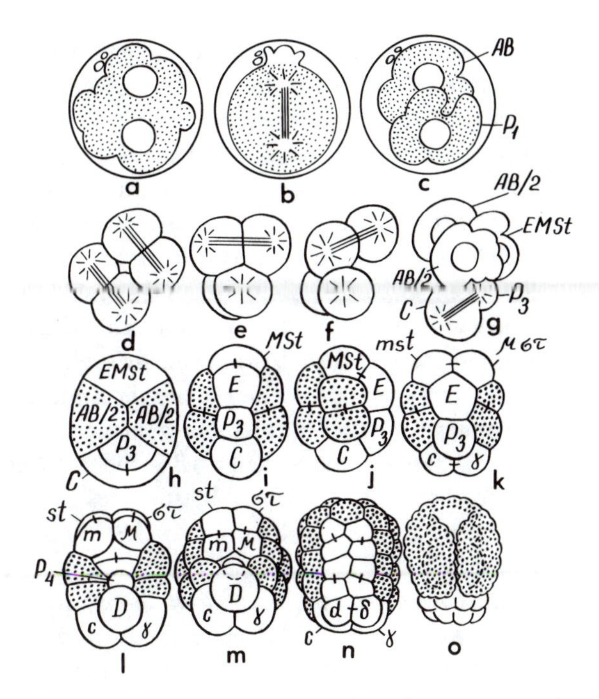

Figure 81. Embryonic development in *Hypodontolaimus inaequalis*. Descendants of blastomere *AB* are dotted. a, cytoplasmic activity of the fertilized ovum; b, first cleavage division; c, cytoplasmic activity of the two-blastomere stage; d, e, and f, different variants of the second division; g, transition to five-blastomere stage; h, five-blastomere stage after diapause; i–l, further cleavage; m and n, first step in gastrulation; o, late embryo.

to blastomere P_1 containing material for the endoderm, mesoderm, stomodeum, posterior ectoderm, and rudiments of the reproductive system. After division, the interphase blastomeres lose their globular shape and begin to exhibit strong cytoplasmic activity. The blastomeres form processes of various shapes and move within the egg envelope (Figure 81c). The second cleavage proceeds by several different means. The spindles of the second division may be oriented in mutually perpendicular directions so that, as the result of division, a tetrahedron results (Figure 81e). In other cases, the spindles of the second division are arranged in one plane (Figure 81d). Finally, the mitotic spindles may be arranged to form a T-shaped solid figure (Figure 81f). After the second division, the blastomeres lose their globular form, begin to move, and at this time it is difficult to distinguish one from another. After some time, one of the blastomeres, descendants of blastomere P_1, which we have designated P_2, divides unevenly into a small cell P_3 and a large cell *C* (Figure 81g). After this, cytoplasmic activity decreases and the blastomeres gradually align in a definite manner: blastomeres *EMSt,* P_3, and *C* lie in one line, and perpendicular to this series are arranged blastomeres descendent from cell *AB*, which have been designated *AB*/2 and *AB*/2 (Figure 81h). At this stage, cleavage temporarily discontinues. Blastomeres of the five-cell embryo become closely applied to one another, and the whole embryo acquires a round shape (Figure 81h). At the five-cell stage, the basic axes and planes of symmetry of the embryo are determined. The plane of the bilateral symmetry passing through blastomeres *EMSt,* P_3, and *C* and along the furrow between blastomeres *AB*/2 and *AB*/2 corresponds to the sagittal plane of the future embryo. Blastomeres *EMSt,* P_3, and *C* are situated on the ventral side; and blastomeres *AB*/2 and *AB*/2, on the dorsal and lateral sides.

Diapause at the five-cell stage in *Hypodontolaimus* continues for 14 to 15 hours, which exceeds the usual time interval between divisions in this species by a factor of sixteen to eighteen. After diapause, the character of cleavage changes abruptly. Blastomeres no longer display cytoplasmic activity, do not move around, and their position is entirely determined by the orientation of the division spindles. All further development is characterized by strict adherence to bilateral symmetry.

On the ventral side of the embryo is formed a chain of cells, *MSt,* *E,* P_3, and *C*, arranged in the sagittal plane of the embryo. Blastomere *MSt* subsequently divides into left and right cells *mst* and $\mu\sigma\tau$, which then divide again and provide blastomeres for the rudiments of the stomodeum (*st* and $\sigma\tau$) and the mesoderm (*m* and μ). Blastomere *E* represents the rudiment for the whole endoderm of the embryo. Blastomere P_3 divides unevenly into anterior cell P_4, representing the sexual rudiment, and a larger posterior cell *D*, which, together with blastomere *C*, becomes part of the posterior ectoderm (Figure 81l). On the dorsal side is formed a distinct layer of primary ectoderm, coming from blastomeres *AB*/2 and *AB*/2. Gastrulation occurs in two stages. In the first stage, blastomeres *m* and μ move in on blastomere *E* from the anterior at the same time that blastomeres P_4 and *D* move in from the posterior. As a result, blastomere *E* is submerged within the embryo. At the second stage, the rapidly multiplying cells of the dorsal ectoderm (descendants of blastomere *AB*) shift ventrad and gradually join along the midventral

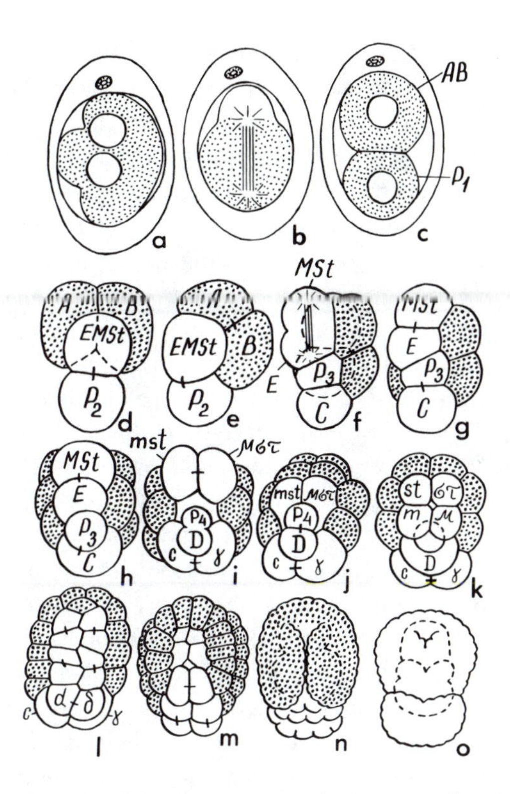

Figure 82. Embryonic development in *Desmodora serpentulus*. Descendants of blastomere *AB* are dotted. a, cytoplasmic activity of the fertilized ovum; b, first cleavage division; c, two-blastomere stage; d, T-shaped stage; e, rhombic stage; f–h, further cleavage; i and j, first step in gastrulation; k–m, late stages of cleavage; n, second-stage gastrulation; o, late embryo.

line, covering the descendant cells of *D* and the sexual rudiment P_4 (Figure 81o). At the anterior end of the embryo there remains an opening lined by cells descending from blastomeres *st* and $\sigma\tau$, which becomes the definitive mouth.

In the order Desmodorida, the fertilized ovum exhibits rapid cytoplasmic activity (Figure 82a). At the time of the first cleavage, near one of the poles of the division spindle, a lobe of transparent yolk-free cytoplasm is formed that goes completely into the anterior end of blastomere *AB* (Figure 82). After the second division, a solid T-shaped figure is formed (Figure 82d), with subsequent formation of a rhomboid figure resulting from a shift of the blastomeres (Figure 82e). At this stage, the definitive plane of bilateral symmetry (sagittal plane) of the embryo is determined, coinciding with the plane of the rhombus. Blastomere *A* occupies an anterior position, blastomere P_2 a posterior position, blastomere *B* corresponds to the dorsal side, and blastomere *EMSt* to the ventral side. On the ventral side of the embryo is formed a chain of blastomeres, *MSt*, *E*, P_3, and *C*, marking the sagittal plane of the embryo. Blastomere *MSt* divides into left and right cells *mst* and $\mu\sigma\tau$, and blastomere *C* into left and right cells *c* and Γ (Figure 82i). The descendants of blastomeres *A* and *B* provide a prominent layer of primary ectoderm. Gastrulation begins by the shift of cells *mst* and $\mu\sigma\tau$ over blastomere *E* (Figure 82j). Blastomere P_3 at this time divides into anterior cell P_4 (reproductive primordium) and posterior cell *D*. Blastomere *E* becomes fully submerged into the embryo (Figure 82j). Subsequently, it provides the beginning for the entire endoderm of the embryo. Blastomeres *mst* and $\mu\sigma\tau$ divide into anterior cells *st* and $\sigma\tau$, providing the precursor of the stomodeum, and posterior cells *m* and μ, which will give rise to the mesoderm. Descendants of blastomeres *C* and *D* represent material for the caudal ectoderm. The second stage of gastrulation occurs as the cells of the primary ectoderm increase their rate of multiplication and expand over the sides toward the midventral line (Figure 82o). The opening preserved at the anterior end of the embryo becomes a definitive mouth.

In the development of nematodes of the order Monhysterida, there is also an early establishment of bilateral symmetry. In the first cleavage, the formation of polar lobes is not observed. The first cleavage is uneven: blastomere *AB* is larger than blastomere P_1 (Figure 83). The interphase blastomeres do not exhibit cytoplasmic activity, and they remain round until the following division. The spindles of the second cleavage are oriented perpendicularly to one another, so that, after the second division, a tetrahedral figure is formed (Figure 83e). Relative to the sides of the body of the future embryo, the tetrahedron is oriented as follows: blastomeres *AB*/2 and *AB*/2 are arranged on the dorsal side, occupying left and right positions; blastomeres *EMSt* and P_2 lie on the abdominal side; blastomere *EMSt* is anterior and blastomere P_2 is posterior (Figure 83). The future sagittal plane passes along blastomeres *EMSt* and P_2 and along the furrow between blastomeres *AB*/2 and *AB*/2. Thus, in *Theristus*, the plane of definitive symmetry is already determined at the four-cell stage. Further cleavage is strictly bilateral and proceeds like that seen

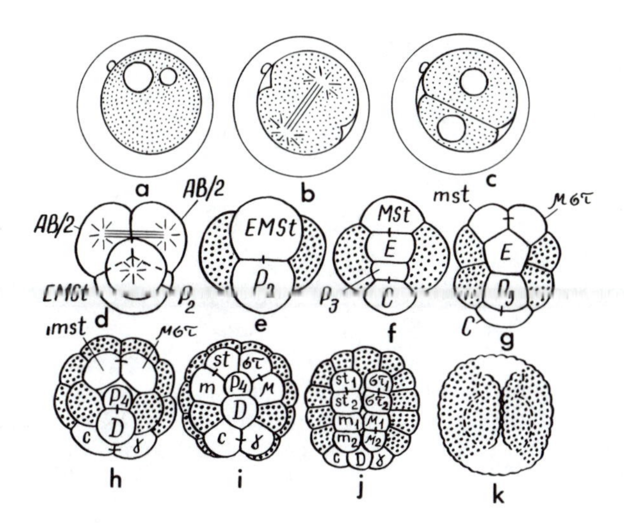

Figure 83. Embryonic development in *Daptonema setosum*. Descendants of blastomere *AB* are dotted. a, fertilized ovum; b, first cleavage division; c, two-blastomere stage; d, second cleavage division; e, four-blastomere stage; f and g, further cleavage; h–j, first stage of gastrulation; k, second stage of gastrulation.

earlier in species of Chromadorida and Desmodorida. Blastomeres *AB*/2 and *AB*/2 provide the beginning for the ectoderm, blastomere *E* forms the entire endoderm, cells *st* and στ are progenitors of the stomodeum, and cells *m* and μ are progenitors of the mesoderm. The sexual rudiment P_4 is isolated after the fourth division. Gastrulation proceeds in two stages (Figure 83).

In the order Araeolaimida, the cleavage process is similar to that observed in the development of Monhysterida. After the second division, a tetrahedron is formed that is oriented relative to the axes and plane of the future embryo as in the previous species. A definitive plane of symmetry in *Axonolaimus* passes through blastomeres *EMSt* and P_2 and along the furrow between blastomeres *AB*/2 and *AB*/2 (Figure 84). Interphase blastomeres do not exhibit any cytoplasmic activity, and all cleavage is strictly orderly and bilaterally symmetrical. The order of isolation of rudiments, potentials of blastomeres, and gastrulation proceeds like that of the three preceding examples.

The development of nematodes of the order Plectida has been described by Drozdovskiy (1977, 1978) from observations of representa-

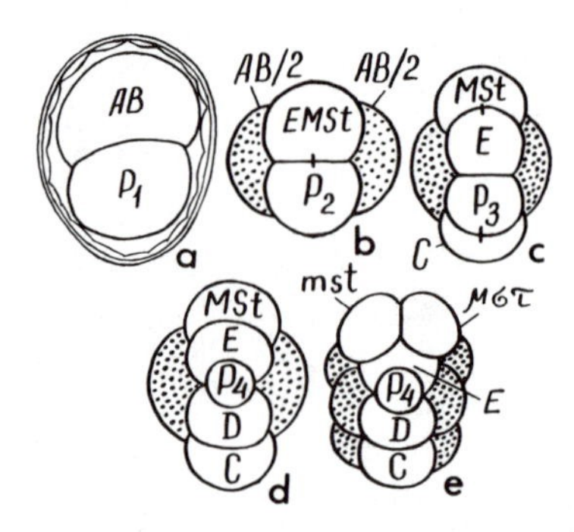

Figure 84. Embryonic development in *Axonolaimus paraspinosus*. Descendants of blastomere *AB* are dotted. a, two-blastomere stage; b, four-blastomere stage; c and d, further cleavage; e, first phase of gastrulation.

tives of the genus *Plectus*. The first cleavage in the ovum of these nematodes is uneven: blastomere *AB* is notably larger than blastomere P_1 (Figure 85, 1 and 7). The furrows of the second division proceed so as to form a T-shaped figure (Figure 85, 2 and 8). However, as the result of the morphogenetic movements of the blastomeres, the T-shaped figure in some species turns into a single plane rhombus and, in others, into a tetrahedron (Figure 85, 6, 4, and 9). Among species with the tetrahedral development, the future sagittal plane of the embryo passes across blastomeres *EMSt* and P_2 and along the furrow between blastomeres *AB*/2 and *AB*/2 (Figure 85). Further development of such embryos is very similar to the development of nematodes of the orders Monhysterida and Araeolaimida.

Among species in which the formation of a rhombus is characteristic, the future sagittal plane coincides with the plane of the rhombus. In the same way as in nematodes of the order Desmodorida, blastomere *A* occupies the anterior position in the rhombus, blastomere *B* corresponds to the dorsal side, blastomere P_2 lies at the posterior end, and blastomere *EMSt* defines the ventral side of the embryo. In further development on the ventral side, a chain is formed with cells *MSt*, *E*, P_3, and *C*, delineating the sagittal plane, and four cells of primary ectoderm are formed on the dorsal side (Figure 85, 5 and 6).

The cleavage of the ovum in various species of the genus *Plectus* is interesting because it combines both variations in the position of the blastomeres from which primary ectoderm will be derived. In one case

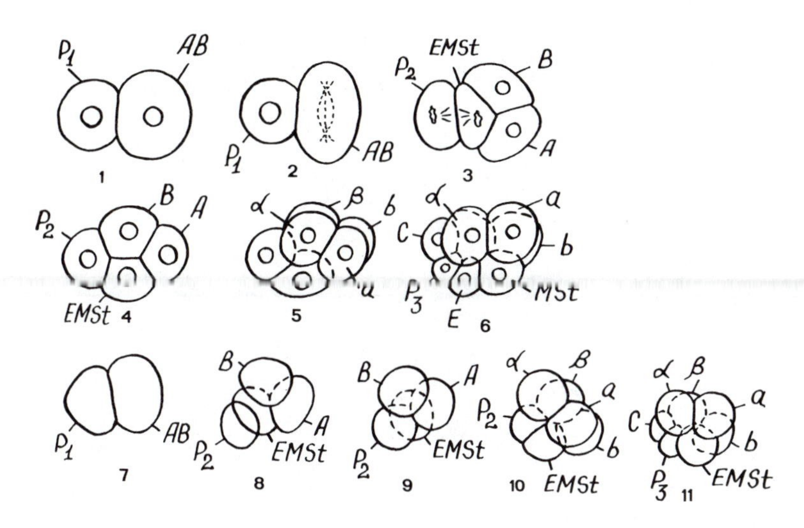

Figure 85. Embryonic development of representatives of the genus *Plectus* through the rhombus figure (1–6) and through the tetrahedron figure (7–11) (after Drozdovskiy 1978).

the primary ectoderm is situated in the anterior and posterior blastomeres (as in nematodes of the order Desmodorida), designated *A* and *B*. In the other case, the primary ectoderm is situated in the left and right blastomeres (as in nematodes of the orders Chromadorida, Monhysterida, and Araeolaimida), accordingly designated *AB*/2 and *AB*/2.

Embryonic Development of the Subclass Rhabditia

The rhombic cleavage noted among representatives of the orders Desmodorida and Plectida is characteristic of the whole subclass Rhabditia. However, the means of transition to the rhombus among representatives of various groups within this subclass are not identical (Figure 86).

Among representatives of the order Ascaridida, the spindles of division in blastomeres *AB* and P_1 are arranged perpendicularly to one another, and as a result of the second division a T-shaped figure is formed (Boveri 1888, 1899; zur Strassen 1896; Zoja 1896; Walton 1918). As a result of morphogenetic movements of the blastomeres, the T-shaped figure, after division, is transformed into a single-plane rhombus (like that described above for Desmodorida and some Plectida). In nematodes of the orders Spirurida and Strongylida, cleavage is described

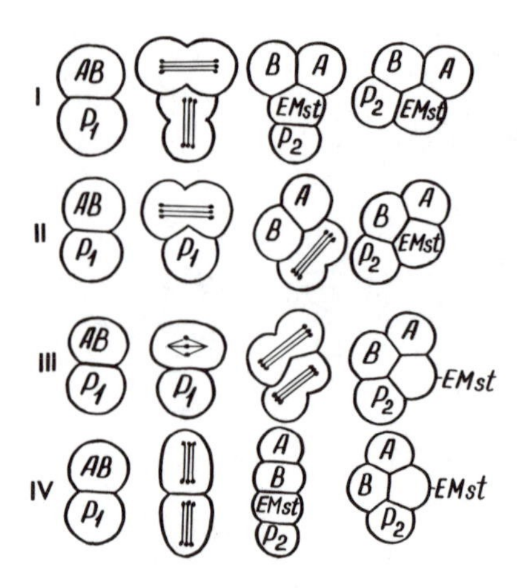

Figure 86. Basic types of cleavage in the subclass Rhabditia: I, T-shaped; II, intermediate; III, parallel; IV, linear.

in which the spindles of the second division are arranged in a T shape. First, blastomere *AB* divides, and three blastomeres form an incomplete T-shaped figure (Figure 87). Blastomere *B* moves to the future dorsal side, and under its pressure, the mitotic spindle in blastomere P_1 turns so that in its division a rhombus figure results. This type of division is called intermediate cleavage (Nigon et al. 1960; Ivanova-Kazas 1975).

Among many representatives of the order Rhabditida, a so-called parallel cleavage has been described (Ziegler 1895; Pal'chikova-Ostroumova 1926; Pai 1928; Seck 1937; Schönberg 1944; Nigon et al. 1960; Badolkhodzhayev 1969, 1970). In this case, the spindle in blastomere *AB* is established perpendicularly to the direction of the spindle of the first cleavage, but already at the metaphase stage, the spindle turns and becomes diagonal. The mitotic spindle in blastomere P_1 is established parallel to the terminal position of the spindle in blastomere *AB* (Figure 86). After this division, the rhombus figure is immediately formed.

Among representatives of the order Tylenchida, ova have a highly elongated form. The spindles of the second cleavage are oriented along the long axis of the ovum, and after the division of the blastomeres, a linear figure is formed (Strubell 1888; zur Strassen 1892; Nagacura 1930; Hattingen 1956; Mulvey 1959; Yuksel 1960; Drozdovskiy 1967, 1968). This type of cleavage, in fact, is called linear. After division, the blastomeres shift to form a figure close to that of a rhombus (Figure 86, IV).

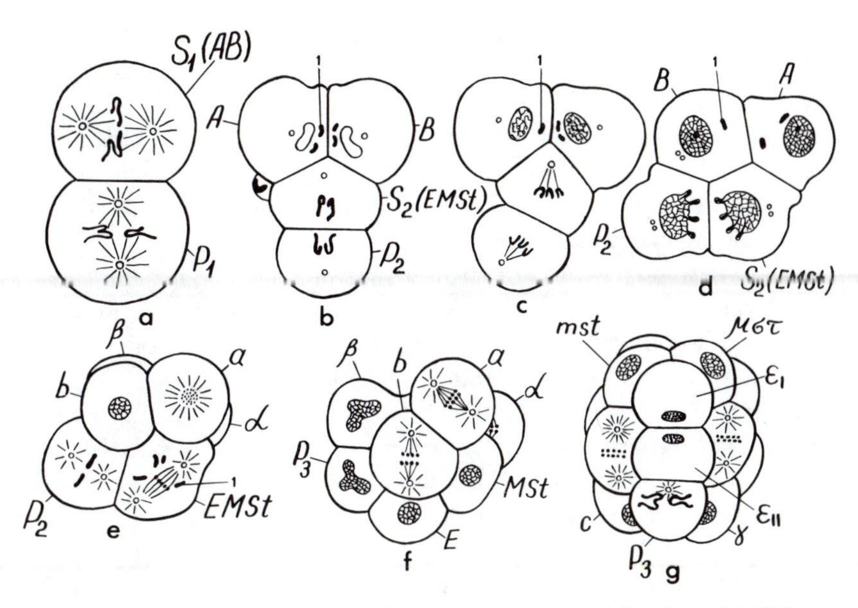

Figure 87. Cleavage in the ovum of *Parascaris equorum* (after Boveri 1899): a, two-blastomere stage; b, T-shaped stage; c, transition to the rhomboid figure; d, rhombic figure; e, third cleavage division; f, eight-blastomere stage, viewed from the left; g, sixteen-blastomere stage, viewed from the abdominal side. 1, pieces of chromatin material cast into the cytoplasm.

Just as among nematodes of the order Desmodorida and some Plectida, the future sagittal plane of the embryo in the development of Rhabditia is determined at the rhombus stage. Blastomere *A* in the rhombus occupies the anterior position, blastomere P_2 the posterior position, blastomere *B* corresponds to the future dorsal side, and the blastomere *EMSt* (*EM*), to the ventral side. Subsequently, on the ventral side of the embryo, there is formed a chain of blastomeres, *MSt* (*M*), *E*, P_3, and *C*, marking the sagittal plane of the embryo, while our blastomeres are formed on the dorsal side: *a*, α, b and β (Figure 87).

Among nematodes of the orders Ascaridida and Spirurida, blastomere *MSt* divides three times and provides the beginning to a pair of stomodeal cells and two pairs of mesodermal cells bordering the blastopore. The blastopore at first has a round form, but then narrows and becomes slitlike. Through the growth of the primary ectoderm (descendants of blastomere *AB*) from lateral to ventral, the blastopore is closed midventrally. The anterior opening provides the mouth, and the posterior one closes. Sometimes the slitlike blastopore is formed by the folding of the flattened embryo along the future anterior-posterior axis, as in *Camallanus* (Martini 1903).

In nematodes of the orders Rhabditida, Strongylida, and Tylenchida, the blastopore closes very early and does not take on the slitlike form. The definitive openings of the digestive tract are placed independently of the blastopore.

The descendants of blastomere *AB* provide the so-called "primary" ectoderm, which forms the larger part of the covering ectoderm of the embryo. Blastomeres *C* and *D* provide the "secondary" ectoderm. In a number of forms, cell *F* separates from blastomere P_4, providing the "tertiary" ectoderm. Views of various authors differ on the fate of secondary and tertiary ectoderm among nematodes of the subclass Rhabditia. Boveri (1899) believed that these cells provide only the covering of the posterior part of the body. Zur Strassen (1896), von Strassen (1959), and Müller (1903) believed that the secondary ectoderm in the later stages of development moves forward along the dorsal side of the embryo and partially displaces the primary ectoderm. We have observed nothing similar in the development of *Gastromermis.* According to Müller's data (1903), part of the mesoderm of the posterior area of the body also has its beginning in the secondary endoderm.

The progenitor of the reproductive system among nematodes of the subclass Rhabditia is isolated after the fourth division (designated P_4) or just after the fifth (designated P_5). A remarkable feature of the isolation of the antecedent of the reproductive system among nematodes of the subclass Rhabditia is the presence in some forms of the diminution of chromatin in the somatic blastomeres. This phenomenon was first discovered in equine Ascaridae by the outstanding embryologist Boveri (1887, 1888, 1899). Later, the diminution of chromatin in somatic blastomeres was found in a number of other forms in subclass Rhabditia (Herla 1894; Meyer 1895; Zoja 1896; Bonnevie 1901; Walton 1918; Fogg 1930; King and Beams 1938; Lin 1954; Moritz 1967). The diminution of chromatin is found also in dicyemids, trematodes, crustaceans, and insects.

In examining the embryonic development of nematodes, we must briefly review the remarkable research done in recent years on *Caenorhabditis elegans* of the order Rhabditida (Ehrenstein and Shierenberg 1980; Sulston et al. 1983). Using various methods, the authors of these works successfully traced the genealogy of cells of the late embryo up to the emergence of a juvenile. The ectoderm of the juvenile consists of 352 cells, coming from blastomeres *AB* and *C*, only 86 going into the hypoderm and the remaining 266 cells constituting the antecedent of the nervous system. The majority of the future nerve cells migrate inward from the so-called "ventral furrow." This furrow goes along the middle line of the ventral side of the embryo at the age of 5.5 to 6.25 hours and probably corresponds to the line of closure of the slit-shaped blastopore in the more primitive nematodes (among representatives of the order Rhabditida, the slit-shaped blastopore is not evident). Some of the neuroblasts of *C. elegans* come from the hypodermis through tangential division. Blastomeres *MSt* and *D* provide the beginning for cells of the primary and secondary mesoderm. From these blastomeres come the ninety-one mesodermal cells of the juvenile (eighty-one cells of somatic musculature, four cells of the specialized muscles of the intestine, four coelomocytes, one cell of undifferentiated mesoderm behind the progenitors of the reproductive system, and one cell forming the slit contacts between the muscle cells of the head end).

The pharynx of *C. elegans* contains seventy-seven cells of which forty-three are epithelial and nerve cells and thirty-four are muscle cells. The epithelial and nerve cells come from descendants of blastomeres *AB* and the muscle cells from descendants of blastomere *MSt*. The intestine of the juvenile consists of twenty cells, coming from the endodermal blastomeres E_1 and E_2. The progenitor of the reproduction system in the juvenile consists of two large cells—descendants of blastomere P_4. Two large cells are related to two small cells—descendants of blastomere *MSt*. From the latter come some structures of the gonads of adult specimens.

Two evolutionary lines can be clearly delineated in the comparative embryology of nematodes. At the base of one of them is the embryonic development of free-living marine nematodes of the order Enoplida. In the cleavage of these forms there is no early establishment of bilateral symmetry nor rigid determination of the fate of blastomeres, such as takes place in the development of most nematodes. Cleavage among marine members of Enoplida is variable. At the four-cell stage, we observe various geometric configurations: tetrahedral, rhombic, and T-shaped. All of these configurations are encountered in the development of nematodes of other groups, but in Enoplida they exist in development in the same species within the context of individual variability. Interphase blastomeres in the development of Enoplida exhibit rapid cytoplasmic activity. A small role in development is played by the movement of blastomeres after division. Individual variability is exhibited also in the existence of some variants in the arrangement of blastomeres in the late stages of cleavage. This can even engender the idea that the cleavage among members of marine Enoplida is indeterminate, but this is not so. The endodermal blastomere in enoplid development is already isolated at the eight-cell stage, and from the fifteen- to sixteen-cell stage, the blastomeres begin gastrulation. Such early determination of the fate of the endodermal blastomere is characteristic only for determinate cleavage. The endodermal blastomere in Enoplida is always tied to the anterior of the two blastomeres of the embryo. This localization of the precursor of the primary endoderm serves as the basis for working out further determinant development.

A primitive feature in enoplid development is the late establishment of the elements of bilateral symmetry. The arrangement of blastomeres at stages of cleavage is more asymmetrical than bilaterally symmetrical. Only from the moment of the elongation of the blastopore is the bilateral symmetry of the embryo no longer in doubt.

In dioctophymid nematodes, the fate of all blastomeres of the embryo is predetermined, beginning at the early stages of cleavage. At the early stages there are two basic configurations—tetrahedron and rhombus—but later both paths of development merge into one. The potential for the ectoderm of the anterior part of the body, the stomodeum, and endoderm lie within the anterior blastomere of the embryo at the two-cell stage. The posterior blastomere of the embryo will give rise to the ectoderm of the posterior part of the body, the mesoderm, and the reproductive system. In contrast to most other nematodes, but similar to nematodes of the order Enoplida, there is no early isolation of the progenitor of the reproductive system in the development among members of Dioctophymida. As in Enoplida, the precursor of the reproductive system lies with the appearance of two cells —right and left—probably coming from blastomeres *c* and Γ. Dioctophymida and Enoplida do not have a median progenitor of the reproductive system characteristic of nematodes of other orders.

Among nematodes of the orders Mononchida and Dorylaimida, cleavage is completely determinant. At the two-blastomere stage, the material for the ectoderm of the anterior part of the body, the stomodeum, and endoderm is localized in the anterior cell. The potential for the ectoderm of the posterior part of the body, the mesoderm, and the reproductive system is localized in the posterior blastomere. In nematodes of these orders, the median progenitor of the reproductive system P_4, appears first, later dividing into anterior and posterior primary sexual cells.

The development of Mermithida also is distinctly determinant from the earliest stages and has a median progenitor cell for the reproductive system.

For all nematodes of the Enoplida whose embryogeny has been studied, it is characteristic to have a pronounced slit shaped blastopore, the division of which leads to the formation of a complete intestine.

Another line of evolution in embryonic development is related to the different arrangements of potentials between the first blastomeres of the embryo. At the two-blastomere stage the potential for the ectoderm of the anterior part of the body and the dorsal side is situated in the anterior blastomere, and the potential for the ectoderm of the posterior part of the body, the endoderm, mesoderm, stomodeum, and reproductive system is situated in the posterior blastomere. This evolutionary branch has not preserved the primitive cleavage we encountered in Enoplida. However, in the development of nematodes of the order Chromadorida are vestiges of such a condition. Up to the five-blastomere stage, the cleavage is variable, the blastomeres are mixed and exhibit rapid cytoplasmic activity. This phenomenon can be regarded as a recapitulation of the original nematode cleavage, which was characterized by variability in the arrangement of blastomeres, their great mobility, and by being weakly determinant. After the five-blastomere stage, the plane of the bilateral symmetry of the embryo is determined, and the cleavage becomes very orderly. For Chromadorida, Desmodorida, Monhysterida, Araeolaimida, Plectida, and all Rhabditia, the presence of the median sexual rudiment is characteristic. It is interesting to note that although the early establishment of bilateral symmetry in cleavage is characteristic for all representatives of this evolutionary branch, the geometry of the figures in which the establishment of bilateral symmetry takes place is varied. In Chromadorida, this is a distinctive five-cell figure; in Monhysterida and Araeolaimida, a tetrahedron; and in Desmodorida, some Plectida, and all Rhabditia, a rhombus. This variation also correlates with differences in the prospective potentials of the blastomeres of the early embryo. In Chromadorida, Monhysterida, Araeolaimida, and part of Plectida, the primary ectoderm is localized in the left and right blastomeres designated *AB*/2, and in Desmodorida, part of

Plectida, and all Rhabditia, it is localized in the anterior and posterior ones designated *A* and *B*.

The growth of the primary ectoderm from the sides and its closing along the middle line of the ventral side in Chromadorida, Desmodorida, Monhysterida, and Araeolaimida can be compared with the closure of the slit-shaped blastopore. A real slit-shaped blastopore is noted in nematodes of the orders Ascaridida and Spirurida. In Rhabditia and Tylenchida, the blastopore closes early and both the mouth and anus break through without a connection with the blastopore.

The data on embryonic development are graphic evidence of the early divergence of the basic branches of nematodes: Enoplida and Chromadorida-Secernentea. The ancestral forms of nematodes were probably characterized by weakly determinant cleavage with variable location for the potential of the various types of embryonic tissue. Among nematodes of the enoplid branch, localization of endoderm was fixed in the anterior of the first two blastomeres of the embryo. Nematodes of the Chromadorida-Secernentea branch have been characterized by the localization of endoderm in the posterior blastomere. This difference in the distribution of potentials between blastomeres in the two basic evolutionary lines of nematodes was first discovered by Drozdovskiy (1975).

The geometric forms of cleavage in nematodes were originally variable, and we encounter this condition among marine Enoplida. In further evolution, various groups of nematodes made basic the variable configurations of blastomeres (tetrahedron, rhombus, and T figures), encountered in Enoplida.

It is interesting that parallel evolution led to the development of similar embryonic characteristics among various groups of nematodes. Thus, independently in both basic branches, there is the formation of the median progenitor of the reproductive system, a uniform number of mesodermal and stomodeal blastomeres and their similar arrangement in the embryo, and so forth.

Independently in both main evolutionary branches of nematodes, there was the formation of bilateral symmetry in cleavage. And, in both main branches, the rhombus figure became firmly established among the higher forms. The plane of the rhombus corresponds to the sagittal plane of the embryo, and the distribution of the potentials of definite structures among blastomeres corresponds completely to the arrangement of these structures in the later embryo.

In analyzing the development of nematodes, we evidently can speak with full justification of the channelization in the evolution of nematode embryogeny. The reasons for this concept are related to the pattern of embryonic development of the juvenile stages of nematodes, which is very similar throughout the class. In general, with consideration of the structure of adult forms as well, nematodes are a surprisingly uniform group with the same arrangement of practically all basic structures. This uniformity, which was described as early as 1865 by Bastian, contrasts sharply with the variability in forms of cleavage, and is the reason for the directional evolution of embryonic development.

Postembryonic Development

Nematodes have direct development. From the ovum develops a juvenile that is similar to and living in the same environment as the adult. The presence of direct development in nematodes is also a consequence of the originally small sizes of representatives of this group. Development with a pelagic larva, which offers significant advantages to the organisms in the sense of widespread settlement and which permits them to use the food resources of pelagic biotopes, entails massive mortality of larvae. Therefore, pelagic development is attainable for relatively large forms that produce a large quantity of ova. Small forms with sexual reproduction, as a rule, have direct development. It is interesting that among the most primitive Bilateria, namely Turbellaria, direct development is characteristic of representatives of all orders (including the most primitive Turbellaria), except for large marine forms of the order Polycladida. Direct development is also characteristic of other groups of small lower Metazoa—Rotifera, Gastrotricha, Kinorhyncha, and others. On the other hand, pelagic development is more characteristic for rather highly organized coelomic Bilateria—annelids, mollusks, echinoderms, and others. Among the latter we encounter one of the most primitive types of ontogeny among Metazoa in general. These are the blastiform stages of development in echinoderms, which acquire a free existence in their aquatic environment.

The postembryonic development of nematodes is accompanied by periodic molting; that is, the discarding of old cuticle, which impedes growth. The number of molts in nematodes is constant within the entire class (i.e., four [Maupas 1990; Filip'jev and Michailova 1924; Pai, 1928; Wessing 1953; Brunold 1954; Frenzen 1954; Chuang 1962; Paramonov

1962; Malakhov 1974a]). The constant number of molts is evidence of great evolutionary stability in nematode ontogeny. In this connection, it is interesting to recall that in postembryonic development among primitive representatives of various evolutionary branches of the Arthropoda, the number of molts is not constant. Rather, the animal grows and periodically sheds during its lifetime, and a constant number of molts characterizes evolutionarily advanced or highly specialized groups.

We encounter four molts in all nematodes, from the primitive free-living marine forms to the specialized parasitic representatives of this class. The original type of postembryonic ontogeny for nematodes is that in which the juvenile of the first stage leaves the ovum and undergoes all four molts in the external environment. This type of development is characteristic of most free-living nematodes of various taxa. Evolutionarily advanced parasitic nematodes of the subclass Rhabditia have more or less undergone embryonization of molting, in that one or two molts takes place within the shell of the ovum. Thus, among most tylenchid phytohelminths, a second-stage juvenile emerges from the ovum (the first molt takes place before the juvenile hatches). The embryonization of development occurs later in many zooparasitic nematodes (for example, in Strongylida) where, in a number of forms, two molts occur within the egg, and it is the third-stage juvenile that hatches.

Very interesting patterns of nematode postembryonic development have been revealed by the quantitative analysis of growth and development of juveniles. Unfortunately, the quantitative study of nematode growth processes has not been given the attention it deserves.

As is known, the general pattern of development in poikilothermic animals is an uneven S-shaped growth curve. The most general growth formula was introduced by von Bertalanffy (1938). This formula proved useful for describing a larger number of groups of animals such as fish, crustaceans, mollusks, and others. In the works of a number of researchers, it was also shown that the entire growth of free-living nematodes of various taxonomic groups is subject to the S-shaped curve (Pai 1928; Wessing 1953; Wieser and Kanwisher 1960; Tietjen and Lee 1973; Malakhov 1974a; Byerly et al. 1976). However, adequate research with regard to the molting or growth of nematode juveniles has been conducted on only a few species. Thus, there are data at present only for *Rhabditia anomala* (Wessing 1953), *Chromadora macrolaimoides* (Tietjen and Lee 1973), and *Pontonema vulgare* (Malakhov 1974a). To generalize these data, it can be stated that the growth curve of these

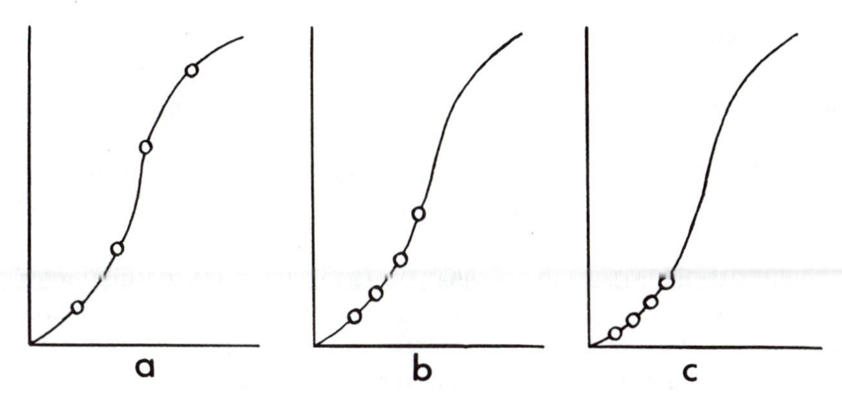

Figure 88. Growth curves of various ecological groups of nematodes. Points indicate molting times. a, free-living nematodes; b, poorly specialized parasites; c, highly specialized parasites.

nematodes has the appearance shown in Figure 88a. The gently sloping parts of the curve correspond to relatively slow rates of growth and fit the young juvenile stages and the sexually mature forms. The most intensive growth fits the segment between the second and fourth molts; that is, it is accomplished by juveniles of the third and fourth stages of development (Figure 88a).

Nematode growth is accompanied not only by change in body size and mass, but also by change in proportion and by significant restructuring of internal organization. Different sections of the body grow at different rates. In the study of the processes of uneven growth in nematodes, the indices of de Man have proven very useful, and they are widely used in the systematics of free-living nematodes and phytohelminths: a = body length/maximum width, b = body length/pharyngeal length, and c = body length/tail length. Indices b and c have special significance. A general rule for all nematodes is that in first- and second-stage juveniles. the most formed sections of the body are the trophico-sensory and caudal sections, while the trophico-genital section is underdeveloped. During the process of postembryonic development, the significance of indices b and c increases, and this reflects the higher growth rate of the trophico-genital section in comparison with the trophico-sensory and caudal section. The most abrupt changes in body proportions occur in the fastest-growing juveniles, stages III and IV, of free-living and phytoparasitic nematodes (Wessing 1953; Yuksel 1960;

Paramonov 1962; Malakhov 1974a). The third and fourth stages of nematode juvenile development, as especially noted by Paramonov (1962), can be fully regarded as the critical period of postembryonic ontogeny.

The dominant development of the trophico-sensory and caudal section in young nematode juveniles has a clearly adaptive significance. At the beginning of life of individual animals, it is more important to have organizational features that promote survival and provide conditions for further growth, while, at the later stages of ontogeny, the most important traits of structure and biology are those responsible for continued survival of the species. In the trophico-sensory and caudal section of the nematode body are situated the organs for finding and seizing food and the organs that play an important role in locomotion, respectively. In the trophico-genital section of the body are localized the progenitors of the reproductive system that are fully developed only in mature animals (fifth stage).

Actually, among juveniles of the first stage, the progenitor of the reproductive system is represented by only a pair of primary germinal cells (Wessing 1953; Yuksel 1960; Chuang 1962; Paramonov 1962; and others). The most rapid development of the reproductive system occurs at the third and fourth stages of development. Rudiments of either male or female gonoducts are formed in third-stage juveniles. In the fourth stage, a functional reproductive system is already completely formed but not yet functional. The females have a rudimentary vagina, and the males have very apparent rudiments of spicules. Thus, change in body proportions, caused by the accelerated growth of the trophico-genital section of the body in third- and fourth-stage juveniles, is reflected in rapid restructuring in the nematode internal anatomy.

It is known that significant changes are undergone in the organization of nematodes with the transition to the parasitic form of life. The postembryonic development of nematode parasites of animals, for example, is complicated by the fact that, in the life cycle of many forms, there is a change of habitat, change of hosts, and complex migration in the host's tissues and organs. To what extent does this affect the rates and patterns of growth of nematode parasites of animals?

It has turned out that a significant share of the growth in nematode parasites of animals takes place in the last (i.e., fifth) stage of development. It is possible to distinguish two groups among nematode parasites of animals (Valovaya and Malakhov 1978). For one of them a characteristic overall growth in body mass in the fifth stage is 50–80%, and the growth curve of these forms takes on the appearance represented in Figure 88b.

For the other group, to which belong such generally known parasites as *Ascaris, Ascaridia,* and *Syngamus,* the amount of growth in the fifth stage is up to 99%. The growth curve of this form is depicted in Figure 88c.

Thus, the tendency to shift basic growth to the fifth and last stage of development can be clearly traced. As paradoxical as it seems, in specialized parasites of animals the most rapid body growth is not accompanied by molting.

The predominant growth in the fifth stage in parasitic nematodes, of course, has adaptive significance. Parasitic nematodes reside in the definitive host (or organ of definitive localization within a single host) only at relatively late stages in its life cycle. In the definitive host are created the most favorable conditions for the growth and development of the parasite. This circumstance is also conditioned by the shift in the period of predominant growth from the mid (third and fourth) stages of development to the last (fifth) stage. The transfer of the basic growth to the last stage of development should be regarded as a specific adaptation to the parasitic form of life.

In this connection, attention must be given to one important circumstance. Essentially, within the confines of the class Nematoda two types of growth are found. In the first instance, nematode growth is accompanied by periodic molts. This growth is characteristic of free-living nematodes and many phytohelminths. Parasites of animals also grow in this way in the first through fourth stages of development. The processes of molting, their mechanisms, and regulation have already been studied to a certain extent and continue to attract the attention of numerous researchers.

The second type of growth is characteristic of the fifth stage of development in a large number of zooparasitic nematodes. In this instance, there are great changes in the structure of the whole organism: the body length increases many times (by a factor of almost 100 in certain species), the cuticle thickens several times, and evidently the hypodermis and musculature undergo some restructuring. These restructurings are not accompanied by molts. The processes of such growth and the rearrangement that takes place in the structure of various organs have not attracted the attention of researchers up to the present time.

Meanwhile, we are dealing with some specific mechanisms. How, for example, can one imagine the increase in length and repeated thickening of the cuticle without molting? What takes place concerning the muscle cells with a hundredfold increase in body length? There are as yet no answers to these questions, and essentially they have not even been asked. These questions are crucial because these mechanisms, still unknown, represent an important acquisition that allows nematode parasites of animals to transfer basic growth to the last (the fifth) stage of development. In future research on the morphology, histology, and ultrastructure of the nematode parasites of animals, attention should be given to the study of mechanisms of this type of growth. The transfer of basic growth to the last stage of postembryonic development is a specific adaptation of nematodes to the parasitic form of life, provided by particular morpho-physiological mechanisms, the understanding of which appears to be of interest not only from theoretical but also practical points of view.

3

Classification of Nematodes and Phylogenetic Relationships within the Class

The classification of nematodes in light of our current perception of the basic trends of evolution within the class has become an increasingly timely problem, and the construction of an all-embracing classification for the entire class Nematoda is a kind of superproblem for nematology. But, even at the start of such an effort we can see what is missing to construct a truly natural system.

Fundamental difficulties in constructing a natural system for nematodes relate to the insufficient study of this large class. Recent rapid expansion in our knowledge about nematodes—the description of new species, genera, families, and even orders—leads to a major revision of the entire classification of the Nematoda. But the structure of nematodes and their development have not yet been studied well enough to construct morphological series within the limits of the whole class.

Other difficulties relate to the specifics of the class Nematoda. Nematodes are an unusually uniform group of animals in the morphological sense. Not only a whole organism, but also many individual systems of organs are constructed according to a single type within the entire class. This uniformity appears even in relatively minor traits of organization, which, in other groups of the animal kingdom, vary within the boundaries of taxonomic groups of low rank. The range of morphological differences between the most primitive and the most evolutionarily advanced nematodes is relatively small.

The common ancestor of all modern nematodes probably already possessed all the common "nematode" traits, and its descendants had a very limited set of morphological combinations for evolution. The evolution of such morphological structures as the hypodermis, sensory organs, pharynx, stoma, and others in most major evolutionary branches of the Nematoda has proceeded in a parallel course, almost all variations being realized within each evolutionary branch. The large role of parallelism in the evolution of morphological structures in nematodes was noticed early by Filip'jev (1921), who wrote: "There is no organ where similar forms in different groups have not developed. . . . There is an impression that an organism in various groups of nematodes is so similar that the same biological reasons cause the same morphological consequences."

Recent publications have been devoted to the classification of individual ecological groups of nematodes, for example, the classification of nematode parasites of vertebrate animals (Skryabin et al. 1949; Yamaguti 1961, 1963; Chabaud 1974), the classification of free-living nematodes (Lorenzen 1981a), or only marine nematodes (Gerlach and Riemann 1973, 1974). Without denying the significance of these works in deepening our knowledge of certain groups of nematodes, let us note that they do not provide a complete idea of the position of taxa in the system of the class as a whole. In this work, we will propose a nematode classification encompassing all ecological groups of the class (Malakhov et al. 1982). Of course, this classification must be regarded as preliminary.

General Classification

The class Nematoda is subdivided into subclasses, but specialists do not have a unified opinion as to subclasses and their respective inclusions.

In a majority of modern taxonomic works on nematology, two subclasses are adopted: Adenophorea (= Aphasmidia) and Secernentea (= Phasmidia). The subdivision of nematodes into two groups was first laid down by Chitwood (1933, 1937), who proposed the names Aphasmidia and Phasmidia. Subsequently it was revealed that the name Phasmidia had been given earlier to an order of insects. Therefore, Chitwood himself (1958) and the majority of other nematologists began to use the names Adenophorea Linstow, 1905, and Secernentea Linstow, 1905. Let us note that Linstow himself (1905, 1909) subdivided nematodes not into two groups, but into four: Secernentes, Resorbentes, Pleuromyarii, and Adenophori. The extent of Adenophori and Secernentes and the concepts of Linstow disagreed substantially with the modern subclasses of Adenophorea and Secernentea.

Chitwood (1933) and Chitwood and Chitwood (1950) gave nematodes the rank of a phylum, and Aphasmidia and Phasmidia were regarded as classes of this phylum. Class Aphasmidia was subdivided by them into the orders Chromadorida and Enoplida, and class Phasmidia into the orders Rhabditida and Spirurida. Roughly the same system was maintained by Pearse (1942), who also divided nematodes into two classes, Aphasmidia and Phasmidia, but he regarded the orders isolated by Chitwood and Chitwood as subclasses, changing the endings as follows: Chromadoria, Enoplia, Rhabditia, and Spiruria.

Andrassy (1974, 1976) divided nematodes into three subclasses: Torquentia (corresponding to Chromadoria), Penetrantia (corresponding to Enoplia), and Secernentia. Drozdovskiy (1975, 1977, 1978, 1980) isolated two subclasses among nematodes—Enoplia and Chromadoria—the latter including, together with Chromadoria proper, all nematodes assigned earlier to the subclass Secernentea.

In the classification of the Nematoda we distinguish three subclasses: Enoplia, Chromadoria, and Rhabditia. The contents of Enoplia and Chromadoria, in our opinion, coincide with Pearse's taxa (1942). The extent of Rhabditia, in our opinion, is greater than that indicated by Pearse. This subclass, in our understanding, includes all nematodes earlier combined in the subclass Secernentea (= Phasmidia). We use the names Enoplia Pearse, 1942 (instead of Penetrantia Andrassy, 1974), Chromadoria Pearse, 1942 (instead of Torquentia Andrassy, 1974), and Rhabditia Pearse, 1942 (instead of Secernentea Linstow, 1905, and Phasmidia Chitwood, 1933), keeping in mind the increasingly deep-rooted trend to confer to taxa of higher rank names derived from the names of representative taxa of lower rank.

The orders included by us in each of the three subclasses are listed below:

Class Nematoda

Subclass Enoplia

Orders: Enoplida
Marimermithida
Mononchida
Dorylaimida
Mermithida
Trichocephalida
Dioctophymida
Subclass Chromadoria
Orders: Chromadorida
Monhysterida
Desmodorida
Araeolaimida
Plectida
Desmoscolecida
Subclass Rhabditia
Orders: Rhabditida
Tylenchida
Strongylida
Oxyurida
Ascaridida
Spirurida

Characteristics of Higher Taxa

Subclass Enoplia Pearse, 1942

Diagnosis: Nematoda. Cuticle with subtle striae on surface layers or smooth with pores and highly impenetrable. Cuticle without sclerotizations. Lateral differentiation very weak or absent. Hypodermis cellular. Musculature polymyarian and of the primary coelomyarian type. Cellular constancy absent in most organs. Cephalic setae in two circles (6+10) or, more rarely, in three circles (6+6+4). Somatic setae usually numerous. Amphids pocket-shaped or, as an exception (Tripyloidina), spiral. Pharynx cylindrical or gradually widening toward base; sometimes divided into narrow anterior half and wide posterior half, and occasionally posterior region of pharynx with bulges. Three or five pharyngeal glands present with either three openings in anterior part of pharynx, or five openings in posterior part of pharynx; less frequently, openings of three glands lie in anterior part of pharynx, and two others in posterior part of pharynx. Midgut polycytose. Parasitic representatives with pharynx sometimes transformed into glandular organ (stichosome), and midgut into reserve organ (trophosome). Cervical gland, when present, saclike with noncuticularized duct opening anterior to nerve ring. Caudal glands present, but reduced in soil and parasitic representatives. Hypodermal glands usually numerous. Male reproductive system with paired testes, and ejaculatory duct equipped with powerful musculature. Supplementary organs present or absent. Female reproductive system monodelphic or didelphic, antidromal or, rarer, homodromal. Cleavage of ovum variable or strictly bilateral. Genetic determinant for endoderm situated in anterior blastomere of two-cell stage. Representatives occur as free-living marine, freshwater, and soil forms, as well as commensals and parasites of plants and animals.

Order Enoplida Chitwood, 1933

Diagnosis: Enoplia. Cuticle subtly annulated (appears smooth with light microscopy observations). Setiform and papilliform cephalic sensilla in two or three circles. If cephalic setiform and of uneven length, then six setae of second circle longer than four setae of third circle. Amphids pocket-shaped or, more rarely, spiral. Labia absent or present with three or six primary labia. Pharynx simple, cylindrical or posterior region of pharynx composed of serial bulbs. Pharynx with three or five pharyngeal glands; glands open with three pores in anterior part of pharynx, or three open in anterior part of pharynx and two in posterior part. Hypodermal glands numerous. Caudal glands present. Cervical gland rarely absent. Female reproductive system didelphic or, rarely, monodelphic. Spicula paired; gubernaculum rarely reduced. Supplements present or absent. Representatives occur as free-living marine, freshwater, and, rarely, soil forms, and very rarely, as commensals of invertebrates. Suborders: Enoplina Chitwood, 1933; Oncholaimina de Coninck, 1965; Tobrilina Tsalolichin, 1976; and Tripyloidina de Coninck, 1965.

Order Marimermithida Rubtzov, 1980

Diagnosis: Enoplia. Cuticle when observed with light microscopy smooth or subtly annulated. Cephalic sensilla papilliform or setiform and in two circles (6+10 or 6+4). Amphids porelike, rarely large. Somatic sensilla reduced. Buccal capsule without specialization and pharynx cylindrical, or buccal capsule reduced to absent, and pharynx reduced. Anus present or absent. Hypodermal glands numerous and well developed. Cervical gland present. Caudal glands reduced. Female reproductive system didelphic. Males known only for *Rhaptothyreus,* with single spiculum. Juveniles parasitic in marine invertebrates; known hosts include representatives of Echinodermata and Priapulida. Taxonomic structure not well defined. Genera: *Marimermis, Trophomera, Rhaptothyreus, Ananus, Acronema,* and *Abos.*

Order Mononchida Jairajpuri, 1969

Diagnosis: Enoplia. Cuticle smooth with light microscopy observation. Cephalic sensilla papilliform, each situated on cephalic protuberance at base of each lip. Amphids large and pocket-shaped. Oral cavity large and strongly cuticularized with one or more onchia. Cylindrical pharynx with five pharyngeal glands opening posterior to nerve ring. Cervical gland weakly developed or reduced. Caudal glands present. Gonoducts paired or single. Spicula paired, gubernaculum present or absent, and supplements present. Inhabitants of fresh water and soil. Superfamilies: Bathyodontoidea Clark, 1961; and Mononchoidea Chitwood, 1937.

Order Dorylaimida Pearse, 1942

Diagnosis: Enoplia. Cuticle very finely striated, but appearing smooth with light microscopy observations. Cephalic sensilla papilliform. Amphids pocket-shaped or porelike. Oral cavity narrow, reduced, and armed with mobile onchostyle. Pharynx differentiated into narrow anterior and wider posterior regions. Each of five pharyngeal glands opens posterior to nerve ring. Midgut with or without prerectum. Cervical gland weakly developed or completely reduced. Caudal glands reduced. Gonoducts paired or, rarely, unpaired. Spicula paired and gubernaculum present or absent. Supplements present. Representatives include free-living soil and freshwater forms and plant parasites. Suborders: Dorylaimina Pearse, 1942; and Diphtherophorina Coomans and Loof, 1970.

Order Mermithida Hyman, 1951

Diagnosis: Enoplia. Cuticle smooth with microscopic pores. Cephalic sensilla papilliform. Amphids pocket-shaped. Oral cavity reduced. Juveniles with onchiostyle. Walls of pharynx form glandular stichosome.

Midgut replaced by trophosome. Cervical gland and caudal glands reduced. Female gonoducts paired. Spicula paired or merged as one. Gubernaculum absent and genital papillae paired. Parasites of members of Arthropoda and, more rarely, other invertebrates (mollusks, turbellarians, nematodes, and hydra). Two superfamilies: Mermithoidea Wülker, 1924; and Tetradonematoidea Artykhovsky and Kharchenko, 1977.

Order Trichocephalida Skryabin and Schulz, 1928

Diagnosis: Enoplia. Body more or less abruptly narrowed anteriorly. Cuticle smooth. Cuticle with rows of slitlike pores—bacillary bands—through which flow secretions of hypodermal glands. Papilliform cephalic sensilla arranged in three circles. Amphids reduced and porelike. Posterior part of pharynx transformed into glandular stichosome. Onchiostyle present, at least in the juvenile stage. Midgut normally developed or partially reduced. Anus present or absent. Cervical gland reduced. Caudal glands absent. Female reproductive system monodelphic. Males with one spiculum or none; gubernaculum absent. Ova plugs at poles. Parasitic in vertebrates, and with or without intermediate host. Three superfamilies: Trichocephaloidea Skryabin and Schulz, 1928; Cystoopsoidea Ssaidov, 1953; and Muspecioidea Chabaud, 1974. The latter superfamily is conditionally included in this order.

Order Dioctophymida Railliet, 1916

Diagnosis: Enoplia. Cuticle smooth. Anterior end transformed into sucker. Cephalic sensilla papilliform; amphids reduced and porelike. Pharynx cylindrical with three pharyngeal glands opening anterior to nerve ring. Onchiostyle present in juveniles. Anus situated at posterior terminus. Cervical gland and caudal glands reduced. Male and female reproductive systems unpaired. Spiculum unpaired and gubernaculum absent. Cloacal opening of males with cuticular bell-shaped bursa. Ova with plugs or thickenings at poles; shell of ova ornamented. Parasites of birds and mammals; intermediate hosts, one (oligochaetes) or two (oligochaetes and fish). Two families: Dioctophymidae Railliet, 1915; and Soboliphymidae Petrow, 1930.

Subclass Chromadoria Pearse, 1942

Diagnosis: Nematoda. Cuticle clearly annulated, often with distinctly sculptured surface and intracuticular sclerotizations. Lateral differentiation of cuticle distinct if present. Hypodermis cellular. Musculature polymyarian or, more rarely, meromyarian. Muscle cells platymyarian. Cell constancy absent or present only in certain organs. Cephalic sensilla arranged in three circles (6+6+4) or, more rarely, in two (6+10). Somatic setae usually numerous. Amphids round, spiral, noose-shaped, or slit-shaped, but never pocket-shaped. Pharynx usually with basal bulb, rarely cylindrical, and usually with three pharyngeal glands; aperture of dorsal gland situated in anterior part of pharynx; apertures of two subventral glands in posterior part of pharynx. Midgut poly- or oligocytose. Cervical gland saclike, or more rarely, paddle-shaped or branched; renette duct rarely sclerotized, and duct aperture situated at various levels of body. Caudal glands present. Hypodermal glands usually numerous. Testes of males paired, musculature of ejaculatory duct poorly developed, and supplementary organs often present. Female reproductive system didelphic or monodelphic, and antidromal or, more rarely, straight. Cleavage of embryo extremely determinant and strictly bilateral; genetic determinant for endoderm situated in posterior blastomere of two-cell stage. Free-living marine, freshwater, and, more rarely, soil forms, and certain species commensals of invertebrates.

Order Chromadorida Chitwood, 1933

Diagnosis: Chromadoria. Cuticle distinctly annulated, with patterns of dots, rods, bands, "basket weave," etc., and usually with lateral differentiation. Cephalic sensilla commonly arranged in three circles; each of first two circles with six papilliform sensilla and third circle with four setiform sensilla; rarely sensilla in two circles, with ten sensilla in second circle. Amphids slitlike or spiral. Oral cavity usually small with wide vestibular opening and mobile labia; walls of cavity with cuticular thickenings. Pharynx usually with distinct bulb. Female usually didelphic, less commonly monodelphic. Spicula paired and gubernaculum and supplements present. Free-living marine, rarely inhabitants of fresh water or soil. Two suborders: Chromadorina Chitwood and Chitwood, 1937; and Cyatholaimina de Coninck, 1965.

Order Monhysterida De Coninck and Schuurmans-Stekhoven, 1933

Diagnosis: Chromadoria. Cuticle with subtle or distinct striae and without sclerotizations and lateral differentiation. Cephalic sensilla arranged in two or three circles, and bundles of cervical setae may be united with cephalic setae. Amphids circular. Stoma small or well developed. Pharynx cylindrical or with basal bulb. Female reproductive system usually monodelphic or, more rarely, didelphic. Spicula paired, gubernaculum present, and supplements rarely present. Marine free-living forms and, more rarely, freshwater and soil forms; several species commensal with invertebrates. Three suborders: Monhysterina de Coninck and Schuurmans-Stekhoven, 1933; Linhomoeina de Coninck, 1965; and Siphonolaimina de Coninck, 1965.

Order Desmodorida De Coninck, 1965

Diagnosis: Chromadoria. Cuticle distinctly annulated, sometimes with appearance of scales or tiles; sclerotization or lateral differentiation absent. Cuticle at head end smooth and may differ from that of body, giving appearance of cephalic armor or helmet (besides Spirinoidea). Cephalic sensilla in three circles, more rarely in two circles; sensilla in first and second circles (or only first) papilliform and sensilla of second and third (often only the third) setiform. Amphids small, round, spiral, or, rarely, hook-shaped. Buccal capsule narrow, poorly developed, and enveloped by pharyngeal tissue. Pharynx with distinct bulb. Female reproductive system didelphic. Spicula paired, sometimes reduced and replaced by gubernaculum (Monoposthiidae). Supplements present or absent. Free-living marine nematodes; rarely freshwater or soil forms. Three suborders: Desmodorina de Coninck, 1965; Draconematina de Coninck, 1965; and Ceramonematina de Coninck, 1965.

Order Araeolaimida De Coninck and Schuurmans-Stekhoven, 1933

Diagnosis: Chromadoria. Cuticle delicately annulated and without sclerotizations or lateral differentiation. Cephalic sensilla situated in three circles, with six papilliform sensilla in each of first two circles and four setiform sensilla in third. Amphids round, sometimes very large, spiral, or noose-shaped. Buccal capsule not developed or cylindrical. Pharynx cylindrical or with basal bulb. Female reproductive system didelphic. Spicula paired, gubernaculum usually present, and precloacal supplements absent. Primarily marine forms, or more rarely, freshwater and soil forms. Families include Araeolaimidae de Coninck and Schuurmans-Stekhoven, 1933; Axonolaimidae Filip'jev, 1918; Diplopeltidae Filip'jev, 1918; and Comesomatidae Filip'jev, 1918.

Order Plectida Malakhov, 1982

Diagnosis: Chromadoria. Cuticle distinctly annulated and without sclerotizations; lateral differentiation sometimes present. Cephalic sensilla arranged in three circles; sensilla papilliform in first two circles and setiform in third. Amphids small and round or simple spiral. Buccal capsule narrow, cylindrical, and often very long. Pharynx with basal bulb and often with corpus and isthmus. Cervical gland with or without branches, but with long, coiled, and sclerotized efferent canal. Female reproductive system didelphic or, rarely, monodelphic. Spicula paired, gubernaculum present or, rarely, absent, and preanal supplements present. Free-living marine nematodes, many freshwater and soil groups, and some saprobionts. Three superfamilies: Plectoidea Örley, 1880; Leptolaimoidea Örley, 1880; and Haliplectoidea Chitwood, 1951.

Order Desmoscolecida Schuurmans-Stekhoven, 1950

Diagnosis: Chromadoria. Body shortened. Cuticle transversely striated with small or very large interstrial rings; rings sometimes encrusted with granules. Cephalic sensilla arranged in three circles; six papilliform sensilla in each of first two circles and four setiform sensilla in third. Amphids round. Somatic setae and special attachment setae present. Buccal capsule small or apparently absent. Pharynx cylindrical. Female reproductive system didelphic. Spicula paired, gubernaculum present, and preanal supplements absent. Most species marine, but some species inhabit fresh water and soil. Three superfamilies: Desmoscolecoidea Schepotieff, 1908a; Meylioidea de Coninck, 1965; and Greeffielloidea Filip'jev, 1929.

Subclass Rhabditia Pearse, 1942

Diagnosis: Nematoda. Cuticle annulated, surface often distinctly sculptured, and lateral differentiation often present. Hypodermis partially or completely syncytial. Body wall musculature polymyarian or meromyarian, and muscle fibers platymyarian or secondary coelomyarian. Cell constancy in some or all organ systems. Cephalic sensilla usually papilliform, arranged in two (6+10) or three (6+6+4) circles, but merger or reduction may decrease apparent number of sensilla. Amphids small and porelike, distinctly moved toward anterior end of body. Somatic sensilla almost completely absent (only deirids or postdeirids persist). Pharynx cylindrical or with basal bulb and often subdivided into corpus, isthmus, and bulb. Three pharyngeal glands usually present; orifice of dorsal gland situated in anterior part of pharynx, and subventral orifices situated in posterior region. Midgut oligocytose or poly- and megacytose. Hypodermal and caudal glands reduced. Phasmids present. Cervical gland branched with long canals in lateral folds of hypodermis; duct of cervical gland sclerotized with orifice at various levels from anterior end of body. Female reproductive system didelphic and homodromus. Males monorchic and ejaculatory duct with strong musculature. Supplements absent or, if present, then unpaired. Paired genital papillae usually present and bursa often present. Cleavage of blastomeres strictly determinant, clearly bilateral; rhombic stage present. Genetic determinant of endoderm situated in posterior blastomere of two-cell stage. Most species parasites of plants or animals, but some saprobiotic soil forms, or, more rarely, aquatic forms.

Order Rhabditida Chitwood, 1933

Diagnosis: Rhabditia. Oral aperture surrounded by six, three, or two secondary labia or, more rarely, labia absent. Buccal capsule five-chambered (cheilostom, prostom, mesostom, metastom, and telostom). Pharynx divided into corpus, isthmus, and bulb. Midgut oligocytose. Three rectal glands present. Musculature meromyarian, and muscle fibers platymyarian. Cervical gland usually H-shaped, sometimes additional excretory system in posterior part of body. Bursa present in males and with bursal papillae; supplements present or absent. Saprobiotic soil forms and, more rarely, aquatic forms; parasites of invertebrates and vertebrates. Three suborders: Cephalobina Andrassy, 1974; Rhabditina Chitwood, 1933; and Drilonematina Chitwood, 1950.

Order Tylenchida Thorne, 1949

Diagnosis: Rhabditia. Secondary labia around oral opening transformed into labiotubercules. Stylet present and derived from prostom, mesostom, metastom, and telostom of ancestral buccal capsule. Guiding apparatus of stylet derived from cheilostom. Pharynx divided into corpus (usually with metacorpal bulb), isthmus, and cardial bulb. Midgut oligocytose. Rectal glands absent. Musculature meromyarian and platymyarian. Cervical gland asymmetrical. Males with caudal alae, or caudal alae reduced; caudal alae usually without papillae. Soil saprobiotic forms, rarely freshwater forms, and many plant parasites or parasites of insects. Two suborders: Aphelenchina Geraert, 1966; and Tylenchina Chitwood, 1933.

Order Strongylida Railliet and Henry, 1913

Diagnosis: Rhabditia. Oral opening encircled by six secondary labia or secondary labia reduced. Radial corona often present. Buccal capsule not divided into five regions, but enveloped by pharyngeal tissue. Pharynx of juveniles divided into corpus, isthmus, and bulb; pharynx of adults pin-shaped. Midgut oligocytose. Musculature mero- or polymyarian, and muscle fibers platymyarian or secondary coelomyarian. Cervical gland H-shaped. Bursa present in males and with bursal papillae; bursa rarely reduced. Parasites of all groups of vertebrates except fish. Life cycle may or may not include intermediate host; if included, then annelids or mollusks. Five superfamilies: Strongyloidea Weinland, 1857; Ancylostomatoidea Looss, 1905; Diaphanocephaloidea Chabaud, 1965; Metastrongyloidea Lane, 1927; and Trichostrongyloidea Gram, 1927.

Order Oxyurida Skryabin, 1923

Diagnosis: Rhabditia. Oral opening surrounded by six, four, three, or two secondary labia. Stoma poorly developed and without separation into five regions. Pharynx usually consists of three sections: corpus, isthmus, and bulb (pseudobulb). Midgut polycytose. Musculature mero- and polymyarian, and muscle cells platymyarian or secondary coelomyarian. Cervical gland H-shaped. Bursa not present in males. Unpaired preanal papillae or suckers often present. Parasites of digestive tract of vertebrates or arthropods. Five superfamilies: Oxyuroidea Railliet, 1916; Cephaloboidea Kloss, 1960; Hystriognathoidea Kloss, 1960; Thelastomatoidea Sanchez, 1947; and Rhigonematoidea Sanchez, 1947.

Order Ascaridida Skryabin and Schulz, 1940

Diagnosis: Rhabditia. Oral opening encircled by three secondary labia, or more rarely, by two, four, or six labia. Buccal capsule poorly developed or, less commonly, well developed, but never with five basic regions. Pharynx cylindrical or slightly broadened posteriorly. Midgut polycytose or myriocytose. Musculature polymyarian, and muscle fibers platymyarian or secondary coelomyarian. Cervical gland H-shaped, but with reduction of one anterior canal, or asymmetrical with single canal in one lateral hypodermal chord. Parasites in digestive tract of vertebrates. Life cycle without intermediate host or with invertebrate intermediate host. Two suborders: Ascaridina Skryabin, 1915; and Heterakina Chitwood, 1971.

Order Spirurida Chitwood, 1933

Diagnosis: Rhabditia. Oral opening surrounded by six rudimentary labia or two large secondary labia. Buccal capsule poorly to well developed, but never with basic structure of five divisions. Pharynx with anterior muscular and posterior glandular regions and cylindrical or pin-shaped. Midgut polycytose or myriocytose. Musculature polymyarian, and muscle fibers secondary coelomyarian. Sexual bursa usually absent and less frequently present. Parasites of vertebrate animals; with, but sometimes without, intermediate host; intermediate host usually arthropods. Three suborders: Spirurina Railliet, 1914; Camallanina Chitwood, 1936; and Filariina Skryabin, 1915.

Phylogenetic Relationships within the Class

The question of phylogenetic affinities within the class Nematoda at present can only be regarded as controversial. The incompleteness of our knowledge permits only the most preliminary comments.

Two phylogenetic lines are distinctly defined in the evolutionary development of nematodes: one of them is nematodes of the subclass Enoplia, and the second, nematodes of the subclasses Chromadoria and Rhabditia (Figure 89). Among nematodes of the subclass Enoplia, the most primitive forms are assigned to the order Enoplida, which includes free-living marine and freshwater nematodes. Forms of the order Enoplida are without cellular constancy and have relatively high regenerative capability. The whole body is abundantly strewn with somatic setae and hypodermal glands. The nervous system is characterized by a complex but weakly centralized structure and by a strong development of the peripheral nervous system. Among some forms, the oral cavity and labia are still completely undeveloped. The pharynx usually is a simple, cylindrical, muscular tube. A primitive characteristic of the order as a whole is the variation in structure of the glandular apparatus of the pharynx: here are encountered forms with five glands opening at the anterior end of the pharynx; forms that have three glands opening close to the anterior end of the pharynx and two in the posterior half; and also forms with three glands. Characteristics of primitiveness are present in the em-

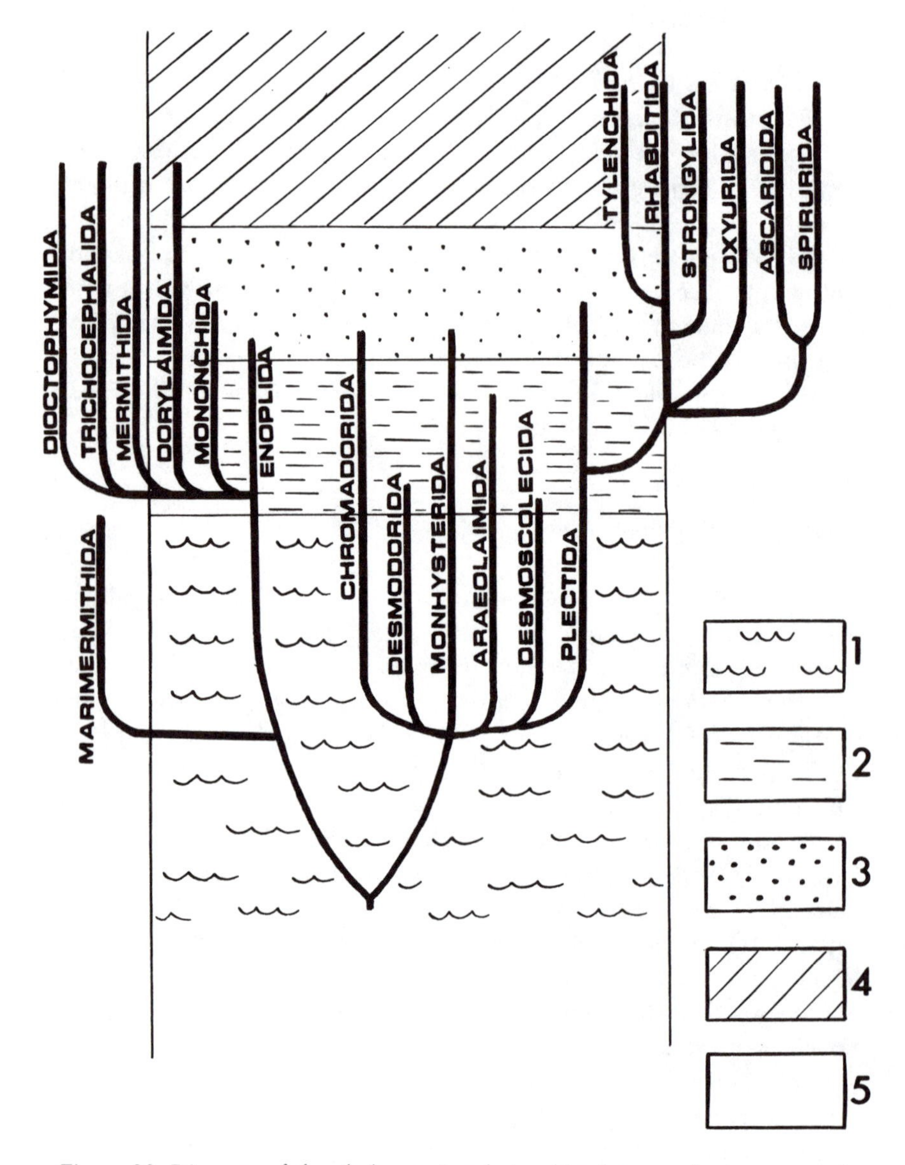

Figure 89. Diagram of the phylogenetic relationships between basic groups of nematodes with consideration of their environment. 1, marine; 2, fresh water; 3, soil; 4, plant parasitism; 5, animal parasitism.

bryonic development of Enoplida. Cleavage of the nematodes of this order is without bilateral symmetry, which is characteristic of early cleavage of other nematodes. Variability in blastomere arrangement suggests that determinant development is not well established, and this condition may be regarded as a characteristic that is ancestral for nematodes. Nonetheless, in Enoplida, localization of endodermal material is fully determinant. It is always related to the anterior first two blastomeres of the embryo, a trait that is further developed among representatives of other orders of the subclass Enoplia.

From ancestral marine Enoplida came also the very primitive group of parasitic nematodes of the order Marimermithida. The juvenile forms of the nematodes of this order are parasites of Priapulida, Echinodermata, Polychaeta, and other marine invertebrates. Characteristic of Marimermithida is a greater or lesser reduction of the gut, and nourishment is accomplished through a porous, highly permeable cuticle. To understand the origin of the parasitism of Marimermithida it is important to note that among lower Enoplida (family Leptosomatidae) are encountered commensals of certain marine invertebrates (sponges and anemones).

A significant part of Enoplida and all Marimermithida are inhabitants of the marine environment. Representatives of two suborders of Enoplida (Tripyloidina and Tobrilina) have adapted to freshwater environments. Two other orders of the subclass Enoplia (i.e., Mononchida and Dorylaimida) are inhabitants of fresh water and soil. Mononchida is a group of purely predatory nematodes, feeding on small soil nematodes. With adaptation to the soil and freshwater environment, there is a simplification of the sensory apparatus. The cephalic sensilla become papilliform, and the primary labia merge with the growth of head tissue—cephalic protuberances—on which the papillae are situated. All Mononchida are predators and have a buccal capsule generally constructed of strongly cuticularized walls bearing one dorsal onchium and sometimes several additional onchia. The cylindrical shape of the pharynx is an ancestral condition. Mononchida have five pharyngeal glands, opening posterior to the nerve ring. The presence of caudal glands is a residual structure persisting from the ancestral aquatic form of life.

Mononchida, in some respects, are an intermediate group between Enoplida and the order Dorylaimida. Members of the latter order are even more specialized to live in the soil (freshwater representatives of this order have probably left the soil environment to secondarily invade aquatic environments). Cephalic sensilla are papilliform, and amphids are pocket-shaped or very small, almost porelike. Caudal glands, the most important attachment apparatus, especially for aquatic nematodes, are reduced. The oral cavity of Dorylaimida is reduced, and it contains a strong onchiostyle—a stylet, which is a transformed onchium (a cuticular process of the buccal wall). The onchiostyle is a fundamental mechanism of the Dorylaimida that helps pierce and devour small animal prey (predatory forms), single-celled algae and mycelia of fungi (phytophages), or plant tissues (phytohelminths).

Mermithida are an order of nematodes that in the juvenile stage are parasites of arthropods or, more rarely, of other invertebrates (including nematodes). Characteristic of Mermithida is the reduction of the digestive tract: part of the pharynx is transformed into a glandular formation—stichosome—and the midgut is transformed into a trophosome (storage organ). It is possible to regard the transformation of the pharynx into a glandular organ as the result of the duplication of the pharyngeal glands. The mermithid cuticle contains numerous pores, and the hypodermis is furnished with a large quantity of glands. Extraintestinal digestion and absorption through the cuticle (Rubtsov 1978) is characteristic of the Mermithida. The basis for these processes is the activity of the hypodermis and stichosome glands and the permeable properties of the porous cuticle. Cephalic sensilla are reduced in members of the Mermithida to papillae; amphids are pocket-shaped and are especially large in males, this being related to the necessity for detecting sexual pheromones.

Phylogenetically the Mermithida probably share a common ancestor with the Dorylaimida. This is supported by many peculiarities of morphology and, in particular, the presence of an onchiostyle in mermithid juveniles and its homologous counterpart in dorylaims—the stylet. The presence of a stylet in mermithid juveniles can be regarded as a recapitulation of the ancestral condition.

The Trichocephalida is an order that includes parasites of invertebrates and, in a number of ways, are close to Mermithida (Rubtsov 1978). The closeness of these two orders is emphasized by the development of the stichosome and the presence of the stylet—an onchiostyle—homologous to that of Dorylaimida. Characteristic of a number of representatives of this order is the development of the bacillary bands—pores in the cuticle through which pass secretions of the hypodermal glands (Wright 1968a, b; Sheffield 1962). On the phylogenetic level, Mermithida and Trichocephalida evolved from ancestral forms of Dorylaimida (Rubtsov 1978).

The position of Dioctophymida in the classification of the Enoplida is problematic. The stylet in juveniles of Dioctophymida seems close to those of Mermithida and Trichocephalida; however, the homology of the stylets of all three groups has not been demonstrated. The glandular apparatus of the pharynx distinguishes Dioctophymida not only from Mermithida and Trichocephalida, but also from Mononchida and Dorylaimida and brings them closer to Enoplida directly. For Dioctophymida it is characteristic that there are three pharyngeal glands, the ducts of which open in the anterior part of the pharynx. The peculiarities of dioctophymid embryonic development also fortify the isolated position of this order. It cannot be excluded that Dioctophymida is an independent branch of Enoplia.

The subclass Enoplia unites a rather broad range of forms: from primitive marine free-living nematodes to deeply specialized parasites of vertebrates, the whole life cycle of which progresses within the vertebrate host. The parasitic representatives of this subclass have a tendency toward an extraintestinal type of nourishment, which is realized most fully in Marimermithida, Mermithida, and, partially, Trichocephalida. A highly permeable cuticle and well-developed hypodermal glands were a preadaptation to this type of feeding.

Parasitism has arisen frequently in various evolutionary branches of Enoplia. Evidently, ancient parasites of marine invertebrates—Marimermithida—arose at a very early stage of evolution from primitive marine enoplids. At a significantly higher level of development of the enoplid branch, the common ancestors of Mermithida, Trichocephalida, and, perhaps independently from the latter two, the ancestors of Dioctophymida split off from the freshwater ancestors of Dorylaimida. Members of the Mermithida remained parasites of freshwater invertebrates. The invertebrate marine intermediate hosts of members of Trichocephalida and Dioctophymida historically were probably the primary host of these forms, and the life cycle of the ancestors of these orders resembled that of the mermithids. Later, the vertebrates, using the invertebrates as food infected with nematodes that were ancestors of Trichocephalida and Dioctophymida, became the final hosts of these parasites. The highly specialized parasites of the order Trichocephalida lost their original primary invertebrate host in the course of evolution and now complete their life cycles within the vertebrate host.

The subclass Chromadoria embraces free-living marine nematodes, and, more rarely, freshwater and soil nematodes. Parasitic forms among representatives of this subclass are not encountered; soil forms are often saprobionts, and some representatives of the order Monhysterida are commensals of marine invertebrates. The fact that among Chromadoria parasites are totally absent has been used as evidence for the very primitive condition of the representatives of this subclass. Among Chromadoria, a number of groups unquestionably possess very primitive characteristics, while others, in the evolution of many systems of organs, have departed far from the primitive prototype.

Overall, Chromadoria contrasts rather sharply with Enoplia. While Enoplia is characterized by a smooth or slightly annulated cuticle and is equipped with pores or deep grooves, Chromadoria is characterized by its members always having a striated or deeply striated cuticle, formed by the external structure of the cuticle with an inner supporting structure, sclerotizations often lying within the cuticle, etc. Canals or deep grooves that cut across the outer layers of the chromadorid cuticle are absent. The musculature among members of Chromadoria belongs to the platymyarian type, although data on this are still limited. Cephalic sensilla of Chromadoria are most often situated in three circles, more rarely in two. Although it is characteristic of Enoplia that the sensilla of the second circle (six setae) are longer than the sensilla of the third circle (four setae), in Chromadoria it is typical to have the four setae of the third circle significantly longer than all other sensilla. Among many forms of the orders Chromadorida, Desmodorida, Desmoscolecida, Araeolaimida, and Plectida, the first and second circles are represented by papillae and only the third one has four setae. There are a few exceptions to this rule, however, among Xyalidae and some Cyatholaimidae. A majority of Enoplia representatives have a pocket-shaped amphid, and only in Tripyloidina have they become circular or spiral. By contrast, members of Chromadoria are not known to have a pocket-shaped amphid. For them, it is characteristic to have amphids that are round, looped, spiral, or slitlike. Primitive groups of Chromadoria are characterized by a simple pharynx, but within the subclass there is a persistent tendency to develop a basal bulb. There are also three pharyngeal glands, the dorsal opening into the anterior part of the pharynx, and the two subventral into the posterior part.

The peculiarities of the glandular apparatus in the pharynx among members of Chromadoria differ distinctly from the situation found among members of Enoplia, which are characterized by variation in the number of glands (3–5–00). If three glands occur in the pharynx of a member of Chromadoria, they open anterior to the nerve ring. However, among some forms of the order Monhysterida, the enoplid type of pharyngeal gland arrangement is found. Thus, in *Anticyclus* and *Paralinhomoeus,* three pharyngeal glands open into the anterior part of the pharynx. In *Eleutherolaimus,* of the same order, the subventral glands each have two openings: one of them is in the anterior part of the pharynx (as in Enoplia), and the other is in the posterior part (as in typical Chromadoria). Possibly this is an indication that the most differentiated pharyngeal glandular apparatus of the Chromadoria, the most differentiated in the morphological sense, was formed from a more primitive glandular apparatus like that of Enoplida.

With regard to embryonic development, Chromadoria, in contrast with Enoplia, shows a different arrangement of potentials between the first blastomeres of the embryo. At the two-blastomere stage in Chromadoria, the endoderm material is tied to the posterior blastomere rather than to the anterior one as in Enoplia. The significance of this in separating the subclasses of nematodes was first indicated by Drozdovskiy (1975). It is interesting that Gastrotricha, which are undoubtedly close to nematodes, have two basic systematic groups within the class that are also distinguished by localization of endoderm in the first blastomeres of the embryo (along with other features). The fact that this feature "operates" within two related classes shows its inherent naturalness.

Ancestral characteristics are dispersed among various orders of Chromadoria. Representatives of the order Monhysterida in some respects can be regarded as a relatively primitive group of Chromadoria. The cuticle among members of Monhysterida is finely striated, lacks sclerotizations, and lateral differentiation is absent. The internal structure of the cuticle is rather complex and resembles that of marine Enoplida. The hypodermis is represented by several types, forming a comparative morphological series similar to that of Enoplida. Cephalic sensilla are arranged in two or three circles. A wide variation in the structure and arrangement of cephalic sensilla is characteristic of Monhysterida. Thus, among representatives of the suborder Linhomoeina there can be four, six, or ten cephalic sensilla (the remaining receptors are papillae). We find an especially complex picture in Sphaerolaimidae and certain species of *Theristus,* where the cephalic sensilla are few and are functionally replaced by tufts of somatic setae.

The circular amphid characteristic of Monhysterida evidently can be regarded as original to the subclass. The sensory endings wind around in a spiral submerged in secretions of the amphidial glands; this is a prototype of the spiral amphid of many Chromadorida. Somatic setae are abundantly strewn over the body in Monhysterida. The pharynx of Monhysterida is often represented by a single cylindrical tube (a majority of forms in the suborder Monhysterina) and, less often, a bulb is developed in the posterior region. Also indicative of primitiveness among several species of this order is the presence of a poorly differentiated enoplid-type arrangement of the pharyngeal glands (see above).

A comparatively impermeable and clearly annulated cuticle is characteristic of nematodes of the orders Desmodorida and Chromadorida. These orders unite the most typical forms of this subclass. For these there is a characteristic development of a real basal bulb, of a pharyngeal gland system typical of the subclass, and a spiral or slitlike amphid. Representatives of these orders are predominantly marine nematodes.

From primitive Monhysterida evolved forms of the orders Araeolaimida and Plectida and, from the latter, also evidently Desmoscolecida. The latter order unites aberrant nematodes that sometimes have been regarded as an intermediate group between nematodes and kinorhynchs. This is not true, of course. The presence of four setae, the conversion of the remaining sensilla into papillae, and the circular amphid allows the supposition that this group branched out from the general Chromadorida-Monhysterida branch together with the ancestors of Plectida. In recent years it has been shown that Desmoscolecida can be inferred from Leptolaimidae through such intermediate forms as *Manunema* and *Anomonema* (Riemann et al. 1971).

The subclass Rhabditia undoubtedly is related phylogenetically to the subclass Chromadoria. These two groups are close in a number of characteristics. The cuticle of Rhabditia, as a rule, is striated, the striae having the same nature as in Chromadoria. The cuticle of Rhabditia is devoid of deep ringed furrows or pores connecting the deep layers of cuticle to the outer environment, and it has little permeability. The musculature of many Rhabditia belongs to the platymyarian type as in Chromadoria. In many Rhabditia there has been a development of secondary coelomyarian musculature, developing ontogenetically and phylogenetically from platymyarian. Both subclasses are close to the general type of pharyngeal gland arrangement. In Rhabditia, there are three glands, the dorsal one opening into the anterior part of the pharynx, and the two subventrals into the posterior part. An important characteristic, close in both subclasses, is the similarity in the fate of the first blastomeres of the embryo. In Rhabditia, as in Chromadoria, endoderm material is situated in the anterior blastomere at the two-blastomere stage. Some authors (Drozdovskiy 1975, 1978), on the basis primarily of embryological similarity, propose the separation of nematodes into only two subclasses, Enoplia and Chromadoria, including Secernentea in the latter. Some characters that distinguish Rhabditia from Chromadoria are not specific. Thus, the conversion of cephalic sensilla into papillae characteristic of Rhabditia, the reduction of the amphid to a porelike state, and the disappearance of caudal glands can be regarded as the result of a transition to the saprobiotic soil and parasitic forms of life. Such transformations are observed among parasitic Enoplia. However, Rhabditia also have a number of specific traits. Thus, in all Rhabditia there has been a development of a branched cervical gland with canals that lie in the lateral chords of the hypodermis. The trend toward complication of the cervical gland is observed also in some Chromadoria, especially in the order Plectida. In *Anonchus,* there is already a branched gland, but its canal is connected to the lateral chords of the hypodermis, and evidently the development of such a cervical gland in Plectida should be regarded as the result of parallelism. In itself, the appearance of the branched cervical gland reflects one means of solving the osmotic problem. All free-living Rhabditia (from which the parasitic representatives of the group evolved) are soil or freshwater nematodes for which the problem of osmotic regulation is acute. However, among freshwater and soil Enoplia and even among most Chromadoria, the problem is solved differently. Among freshwater Enoplia, not only is there a lack of hypertrophy of the cervical gland, but, on the contrary, there is a decrease in size of the cervical gland and even its complete reduction together with the caudal glands. Probably, freshwater and soil Enoplia "solve" the osmotic problem by some other means of equalizing osmotic pressure. Freshwater and soil Plectida have developed numerous hypodermal glands that participate in osmotic regulation. In Rhabditia there has been obvious reduction of the somatic hypodermal glands, and increased branching of the cervical gland has developed. Among some Rhabditia, in addition, there is still a branched gland in the posterior half of the body.

In addition to the hypertrophied cervical gland, Rhabditia have preserved another pair of glands that are related to phasmids. Phasmids are sensory glandular organs, undoubtedly homologous to the somatic sensory glandular organs of Chromadoria and Enoplia. However, in Chromadoria and Enoplia the number of sensory glandular organs is very large, and their arrangement varies among closely related forms and often also within a given species. For all Rhabditia, only one pair of sensory-glandular organs is characteristic (i.e., phasmids situated in the tail section). Constancy in the arrangement and structure of phasmids within the taxonomic limits of all Rhabditia make them the most important characteristic of the subclass.

There is no doubt that the taxonomic distance separating Rhabditia and Chromadoria is significantly less than the gap that separates Chromadoria from Enoplia. Rhabditia, it may be supposed, comes from a group of freshwater Chromadoria, probably from freshwater (or soil) Plectida. As recognized by a number of authors within the subclass Rhabditia, two basic evolutionary branches can be distinguished: one of them unites nematodes of the spiroascarid branch and the other, nematodes of the rhabditostrongylid branch (Skryabin 1941, 1946; Pearse 1942; Chabaud 1957; Cameron 1962; Sprent 1962; Inglis 1965; Ivashkin 1978).

Nematodes phylogenetically related to Rhabditida (Tylenchida, Strongylida, and Oxyurida) undoubtedly entered into parasitism in animals and plants from soil saprobiosis, where representatives of the order Rhabditida even today form a rich faunistic complex. The saprobiotic forms that inhabit decaying vegetation and animal tissues live in an environment of impoverished oxygen supply but rich in easily assimilable organic substances, and essentially they appear to be preadapted to parasitism. From habitation in dead tissues of animals and plants, it is only a step to true parasitism—existence in a live organism and derivation of food from it. Among forms of the order Rhabditida, there are a number of forms at the stage of becoming parasites (textbook examples are *Rhabdias buffonis* and *Strongyloides stercoralis*). Nematode phytohelminths of the order Tylenchida undoubtedly went from saprobiosis to parasitism in plants.

Nematodes of the orders Strongylida and Oxyurida also entered into parasitism by way of soil saprobiosis. The hosts of Strongylida for the most part are land vertebrates. As the most primitive, the life cycles of such Strongylida should be reexamined; the juvenile leaves the ovum alive, molts in the external environment (soil), and then percutaneously or perorally penetrates the host organism. The life cycles of this type are fully comparable with the life cycles of the semiparasitic Rhabditida and can illustrate the routes of establishing parasitism in this group. A variety of routes for penetrating a host are related to the soil-saprobiotic origin of strongylid parasitism. Evidently, the ancient land vertebrates, coming into contact with saprobiotic breeding grounds in a literal sense, either swallowed the juveniles of strongylid ancestors, or the latter penetrated the relatively thin skin covering.

Members of Oxyurida are parasitic to a wide spectrum of hosts, among which are various land vertebrates and a wide variety of land invertebrates. Only a few species (probably secondarily) passed into parasitism among freshwater fish. The variation in hosts of the Oxyurida is evidently explained by the simultaneous transition of oxyurid ancestors to parasitism in the intestines of various animals, both vertebrates and invertebrates, that were connected with the saprobiotic breeding places.

Evidently, if we recognize the formation of parasitism in the rhabditostrongylid branch through soil saprobiosis, then the time of the formation of parasitism in this group can be assigned to a period no earlier than when the saprobiotic breeding grounds themselves were formed; that is, no earlier than the appearance of abundant flora and fauna on land and no earlier than the appearance of soil as a special specific environment for habitation. Consequently, the epoch of parasitism formation in this evolutionary branch can be assigned to the period when land plants and animals appeared; that is, the Devonian period, probably even toward the end of that period (no earlier than 350 million years ago). Parasitism in this group arose continually and in waves: each new group of animal hosts appearing on earth, along with parasitic nematodes they inherited from their ancestors, also acquired new parasites from the saprobiotic breeding grounds. Groups such as the Strongylida and the Oxyurida embrace already rather specialized nematodes that long ago made the transition to parasitism. But the process of newly evolving parasitism in the rhabditostrongylid branch did not end. Even in our time, many forms of the order Rhabditida are making the transition to parasitism. It must be supposed that, in time, in the process of divergence and further specialization toward parasitism, these forms could provide a beginning for large new taxa of parasitic nematodes.

In distinction from the rhabditostrongylid evolutionary branch of development, which are parasites basically of land animals, the spiroascarids are parasites not only among land animals, but also among a wide circle of primarily aquatic hosts that belong, moreover, to the most primitive groups of vertebrates (round-mouthed and cartilaginous fish, cartilaginous ganoids, and bony fish). Primary aquatic hosts are especially characteristic to such groups as Cucullanidae, Camallanidae, and Anisakoidea. Among a number of other groups that are parasites of land vertebrates, connection with an aquatic environment can be maintained through an aquatic invertebrate intermediate host.

Juveniles of members of Cucullanidae, Camallanidae, and Anisakoidea enter the outer (aquatic) environment from the ovum, lead a free form of life, and molt. The life cycle of Anisakoidea and Camallanidae progresses with the participation of an intermediate host, a role provided by representatives of crustaceans or annulated worms. Infection takes place perorally with the second-stage juveniles. The juveniles perforate the wall of the intestinal tract of their intermediate host, enter the body cavity where they molt, and transform into third-stage juveniles. Further development requires that the invertebrate host be swallowed by the final vertebrate host.

The study of the life cycle among members of the Cucullanidae has been neglected. In recent years, data have appeared that indicate the presence, at least among some Cucullanidae, of life cycles with a single vertebrate host. Thus, in the case of *Truttodacnites stelmoides* (a parasite of lampreys), it has been shown successfully in experiments that the second-stage free-living juveniles are swallowed by the lamprey juveniles—sand diggers—that filter the bottom layers of water. The juvenile nematodes first enter the liver, molt, and then reenter the intestinal lumen, where they reach sexual maturity (Pybus et al. 1978).

The same pattern is found in the life cycle of *Cucullanus cirratus,* a parasite of cod (with the only difference that the histotrophic phase of juvenile development occurs in the mucous covering of the stomach) (Popova and Valovaja 1978). The life cycle of another nematode of the same genus, *C. chabauidi,* progresses without the involvement of an intermediate host.

There are two points of view of the way that parasitism forms among spiroascaridid nematodes. One of them regards the intermediate invertebrate host as the primary one and its presence as a recapitulation of its ancestral state (Ivashkin 1978). Unfortunately, up to the present time, it has not been known for certain whether or not life cycles of nematodes of the orders Spirurida and Ascaridida take place entirely within invertebrate hosts (as among some Rhabditida and many Oxyurida).

Another point of view proposes that the oldest hosts of spiroascaridid nematodes were vertebrates or, more precisely, primitive aquatic ancestors of vertebrates—Agnatha. It is known that the oldest Agnathi appeared in the Ordovician Period with a great variety of species in marine and freshwater environments. It was in fresh water in the Silurian Period that the ancestors of Gnathostomata were isolated from the early Agnatha (probably from Pteraspidomorpha). Agnatha were primarily filter feeders or benthic deposit feeders. The absence of jaws prevented this group from entering into active predation. The body of early Agnatha (Ostracodermi) was clad completely with large or small shields. All Ostracodermi died out in the Devonian. The only modern group of Agnatha, the Cyclostomata, are semiparasites whose juveniles, however, lead the life of filter feeders; this undoubtedly is a recapitulation of the life of Cyclostomata ancestors.

In the filtering process, the earliest Agnatha could easily ingest spiroascaridid juveniles living in the benthic sediments. With the beginning of the parasitic phase, the spiroascaridid at first localized in the epithelial walls of the intestinal tract because these small and poorly specialized parasites did not yet possess the ability to maintain themselves against the flow of food masses through the gut lumen. Evidently, this is related to the primary localization of primitive spiroascaridid juveniles in the wall of the intestinal tract. Not all spiroascaridids subsequently remained parasites in the gut lumen. From the gut wall, nematodes could easily access the bloodstream and from there various tissues and organs. This route is widely used by many nematodes of this evolutionary branch, but it is especially characteristic of certain Spirurida.

The appearance of mandibles allowed vertebrates to become predators; this made victims of a large circle of invertebrates in the environs of the early Cyclostomata. In itself, the acquisition of mandibles was a very great aromorphosis in the organization of vertebrates and facilitated their biological progress, but simultaneously this also indicated that there was a new route for infestation by parasitic worms. Invertebrates, as food for Cyclostomata, of course also spontaneously became infected by juveniles of spiroascaridid nematodes. Turning up in the intestine of an invertebrate, the juvenile spiroascaridid nematode penetrated the gut wall in an attempt to establish itself as an intestinal parasite, and inadvertently made its way into the body cavity, where it could not complete its development. The predatory Cyclostomata, swallowing the invertebrates that were simultaneously infected by juvenile nematodes, became their definitive hosts. Reinforcement of the route of infestation through an intermediate host led to the formation of heteroxic life cycles that were typical for most spiroascaridids.

If one accepts this point of view, then the epoch of the formation of parasitism in the spiroascaridid branch of Rhabditia is no more recent than early Ordovician (about 500 million years ago), and consequently spiroascaridids are earlier in the geological sense than the parasitic Rhabditostrongylida.

The primary route of penetration of a spiroascaridid into a host organism is the peroral route. For rhabditostrongylids, there are two primary routes: peroral and percutaneous. It can be supposed that early spiroascaridids could not penetrate the skin of their primary hosts (early aquatic vertebrates) because the latter had dense integument. And, really, the body of early aquatic vertebrates, as a rule, was encased in a test of scales or bony plates. The latter are especially characteristic of the jawless scaled Ostracodermi and early Cyclostomata—the testaceous fish Placodermi.

The transition to parasitism in the spiroascaridid branch took place once and the primary hosts of this nematode group were aquatic (freshwater) ancestors of vertebrates. From that time on the entire evolution of the spiroascaridids was connected with the evolution of their hosts. As is known, the formation of vertebrates progressed in fresh water, and only from the Middle Devonian did vertebrates enter the sea. In this connection, it is important to note that the evolution of the ancestors of Rhabditia also took place in fresh water. This corresponds well with the distinct development of secretory glands among representatives of this subclass; this probably reflects one of the ways of solving the osmotic problem (see above). Both basic branches of Rhabditia had their beginning in free-living freshwater ancestors. The ancestors of spiroascaridids very early became parasites of the freshwater ancestors of vertebrates. The antecedents of the nematodes of the rhabditostrongylid branch took over the habitats of the soil saprobionts and entered into parasitism among various groups of vertebrates and invertebrates coming into contact with these saprobiotic habitats. The transition to parasitism in the rhabditostrongylid branch has not yet ended, and in the future, the saprobionts still will serve as a source of new taxa of parasitic nematodes.

Along with morphological parallelism and partially causing it, a phenomenon of ecological parallelism also can be observed among nematodes; that is, parallel conquest of similar habitats by representatives of different nematode groups.

The most primitive groups of the subclasses Enoplia and Chromadoria are represented by free-living marine nematodes. At the same time, free-living marine nematodes represent the most diverse and speciose ecological grouping of nematodes. Different branches of both subclasses have provided the origin of freshwater forms from which, in turn, also came the soil species. Among such orders as Enoplida, Chromadorida, Monhysterida, and Plectida there are all three forms: marine, freshwater, and soil. Some orders are represented almost exclusively by freshwater and soil forms: Mononchida, Dorylaimida, and Rhabditida. The soil forms provided the origin of plant parasites.

The transition of various groups of nematodes to parasitism in animal organisms was also accomplished in parallel. Among early members of the marine order Enoplida were to be found the ancestral Marimermithida, parasites of marine invertebrates. From freshwater and soil forms close to Dorylaimida evidently came parasites of invertebrates, Mermithida; and parasites of vertebrates, Trichocephalida and Dioctophymida. Antecedents of the spiroascaridids and some branches of the rhabditostrongylids passed in parallel into a zooparasitic mode of life. Schematically, the phylogenetic relationships of nematodes and the sequence of environmental conquests are represented in Figure 89.

4

Origin of Nematodes and Origins of the Nematode Biological Process

The origin of nematodes and their position in the classification of the animal kingdom is one of the most complicated problems of modern zoology. Many different points of view have been expressed, but most of them now have only historical interest. Thus, the theories of nematode origin from Echinodermata (Bastian 1866), Chaetognatha (Schneider 1866), and Annelida (Greeff 1869) can be rejected unconditionally. Also, appropriately rejected is the once widely held idea that nematodes came from neotenous larvae of Diptera (Rauther 1909; Seurat 1920; Baylis 1924, 1938). Only one point of view has withstood the test of time, and if not generally accepted, it is still widespread. This is the concept of a close affinity between Nematoda and Gastrotricha and the origin of both groups from a common ancestor (Bütschli 1876; Martini 1913; Paramonov 1937, 1962; Beklemishev 1944, 1964a, b; Hyman 1951; Steinböck 1958).

Contemporary Information on the Structure of Gastrotrichs

Answering the question of the phylogenetic relationships between nematodes and gastrotrichs has been made difficult by one circumstance.

Gastrotrichs are a relatively small, not much studied group of animals, the least studied of them until recently being the primitive marine gastrotrichs of the superfamily Macrodasyoidea. At the same time, among nematodes, rather strongly specialized parasitic forms were well studied. Naturally, a comparison between specialized representatives of both groups was difficult.

In recent years, the situation has changed. Several detailed investigations have been published that are devoted to the ultrastructural anatomy, the embryological development, and the biology of the primitive Macrodasyoidea (Teuchert 1968, 1972, 1974, 1975, 1976a, b, 1977a, b). We also have at present a definite set of information on the anatomy, ultrastructure, and embryonic development of free-living marine nematodes. This provides an opportunity to attempt a comparative analysis of the two groups at a new level.

The marine gastrotrichs of the order Macrodasyoidea inhabit capillary spaces between soil particles. Their body length is very extended. In cross section the gastrotrich body is clad in a thin cuticle of a fibrillar structure. The overall cuticle thickness is so negligible that it also covers the bases of the cilia without inhibiting their movements. The gastrotrich epidermis is represented by cellular epithelium of the submerged type. On the ventral side of the body, epidermal cells have cilia or flagellae (Teuchert 1977a). The epidermis among members of Macrodasyoidea is rich in glandular formations of two basic types. The first type is represented by single-cell glands that exude a mucous secretion (mucopolysaccharide). This mucus helps the body glide between the sand particles and protects it from harm. The other type of glandular formation is the so-called attachment or grasping tubules (Figure 90).

The attachment tubules are a kind of sensory-glandular organ with double functions. Each has three cells, which are support, glandular, and sensory. All three cells have long protuberances clad in an overall cuticle. The protuberance of the so-called support cell contains numerous microtubules, and close to the apex there are accumulations of round granules of secretion that exude through an opening in the apex of the protuberance. The glandular cell contains large granules of secretion (albuminous) that accumulate in an ampulla at the end of the protuberance of the cell, and which may then be discharged through an opening. The sensory cell has a protuberance reaching half the length of the whole attachment tubule, and a sensory cilium extends from the tip of the protuberance. Axons from the sensory cells

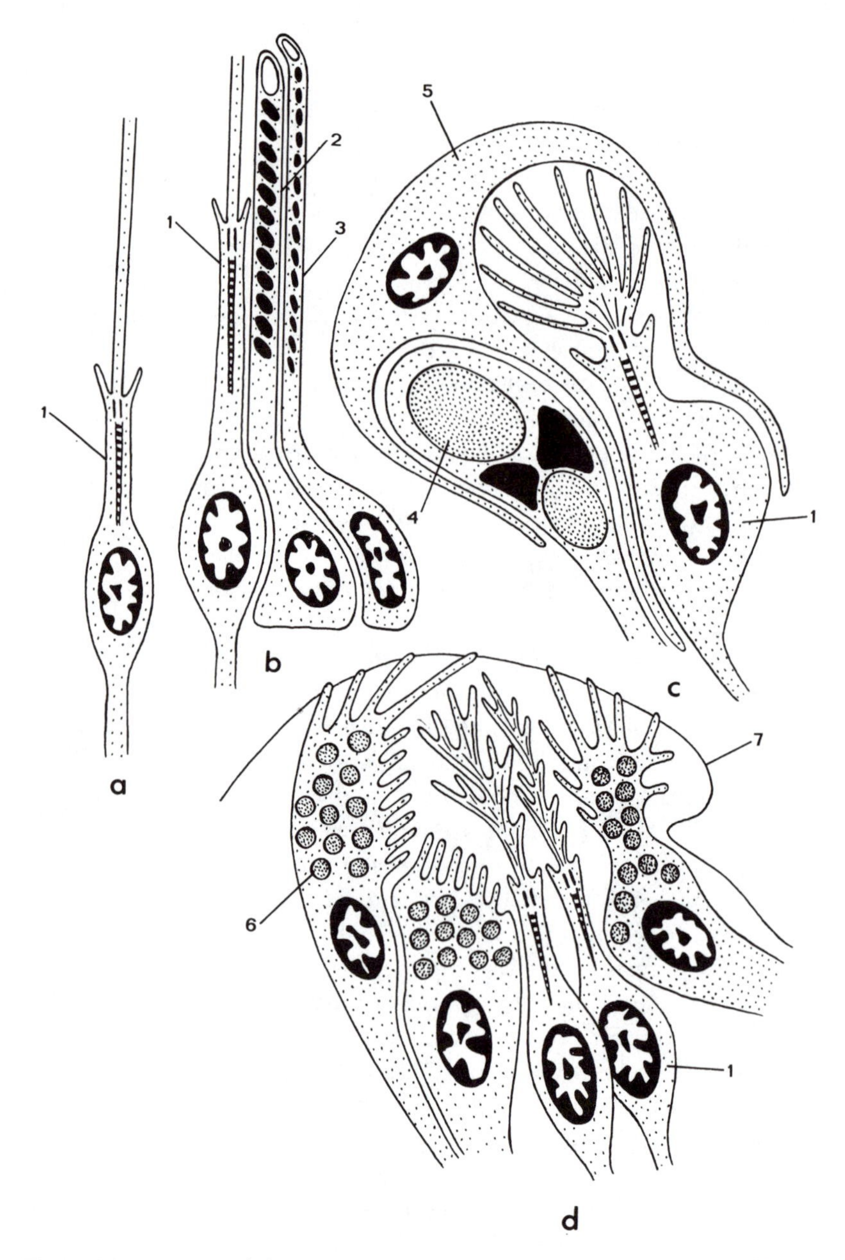

Figure 90. Diagram of the sensory organs of Macrodasyoidea (Gastrotricha) (after Teuchert 1977a): a, sensory hair; b, attachment tubule; c, anterior pair of cephalic sensory organs; d, posterior pair of cephalic sensory organs. 1, sensory cell; 2, glandular cell of the attachment tubule; 3, support cell; 4, glandular cell of the anterior pair of cephalic sensory organs; 5, integumentary cell; 6, glandular cell of the posterior pair of sensory organs; 7, cuticle.

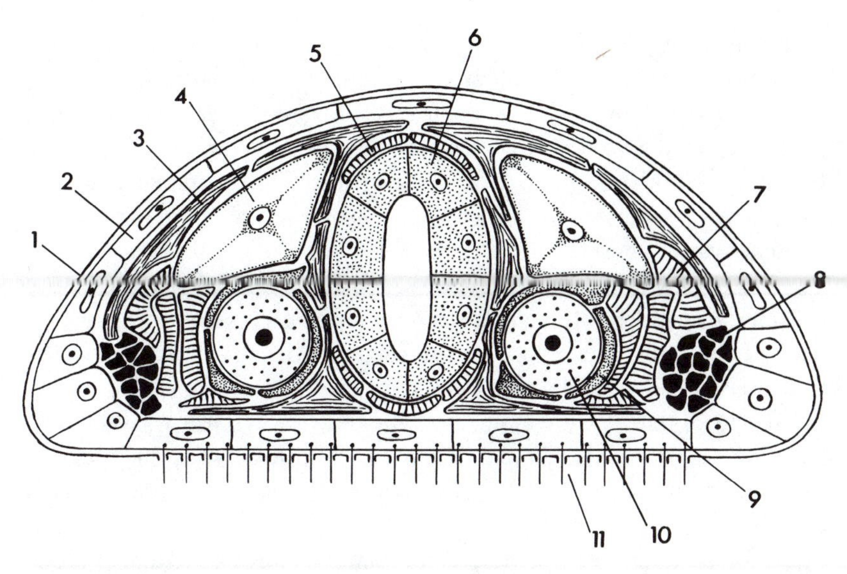

Figure 91. Diagram of a cross section of the body of a member of the order Macrodasyoidea (Gastrotricha). 1, cuticle; 2, epithelium; 3, circular musculature; 4, Y-organ cells; 5, longitudinal muscle cell adjacent to gut; 6, intestine; 7, longitudinal muscles of body wall; 8, ventrolateral nerves; 9, cells of gonoduct wall; 10, gonads; 11, ventral ciliary strip.

of the attachment tubules extend to the ventrolateral trunk (Teuchert 1977a).

The longitudinal musculature is constructed from several basic muscle bands, of which the best developed are the ventrolateral muscle bands. Each dorsolateral muscle band is represented in section by one whole cell. Individual longitudinal muscle cells are arranged along the periphery of the digestive tract from the ventral and dorsal sides (ventral and dorsal muscle bands).

Members of Macrodasyoidea lack a dense layer of circular musculature. Muscle cells oriented perpendicularly to the longitudinal axis of the body join the intestine on the dorsal and ventral sides, and laterally they surround the gonads and Y-organ. Dorsoventrally oriented muscle cells pass between the intestine and Y-organ (Figure 91).

The gastrotrich nervous system consists of the following basic components. The central structure of the nervous system is the circumpharyngeal nerve ring, which surrounds the anterior part of the pharynx.

The ring, which is thicker on the dorsal side than on the ventral side, consists primarily of nerve fibers with the nerve cell bodies situated anterior and posterior to the nerve ring. Ventrolateral nerve trunks begin at the nerve ring with double rootlets. The ventrolateral trunks, which include nerve fibers and individual neurons, fulfill a mixed function. They receive processes of the sensory neurons and also the innervation branches of the muscle cells (Teuchert 1977a).

Besides the attachment tubules, which are organs of mixed function in Macrodasyoidea, there are also more specialized sensory organs, especially the numerous sensory hairs (Figure 90a) that are strewn over the anterior end of the body. They consist of sensory cells that adjoin the inside of the integumental epithelium, with a sensory cilium surrounded by stereocilia. Similar sensory cells can be situated singly or together in the sensory bulbs from which extend several independent flagella.

Two pairs of very specialized sensory structures are situated along the sides of the head end. The anterior pair of organs have been formed by one or two sensory cells surrounded by integumental cells. The receptor apparatus of the sensory cells consists of a dendritic process at the base of which lies the basal body of the cilium. A short distance from the basal body, the dendritic process branches into tufts of microvilli, each of which contains one microtubule. The cavity in which the microvilli are situated is formed by integumental cells. The whole organ is covered by cuticle. Thus, the receptor apparatus of the anterior sensory organs does not come into contact with the external environment (Figure 90c).

The posterior sensory organs each contain ten to twelve sensory cells. The dendrites branch into the cavity, which is fringed with goblet-shaped secretory glands. Externally the organs are covered by thin cuticle bearing pores through which microvilli of the goblet-shaped cells have contact with the outside environment. Processes from the receptor cells are not connected with the outside environment (Figure 90d).

The function of the sensory hairs most likely is the reception of mechanical stimuli. Mechanical reception is also evidently characteristic of the sensory cells of the attachment tubules. The reception type in the anterior and posterior sensory organs is problematical. Teuchert (1976b), after investigating these organs at the ultrastructural level, proposed a photoreceptor function for the anterior sensory organs and a chemoreceptor function for the posterior sensory organs. It should be noted, however, that the morphology of the anterior and posterior sensory organs provides very little to support such an interpretation.

The body cavity of members of the gastrotrich superfamily Macrodasyoidea is not developed. A special Y-organ is situated to the right and left of the gut (Remane 1935, 1936). As shown by ultrastructural research, the Y-organ represents two rows of cells with strongly developed vacuoles that provide high turgor. With the light microscope, these vacuoles are perceived as elements of a schizocoel, but at present their internal cellular nature is not fully explained (Teuchert 1977b). The Y-organ plays an important role in the locomotion of Gastrotricha. In some ways, the Y-organ, constructed from strongly vacuolized cells, is analogous to the chord of lower Chordata.

The intestine of Gastrotricha is divided into three basic sections: an extended muscular pharynx, a midgut, and a short posterior intestine. The pharynx is a muscular tube with a triradiate lumen. In Macrodasyoidea the radii of the lumen are oriented differently than in nematodes: one radius is dorsal and the other two subventral. In cross section, the pharyngeal wall consists of fourteen to sixteen cells in the cytoplasm of which are cross-striated fibers. The pharynx is innervated by a pharyngeal nerve plexus that lies next to the pharyngeal wall. A specific peculiarity of the Macrodasyoidea pharynx consists of special pharyngeal pores that unite the cavity of the pharynx with the external environment (Figure 92). The midgut contains two basic types of cells: absorptive (some of which evidently are capable of digestion within the cell) and secretory. The posterior intestine is short and ends at the ventral anus.

The three-part digestive tract largely determines the subdivision of the whole gastrotrich body into three basic sections: the trophico-sensory section, containing organs for capturing food, sensory organs, and central sections of the nervous system; the trophico-genital section, in which organs of food digestion and the reproductive system are localized; and a short caudal section, which does not contain derivatives of the digestive tract, but has specialized caudal attachment tubules. The regions of the gastrotrich body are defined by the regions of the digestive system (Figure 92).

The excretory system in members of Macrodasyoidea is represented by paired protonephredia situated in the trophico-genital section (Teuchert 1972).

Macrodasyoidea are hermaphrodites. Paired testicles produce flagellar spermatozoids, which are unusual in that they have a very long

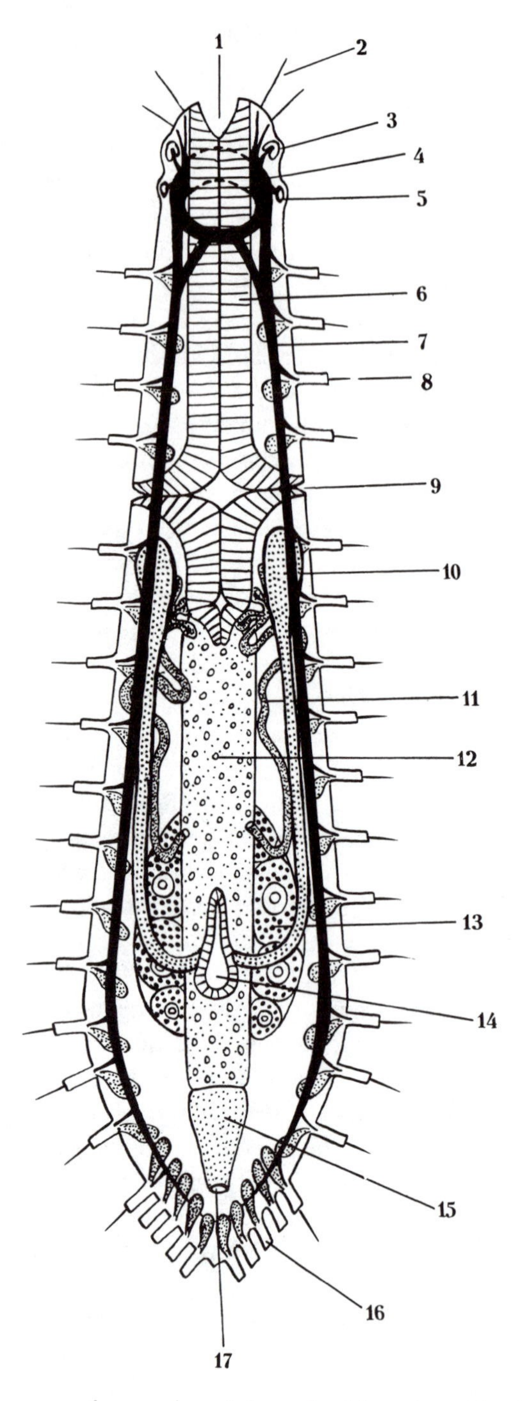

Figure 92. Diagram of a member of the order Macrodasyoidea (Gastrotricha). 1, mouth; 2, sensory hairs; 3, anterior pair of cephalic sensory organs; 4, circumpharyngeal nerve ring; 5, posterior pair of cephalic sensory organs; 6, pharynx; 7, ventrolateral nerves; 8, attachment tubules; 9, pharyngeal pores; 10, testes; 11, protonephridia; 12, midgut; 13, ovaries; 14, penis; 15, posterior gut; 16, caudal attachment tubules; 17, anus.

acrosome, a spirally twisted nucleus with a giant, spiral mitochondrium and a relatively short flagellum. Fertilization in Macrodasyoidea is internal, but the copulatory organ is absent or represented by a simple muscular penis. Ovaries are paired, but proximally merged. The ova are deposited into the external environment through the female gonopore, which serves both for copulation and oviposition. In other species, the ova are simply expelled through an aperture in the body wall. Macrodasyoidea have low fecundity, and the large ova are laid singly.

Chaetonotoidea, the most specialized group of Gastrotricha, comprises mostly freshwater forms. Marine species of this group evidently returned secondarily to the sea from fresh water. These are very small organisms (50 to 100 m) with a compact body armored on the dorsal side with a strong cuticular test. The cuticle among members of Chaetonotoidea is constructed more complexly than that of Macrodasyoidea (see, for example, Rieger and Rieger 1977), but performs only protective functions.

The Chaetonotoidea nervous system is the same as that of Macrodasyoidea. The attachment tubules are present only at the posterior end of the body. There is no Y-organ, and the schizocoel is represented by narrow spaces between organs. The digestive system in Chaetonotoidea is represented by a muscular pharynx, a midgut, and posterior intestine, which opens by way of a dorsal anus. The radii of the pharyngeal lumen in Chaetonotoidea are not oriented like those of Macrodasyoidea, but rather like those of nematodes, with one radius directed ventrally and the other two subdorsally.

Many species of Chaetonotoidea are represented only by parthenogenetic females. Gastrotrich ova are ellipsoid with the longitudinal axis corresponding to the future anterior-posterior axis of the embryo. Polar bodies in ova of members of Chaetonotoidea are shed at the equator of the ovum, and, in Macrodasyoidea, they are shed close to the anterior end of the ovum. The first cleavage is always equatorial and development is determinant. In Gastrotricha, as in nematodes, there are two variants of development. In Chaetonotoidea, insofar as can be judged by the data of Beauchamp (1929) and Sacks (1955), the anterior blastomere of the two-cell stage contains ectodermal material of the anterior part of the embryo, the stomodeum, and endoderm, and the posterior part, ectoderm of the posterior part of the body, the mesoderm, and reproductive primordia. In the development among members of Macrodasyoidea, which has been investigated in detail by Teuchert (1968),

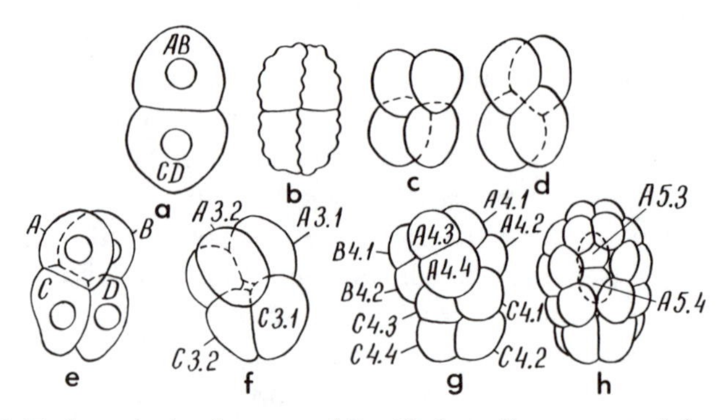

Figure 93. Embryonic development of *Lepidodermella squamata* (after Sacks 1955): a, two-blastomere stage; b, four-blastomere stage; c and d, rotation of blastomeres; e, four-blastomere stage before the third division; f, eight-blastomere stage; g, sixteen-blastomere stage; h, beginning of gastrulation.

the anterior of the two blastomeres of the embryo contains material for the mesoderm, stomodeum, and part of the ectoderm, and the posterior one contains material for the entoderm, the basic part of the ectoderm, and part of the mesoderm.

Division in Macrodasyoidea and Chaetonotoidea are different. In members of Chaetonotoidea, just after the second division, there arises a flat four-blastomere figure (Figure 93b). After division, the sister pairs of blastomeres shift to a 45° angle to one another, and an asymmetrical tetrahedral figure results (Figure 93c and d). After the third division there results an arrangement in which the anterior four blastomeres (descendants of anterior blastomere *AB*) lie precisely in line with the intercellular facets between the four posterior blastomeres (descendants of blastomere *CD*). This figure is essentially still radially symmetrical. The only sign of bilateral symmetry is the somewhat smaller size of the cells on the dorsal side of the embryo. The cleavage plane of the fourth division of all blastomeres is perpendicular to the anterior-posterior axis of the embryo. At the sixteen-blastomere stage, the embryo bends to the dorsal side in the sagittal plane and acquires a bilaterally symmetrical structure. The fifth and subsequent divisions progress differently in various blastomeres of the embryo, and their orientation corresponds to bilateral symmetry (Figure 93g and h).

Gastrulation begins after the fifth division with the submergence of two cells (descendants of one of the blastomeres of the sixteen-cell embryo) into the embryo. The blastopore gradually narrows and becomes the oral opening. The anus is formed at the posterior end of the embryo. The slit-shaped blastopore in Chaetonotoidea is not observed. However, between the mouth and anus, in a number of forms, there lies a groove that ties the two openings together. Two large cells, lying laterally in the posterior third of the embryo, are the precursors of the reproductive system (Beauchamp 1929).

The development of Macrodasyoidea is characterized by a more even formation of bilateral symmetry than that in Chaetonotoidea (Figure 94). After the second division in development among members of Macrodasyoidea, a tetrahedral figure appears in which blastomeres *A* and *B* occupy the anterior end of the embryo, and blastomeres *C* and *D*, the posterior. After the completion of the division, blastomere *C* shifts to the dorsal side of the embryo, and the embryo acquires a bilaterally symmetrical structure (Figure 94). In the transition to the eight-cell stage, blastomeres *A*, *B*, and *C* divide in a longitudinal direction, and blastomere *D*, in a transverse direction (Figure 94). One of the descendants of blastomere *D*, designated blastomere *E*, moves from the posterior end to the ventral side of the embryo (Figure 94j and l). This blastomere provides all of the endoderm. The rudiment of the endoderm in Macrodasyoidea is thus apparent beginning with the eight-blastomere stage. Further division is strictly bilateral. Blastomere *C* provides the beginning for the basic part of the ectoderm. Blastomeres descendant from *A*, *B*, and *D* of the four-cell stage and arranged around blastomere *E* provide the beginning of the mesoderm. Gastrulation begins at the thirty-cell stage and proceeds primarily by epiboly. The blastopore is at first round and then elongates to become slitlike (Figure 94r). The slitlike blastopore closes from posterior to anterior. The precursor of the reproductive system is already multicellular by the late stage of development.

Thus, although adult Macrodasyoidea preserve a number of primitive traits, their cleavage is more advanced than that of Chaetonotoidea, the latter characterized by early establishment of bilateral symmetry and early (as early as the eight-cell stage) isolation of ectoderm. On the other hand, Macrodasyoidea preserve a well-expressed slitlike blastopore; this evidently should be regarded as a primitive characteristic of their organization.

Macrodasyoidea are inhabitants of narrow capillary spaces of marine sediments. They have an elongated, thin body suited to moving through interstitial spaces. The basis for locomotion in Macrodasyoidea

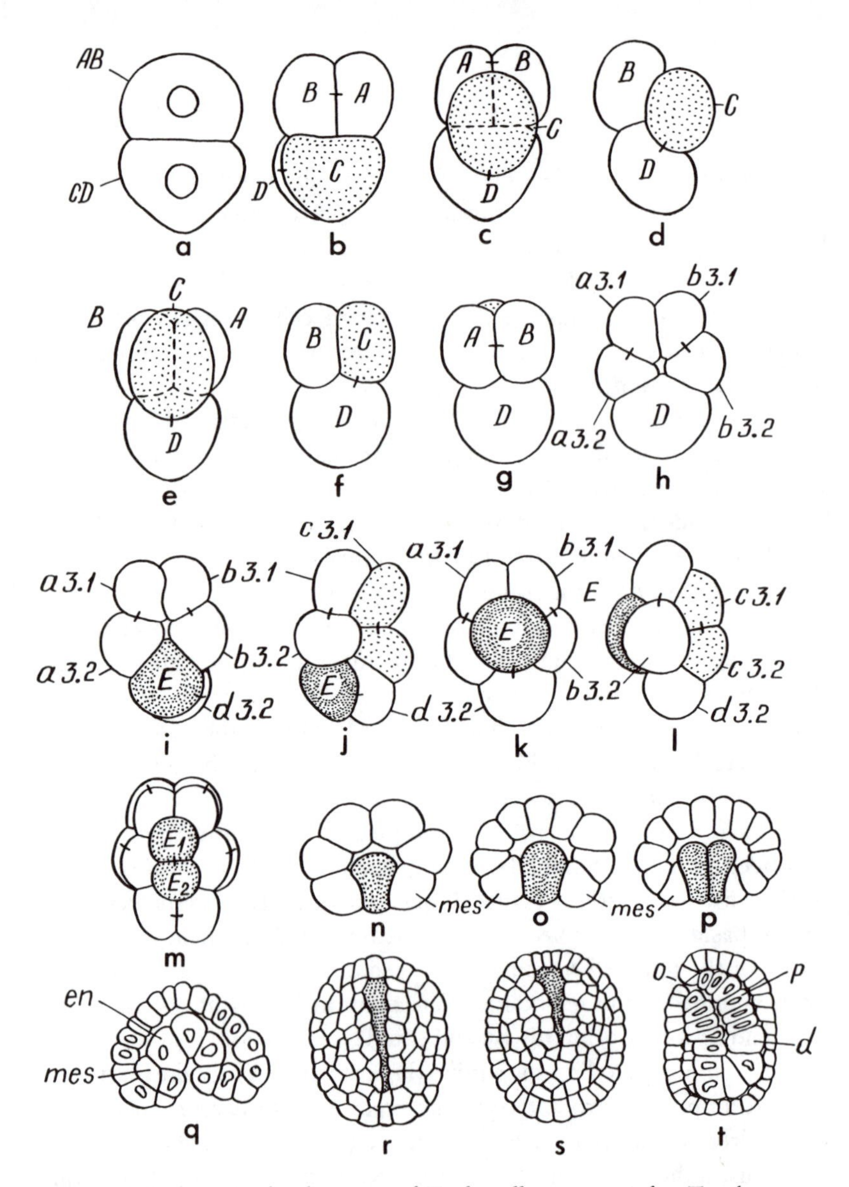

Figure 94. Embryonic development of *Turbanella cornuta* (after Teuchert 1968). (b, c, and e viewed from the dorsal side; d, f, j, and l in lateral view; g, h, i, k, and m viewed from the ventral side). a, two-blastomere stage; b, four-blastomere stage; c and d, movement of blastomere C to the dorsal side; e–g, four-blastomere stage after the third cleavage; h and i, third cleavage; j, eight-blastomere stage just after the third cleavage; k and l, movement of blastomere *E* to the center of the ventral side; m, sixteen-blastomere stage (beginning of gastrulation); n–p, cross section of embryo at different stages of gastrulation; q, cross section of embryo at the slitlike stage of the blastopore; r, slitlike blastopore; s, closure of slitlike blastopore; t, sagittal section of late embryo. d = midgut, en = endoderm, mes = mesoderm, o = mouth, p = pharynx.

is the shortening of the longitudinal musculature, cilia playing only an auxiliary role in locomotion. Gastrotrichs attach themselves to sediment particles with the aid of adhesive organs and pull the body toward the point of attachment. With musculature among members of Macrodasyoidea being largely represented by longitudinal muscles, antagonist muscles are necessary to effectively accomplish muscular locomotion. Various structures among different gastrotrichs serve as these antagonists (Teuchert 1977b). In *Turbanella* the antagonist is the Y organ, in *Macrodasys,* the vacuolized cells of the epithelium; and, in *Chordodasys,* a band of chordlike vacuolized cells in the dorsal wall of the midgut. The cuticle among members of Macrodasyoidea is delicate and never plays a locomotor role. For small Chaetonotoidea, the basic means of locomotion is the operation of the cilia on the abdominal side of the body.

Comparison of Nematode and Gastrotrich Structure

A very detailed comparison of nematode and gastrotrich organization has been published by Paramonov (1937 and 1962), and many points of his analysis have not been given their due. According to the sum of comparative anatomy indices, nematodes and gastrotrichs are undoubtedly two close groups, gastrotrichs having the more ancestral morphological organization. The overall organizational symmetry among all members of Gastrotricha is bilateral, and virtually all organ systems of gastrotrichs are bilaterally symmetrical. With movement, the ventral side, also bilaterally symmetrical, is turned to the substratum and serves as the animal's crawling surface.

Nematodes partially lose their bilateral symmetry and secondarily develop radial symmetry in a number of organ systems. The secondary position of nematode radial symmetry can be shown in the following manner. First, the radial symmetry characteristic of certain organ systems of adult nematodes is formed during the late stages of ontogeny, superimposed on the basically bilaterally symmetrical stages of embryonic development. Second, radial symmetry affects the surface structures of the nematode body: the cutaneous-muscular sac, pharynx, sensory organs of the anterior end, and peripheral nerves. Deep-lying organ systems are bilaterally symmetrical, such as the nervous system as a whole, digestive glands, digestive system as a whole (if the position of the anus is taken into account), and the excretory and reproductive systems. Third, radial symmetry in various nematode organ systems is not congruent. Thus, the sensory organs of the anterior end and pharynx are subordinate to triradial symmetry, and the cutaneous-muscular sac is governed by biradial symmetry. The combination tri- and biradial symmetries is congruent with bilateral symmetry. The only general plane of symmetry is the animal's sagittal plane. This condition is not accidental, of course. It shows that nematode symmetry is fundamentally bilateral.

The formation of radial symmetry in the nematode body wall is undoubtedly related to the peculiarities of the biomechanical movement of nematodes and the loss of the ventral crawling surface. The nematode body becomes round in cross section and the ventral side loses external differentiation. Further development of undulatory locomotion leads to lateral crawling surfaces.

The body walls of nematodes and gastrotrichs are constructed in a similar pattern. Gastrotrichs and nematodes are clad in cuticle, under which lies an epithelium. Special significance is acquired in both groups by the longitudinal body-wall musculature, which is divided into separate muscle bands in both gastrotrichs and nematodes.

At the same time, the body wall in gastrotrichs is significantly more ancestral than that of nematodes. The gastrotrich cuticle is very delicate, very simply constructed, and has a purely protective function. The rather delicate cuticle of the dorsal side of Chaetonotoidea has a simple internal structure and represents a purely protective formation. Nematode cuticle, besides the protective function, becomes the antagonist of the longitudinal musculature, a most important component of the locomotory system. In this connection, the nematode cuticle acquires a very complicated internal structure that surpasses all that is known about the cuticular formations of invertebrates.

The nematode hypodermis undoubtedly represents a more highly differentiated tissue structure than the submerged epithelium in gastrotrichs. The cells of the nematode hypodermis are sharply differentiated into a thin plate adjoining the cuticle and a cytoplasmic process making up the hypodermal chords. The differentiation of nematode hypodermis into interchordal and chordal components is related to the necessity for tight adhesion between the cuticle and musculature. The cells of the epidermis on the ventral side of gastrotrichs have flagella or cilia, which are completely absent in the hypodermis of nematodes. Muscle fibers on the ventral side of gastrotrichs and nematodes are obliquely striated. The musculature of the nematode body wall is represented by only one layer of longitudinal muscle. But the layer of longitudinal muscle in nematodes is significantly more integrated (essentially there are in all two basic muscle bands, dorsal and ventral, that are mutually antagonistic) than the longitudinal muscles of gastrotrichs. Along with the longitudinal muscles of gastrotrichs, there are preserved elements of circular musculature as well as dorsoventral muscles that allow them to change the width and thickness of the body. Very complete organization of the cutaneous muscular sac allows them to use highly effective undulatory locomotion. The movements among gastrotrichs of Macrodasyoidea are also to a significant degree based on shortening the longitudinal musculature. In gastrotrichs, the antagonists of the longitudinal muscles are various structures—vacuolized cells of the epidermis, the Y-organ, and chordlike formations in the midgut—but all of these structures are incapable of providing as much completely undulatory locomotion as we encounter in nematodes. There is an impression that the ancestors of nematodes and gastrotrichs "searched for" a more effective system of muscular locomotion, with the most successful having been the variant realized in nematodes.

The central organ of the nervous system in nematodes and gastrotrichs is the circumpharyngeal nerve ring and associated nerve cells. There are no real nerve ganglia in either nematodes or gastrotrichs. The homolog of the ventrolateral nerve trunks common among gastrotrichs is the ventral nerve trunk in nematodes. The ventral nerve trunk in nematodes is the structure in the nematode nervous system strongest and richest in nerve cells. The paired nature of the nematode ventral nerve trunk is clear. It proceeds from the circumpharyngeal nerve ring in paired rootlets, which again divide before the anal opening. Other longitudinal nerves in the nematode body are related to the necessity of providing equal innervation to the radially symmetrical body wall. Let us recall in this connection that supplementary longitudinal nerves in the nematode body are largely composed of nerve fibers coming from the ventral trunk.

The radial symmetry of the sensory nerves of the anterior end of nematodes is a result of the radial symmetry in the arrangement of sensory receptors. In gastrotrichs, both the body wall and the sensory structures maintain bilateral symmetry, and this also results in bilateral symmetry of the nervous system.

The sensory organs of nematodes and gastrotrichs reveal more similarity with one another than do the less specialized sensory organs we compare. The general prototype for sensory organs in nematodes and gastrotrichs is the sensory-glandular organ. In gastrotrichs, this prototype in its least specialized form is the attachment tubule, and in nematodes it is the somatic seta with associated glands. It is interesting that nematode somatic setae may acquire a grasping and attachment function (Desmoscolecida and Draconematidae). This is a good example of parallelism based on common morphology.

Nematode amphids are rather specialized organs of chemoreception, and their structure does not reveal special similarity with the organization of anterior or posterior sensory organs of the gastrotrich head end. Let us note, however, that the anterior and posterior sensory organs in gastrotrichs, just as in nematode amphids, maintain their sensory-glandular nature.

The digestive systems of nematodes and gastrotrichs are constructed in the same pattern. The digestive tube in both groups consists of a long muscular pharynx or esophagus that opens with a terminal mouth, a midgut, and a short posterior intestine. The partitions of the digestive system determine the subdivision of the body into three basic sections: trophico-sensory, trophico-genital, and caudal. The capturing of food, sensory structures, and the central sections of the nervous system are related to the trophico-sensory sections. The organs of food digestion and the reproductive system are situated in the trophico-genital section. The caudal section contains specialized attachment glands.

The food-grasping mechanism in both groups is essentially identical: a muscular pharynx/esophagus operates like a suction organ. As might be expected, the pharyngeal lumen in both groups has a triradiate shape, which, as we have seen, operates effectively as a muscular pump. However, in gastrotrichs and nematodes, and even among various groups of gastrotrichs, the triradiate shape of the lumen has been formed in parallel. Thus, in Macrodasyoidea, one radius of the lumen is dorsal and the other two subventral, and in Chaetonotoidea the situation is reversed. In nematodes, the orientation of the radii of the lumen is similar to that of Chaetonotoidea: one radius is ventral, and two are subdorsal. The different food specialization of nematodes is expressed in the differences in types of oral cavity structure.

Both nematodes and gastrotrichs are characterized by the reduction of the parenchyma. The last remnants of a parenchyma, but already rather specialized, are the cells of the Y-organ in gastrotrichs and the parietal cells of nematodes. The body cavity is still almost undeveloped in primitive nematodes. In more advanced forms there is a widened schizocoel, filled with liquid, that facilitates uniform transmission of pressure on the cuticle and serves as the internal environment of the organism.

Gastrotrichs have an oligomerous protonephridial excretory system. The protonephridia are quite reduced in nematodes, and the secretory function is transferred wholly to numerous integumental glands and phagocytic cells.

Gastrotrichs, like nematodes, have internal fertilization. The gastrotrich reproductive system is at a significantly lower level of morphological differentiation than is the reproductive system of nematodes. Paired gonoducts in gastrotrichs are arranged symmetrically relative to the animal's sagittal plane. Male copulatory organs are absent or are represented by a primitive penis. The female gonophore may be absent, in which case the deposition of ova takes place simply through holes in the body wall. The nematode reproductive system is characterized by morphological simplicity but is not ancestral. Paired gonoducts are usually opposed, one directed anteriorly and the other posteriorly, and this is related to the sharply diminished diameter of the nematode body. Sometimes one of the ducts is reduced. There are specialized copulatory organs, spicules, not homologous to the penis of the gastrotrichs. Nematodes always have two sexes, if one leaves out the cases of secondary, protandrous hermaphroditism.

For a long time, and continuing to the present, the idea has been expressed that the phylogenetic closeness of nematodes and gastrotrichs has not been supported by embryological data. New data on the development of nematodes and gastrotrichs that have largely appeared in the last two decades permit a new approach to an embryological comparison of the two taxa.

In both groups the ova are characterized by uniform distribution of the yolk, and as a rule, they are elongated along the anterior-posterior axis of the embryo. The spindle of the first cleavage is also established along the long axis of the ovum. After the second cleavage in gastrotrichs, blastomeres assume a tetrahedral configuration, involving a shift from a plane figure to the tetrahedral in the case of the Chaetonotoidea. The geometry of early subdivision in gastrotrichs is very close to that among members of the marine Enoplida. Thus, as in nematodes, gastrotrichs have two basic variations in their determinant cleavage: in one case, endodermal material is situated in the anterior blastomere of the two-cell stage of the embryo (nematodes of the Enoplida branch and Chaetonotoidea), and in the other case it is situated in the posterior blastomere (nematodes of the Chromadorida-Secernentea branch and Macrodasyoidea). In both groups there is an early tendency to establish bilateral symmetry in subdivision. The most primitive type of development, with relatively late establishment of bilateral symmetry among gastrotrichs, is characteristic of Chaetonotoidea and, among nematodes, Enoplida. Let us note that development among members of Enoplida is an example of the most primitive type among both nematodes and gastrotrichs and, evidently, is close to the original type of embryogeny characteristic of the ancestors of both nematodes and gastrotrichs.

What group of gastrotrichs is closest to the ancestors of nematodes, and at what level did the historical fate of nematodes and gastrotrichs diverge? According to their general habitat, use of muscular locomotion, and development of sensory-glandular organs, marine gastrotrichs of the order Macrodasyoidea are close to nematodes. But they differ sharply from nematodes in orientation of the radii of the pharyngeal lumen. On the other hand, the pharyngeal structure characteristic of specialized Chaetonotoidea is just like that of nematodes. According to the type of determinant development in early cleavage, nematodes of the Enoplida branch resemble those of Chaetonotoidea, but nematodes of the Chromadorida-Secernentea branch are more like members of Macrodasyoidea. Does this perhaps mean that each of the two groups of nematodes developed from its own group of gastrotrichs? No. It means that neither Chaetonotoidea or Macrodasyoidea were ancestors of nematodes. Nematode ancestors were a third group of gastrotrichlike organisms, the remains of which we do not find among modern fauna precisely because this group has evolved into nematodes. These organisms had a set of traits that allowed those transformations that led to the formation of nematode organization and the absence of which among Chaetonotoidea and Macrodasyoidea kept them from becoming more than gastrotrichs.

The historic fates of nematodes and gastrotrichs diverged at the level of ancestors that did not have a three-sided suctorial pharynx (but probably a simple muscular pharynx) and had a primitive, undeveloped subdivision with a temporary localization of rudiments. In the three evolutionary branches of this group, the suctorial pharynx was formed independently; the radii in the lumen of members of Macrodasyoidea were directed dorsally and subventrally; and among the ancestral group of nematodes they were directed ventrally and subdorsally, as was also the case among Chaetonotoidea. In the following stage, types of determinant development of the ovum also formed in parallel: in Macrodasyoidea, the fate of the endoderm was situated in the posterior blastomere of the two-cell stage, and in nematodes of the Chromadorida-Secernentea branch, in Chaetonotoidea, and in nematodes of the Enoplida branch, the fate of the endoderm was situated in the anterior blastomere. The same parallel transformations can also be found in other characteristics of structure and development.

Thus, the historical fate of nematodes and gastrotrichs diverged very early, at the level when their common ancestor still did not have many of those traits that we consider characteristic of nematodes and gastrotrichs. One cannot help recalling the theorem expressed in a semifacetious form by Beklemishev (1976), that "the point of divergence of each pair of evolutionary branches may be put below any level of primitiveness set in advance!"

Biological Advancement of Nematodes

Gastrotrichs are a relatively small group of animals, numbering less than 300 species according to recent data. Nematodes are one of the very largest groups in the animal kingdom, with a vast range of exploited environments. The number of nematode species according to estimates by various authors is between 100,000 and 1 million. With regard to numbers of individuals, nematodes undoubtedly exceed all other multicellular animals. The roots of such clearly expressed nematode biological progress consist of deeply aromorphous transformations in their organization. The most important aromorphoses in the nematode system of organs were once thoroughly analyzed by Paramonov (1962), and we can add little to that analysis.

The most important aromorphoses in nematodes was the acquisition of polarized tissues. The original type of tissue organization in Metazoa was mixed parenchyma (Beklemishev 1925), which was widespread among lower Bilateria. Along with evolution, an increasing number of tissues "split" from the mixed parenchyma, but among parenchymatose Bilateria (flatworms and nemertines), parenchyma is preserved along with highly differentiated tissues. Among nematodes, parenchyma is completely reduced, and all tissues are constructed of highly specialized, polarized cells. One of the reasons for the reduction of parenchyma and specialization of tissues in nematodes was the reduction in nematode size. The reduction in body size "did not leave space" for parenchyma to fill the spaces between organs. Actually, among lower nematodes, the organs are tightly squeezed together and the small spaces between them are filled with a noncellular substance. The reduction of the body to microscopic size characteristic of most free-living nematodes meant organs of few cells. Actually, among relatively primitive forms, some organ systems are constructed of only a few cells. Among rather specialized soil forms (genus *Rhabditis*) the total number of cells in the whole organism is about 700. The reduction in number of cells required the intensification of function, and this fostered highly specialized cells and tissues.

The development of relatively thick and complex cuticle should also be regarded as an aromorphous transformation in nematode organization. The cuticularization of integuments is a rather common phenomenon in the animal kingdom, but this does not often lead to such important changes in the organization of an animal as it did among nematodes. Along with effective protective properties, the nematode cuticle provides for strong, undulatory locomotion for these animals. In no other group of animals do the cuticular coverings take such an active part in the animal's locomotion, and in no other group does the cuticle acquire such a complex structure.

The acquisition of a strong cuticle caused reduction of the ciliated integument and stimulated the development of a purely muscular type of movement. The muscle structure deserves special attention among aromorphous transformations in the organization of nematodes. Using the example of the musculature, it is especially obvious that the aromorphous transformations affect all levels of the animal's organization (Severtsev 1967). The parenchymatose Bilateria have smooth musculature organized in a muscular envelope containing both circular and longitudinal muscles. In nematodes, the muscular system is constructed of very specialized cells, the ultrastructure of which demonstrates an organization of actin-myosin filaments that is in accord with that of obliquely striated musculature. The cell body itself is clearly differentiated into a contractile and plasmatic part and an innervation branch. The longitudinal muscle cells lie in one layer, the general architecture of the muscle layer corresponding to biradial symmetry that responds to the peculiarities of the biomechanics of nematode movements.

The organizational features of the cuticle, hypodermis, and musculature provide for effective nematode locomotion based on bending of the body. Effective means of movement, undoubtedly more than other factors, has facilitated the biological progress of nematodes.

In contrast to the parenchymatose Bilateria, nematodes have a complete gut. The nematode digestive system is characterized by a high degree of physiological advancement. The nematode pharynx is a very effective muscular pump. The pharynx wall is formed by muscle cells, the organization of which is cross-striated musculature. The aromorphous significance of the pharynx follows from the fact that this type of organization in the anterior section of the digestive tube is preserved unchanged in all types of nematode feeding. This shows that the acquisition of this type of pharynx was not a partial adaptation but an adaptation of general significance. The combination of the activity of the pharyngeal glands and the secretory and absorbing actions of the midgut cells provides quick and complete digestion of food almost entirely within the gut lumen. The short midgut is constructed of highly specialized cells with a developed brushlike border. In histological structure, the wall of the midgut is a more highly differentiated intestinal epithelium. The midgut has no bends, supplemental glands, or processes (with a few exceptions), and so forth; consequently, it has a sufficiently high degree of physiological activity to provide for the final stages of digestion and complete absorption.

The aromorphous transformation also affected the nervous system and sensory receptors, but we still know too little about these organ systems to assess them in this context. The nematode nervous system developed in a distinctive way, but as a whole, despite the preservation of very primitive features, it represents a highly differentiated organization. The longitudinal nerves of nematodes are clearly differentiated into sensory and motor nerves, and this is expressed even at the histochemical level. Among some free-living soil nematodes, the nervous system, which has a total of 200 neurons, serves a large variety of physiological and behavioral reactions. The nematode sensory organs, especially among free-living forms, are varied, and according to level of development are comparable with those of Arthropoda—the most highly developed group of invertebrates.

The schizocoel has already been formed within the class Nematoda. The presence of the schizocoel has a positive effect on nematode locomotion because it aids the uniform transmission of internal pressure to the cuticle. Among relatively large nematodes, the schizocoel takes on the functions of distribution. All this determines the participation of the schizocoel in the overall process of aromorphous transformation in nematode origin and evolution.

The aromorphous evolution, which means the acquisition of general rather than particular adaptations, leading to an increased intensity in the exchange of substances, an increase in the degree of independence from the external environment, and permitting expanded habitat exploitation, proceeds not as an abstract general progress but as an adaptation to a fully specific and real habitat.

The evolution and development of nematode structure took place in the interstitial spaces of marine sediment where we have up to now encountered the richest and most varied nematode fauna. The closest relatives of nematodes, members of the gastrotrich order Macrodasyoidea, also live in the interstitial spaces. Living within the capillary spaces makes a definite demand on the structure of the organisms. The body must have a long and narrow form and a small diameter so that the animal may pass through the interstitial spaces between sediment particles. The body surface should be protected by a dense cuticle so as to defend against mechanical damage by sedimentary particles. The animal should be capable of bending the body so as to follow the bends in the capillary routes. The organisms that live in such conditions should perceive equally from all sides sensory information largely of a mechanical and chemical nature. It is easy to see that the structure of nematodes is an ideal response to just such a biotope. The roots of nematode biological progress consist of the fact that, having adapted to a specific ecological niche, they simultaneously acquired a set of organizational traits that allowed them to conquer a new type of environment.

In speaking about evolutionary paths in the animal kingdom, we have become accustomed to thinking of entering new environments, like emerging from aquatic to terrestrial habitats, conquering air, and so forth. Nematodes conquered a new environment—narrow capillary spaces. Biotopes of such a type are widely dispersed among widely varied biocenoses. We find them in marine and freshwater sediments, in soil, in intercellular spaces of plants, and, to some degree, in animal tissues. Everywhere, we find rich nematode fauna. The reason for their surprising morphological uniformity and their colossal eurybionticism is that we find narrow capillary spaces in the most varied associations that are most suitable for nematode habitation. One of the first researchers on free-living nematodes, Bastian (1865), was struck by how little parasitic nematodes differed from free-living representatives of the same class and how omnipresent nematodes were.

Among nematodes, of course, there are many forms that inhabit other environments and have aberrant morphology and large size. But most nematodes are small spindle-shaped or threadlike organisms, the entire morphology of which is best adapted to habitation in narrow spaces. It is these small mobile worms that have penetrated into all natural associations, have provided gigantic variation in species, and have attained colossal numbers, permitting them to play a most important role in the transformation of matter and energy in the biosphere.

5

Classification of the Pseudocoelomates

Pagenstecher (1875), evaluating the status of zoology at that time, came to the conclusion that the phylum Vermes, then accepted among most zoological leaders, became a group that included everything that did not belong in other phyla of the animal kingdom. A hundred years of zoological development has substantially changed the classification of animals. The old phylum Vermes, which was a kind of zoological "dumping ground" and was not a natural group of organisms, was abolished. After elimination of this collective worm taxon, however, the remainder was united in the phylum Aschelminthes, or primary-cavity worms [pseudocoelomates]. The unnaturalness of the union of clearly heterogeneous groups into the phylum Aschelminthes was intuitively felt by many zoologists. Thus, for example, Chitwood and Chitwood (1950), in discussing the origin of nematodes, wrote the following: "The most notable characteristic of all the discussions about the kindred ties of nematodes from the time of Huxley to the present was that the concept of Nemathelminthes . . . has never been accepted by anatomists, although, judging by appearances, it has entered into zoological literature." It is not by chance, therefore, that repeated attempts have been made to divide Aschelminthes into natural groups. Let us review a few of them.

Grasse et al. (1961) divided Aschelminthes into the phyla Nemathelminthes (with classes Nematoda and Nematomorpha), Nematorhyncha (with classes Kinorhyncha and Gastrotricha), and Rotifera. These authors regarded Priapulida and Acanthocephala as groups of indeterminate taxonomic position. Brien (1961) left Nematoda, Nematomorpha, and Acanthocephala in Nemathelminthes, and he separated other groups into independent taxa. Andrassy (1976) more recently favored uniting only two classes in Nemathelminthes (i.e., Nematoda and Nematomorpha). Lang (1953) approached the subdivision of Aschelminthes differently, dividing all pseudocoelomate worms into two groups according to the presence or absence of a proboscis: Rhynchohelminthes (classes Priapulida, Acanthocephala, and Kinorhyncha) and Arhynchohelminthes (classes Nematoda, Rotatoria, Nematomorpha, and Gastrotricha).

Acanthocephala (Skryabin and Schulz 1931; Hyman 1951; Petrochenko 1956) and Priapulida (Kaestner 1965) have been separated rather often into independent phyla. Finally, some authors, discouraged in attempts to find kindred relationships among the separate classes of pseudocoelomate worms, raised the ranks of all the classes to independent phyla (Chitwood and Chitwood 1950). With respect to the latter, one can only note that raising the ranks of all classes to phyla does not solve the problem of the position of this classification in the animal kingdom and only overloads an otherwise complicated zoological classification.

These attempts to separate pseudocoelomate worms have not been widely recognized. One of the reasons for this is that, in dividing a group of classes into independent phyla, the organizational characteristics of adult forms are considered, as a rule, but embryological traits are often completely ignored. Thus, the resemblance of nematodes and hairworms accepted in all the above-cited attempts is based only on the structure of adult forms (similar body form, presence of elastic cuticle, and preservation of a singular longitudinal musculature), while profound differences in the embryonic and larval development were not considered at all.

In comparing the organization of various groups in the animal kingdom, we should certainly remember that in modern comparative morphology the organism is regarded as a morpho-process, and consequently, all stages of the life cycle are equally the objects of comparison, and no stage is less important than the others.

Considering all of the above, we come to the analysis of the organization of various groups of pseudocoelomate worms, beginning with the examination of the early stages of individual development.

Rotifera

The embryonic development of Rotifera has been studied rather thoroughly (Salensky 1872; Tessin 1886; Jennings 1896; Tannreuther 1920; Nachtwey 1925; Beauchamp 1956; Pray 1965; Lechner 1966), but there is still no generally accepted interpretation of the embryogeny of these animals.

Rotifer ova have an elongated shape. The animal-vegetal axis, insofar as can be determined by the position of the polar bodies, is perpendicular to the long axis of the envelope (Figure 95). In the development of rotifers, there are rather complex changes in the position of the embryo within the ovum envelope; these were recently analyzed by Ioffe (1979). The first four blastomeres are arranged as in a quartet spiral division, blastomere *D* being significantly larger than the rest of the cells of the basic quartet (Figure 95c and d). During the separation of the first quartet of micromeres, the embryo rotates 45° from its original position within the ovum envelope (Figure 95). This rotation continues through further division, so that, at the sixteen-cell stage the micromeres, with associated polar bodies, come close to the other end. The animal-vegetal axis of the embryo at this stage coincides with the long axis of the envelope and is perpendicular to the original position of the animal-vegetal axis of the ovum. Subsequently, the large blastomere—a descendant of blastomere *D* of the "basic quartet"—submerges inside at the vegetal pole of the embryo (which is now situated near one of the narrow ends of the envelope). Here begins the submergence of the endodermal macromeres, descendants of blastomeres *A*, *B*, and *C*. Blastomere *D* does not take part in the formation of the endoderm because the cells corresponding to the endodermal macromeres of this quadrant deteriorate (Lechner 1966). The blastopore forms at the vegetal pole of the embryo (near one of the narrow ends of the envelope). The rotifer embryo at this stage is subordinate to monoxenous heteropolar symmetry. The axis of the symmetry is the animal-vegetal axis of the embryo, marked at one end by a blastomere and, at the other, by macromeres with a polar body squeezed between them (Figure 95j). In the next stages begins a mutual exchange in the position of the blastomere and animal pole of the embryo. The polar body, marking the animal pole of the embryo, moves along the future dorsal side of the embryo to the blastopore (Figure 95k–n). This bringing together of substance from the animal pole and blastopore is accomplished presumably by the spreading out of the material

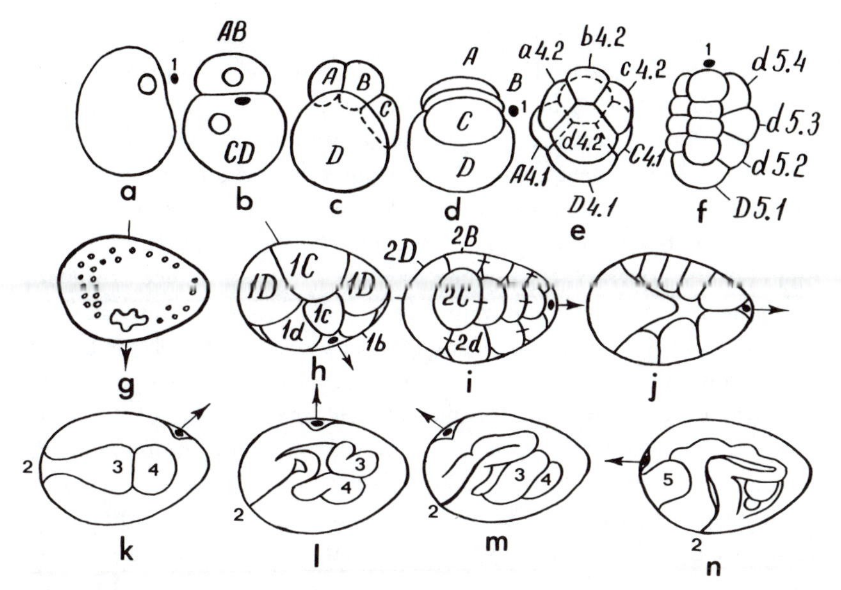

Figure 95. Embryonic development of a member of *Asplanchna* (Rotifera) (after Tannreuther 1920 [a–f]; after Ioffe 1979 [g–n]): a, ovum; b, two-blastomere stage; c and d, two aspects of the four-cell stage; e, eight-blastomere stage; f, sixteen-blastomere stage; g–i, rotation of the embryo in the ovum covering during the process of division; j, beginning of submersion of the primordial sex cells; k–n, union of material from the animal pole and the blastopore. 1, polar body; 2, blastopore; 3, primordial sex cells; 4, yolk material; 5, stomodaeum.

from the *D* quadrant of the embryo. It is true that Lechner (1966), after using ultraviolet light to study blastomeres in the division of *Asplanchna,* discovered that blastomere *B* produced defective cells on the dorsal side of the embryo; this permitted him to conclude that tissue of the dorsal side originates from blastomere *B*. However, Ioffe (1979) called attention to the fact that such illumination disturbs the process of forming the gullet and cerebral ganglion, the construction of which is derived from descendants of blastomere *B*. Material from blastomere *B* remains on the surface of the dorsal side of the embryo and does not go inside, leaving a place for descendants of blastomere *D,* as takes place in normal development.

The blastopore of rotifers goes entirely into the formation of the oral opening. The anus (more precisely, the cloacal opening) is formed in the later stages of development and externally relative to the blastopore. Despite the external distinctiveness of rotifer embryogeny, the symmetry of their development is very close to that of lower Spiralia.

The rotation of the rotifer embryo within the ovum envelope during cleavage does not have much pro-morphological significance. This rotation is caused by the necessity to accommodate the constant extension in the animal-vegetal direction of the embryo within the narrow envelope. On the contrary, the coming together of the substance of the animal pole and the blastopore reflects a real change in the respective positions of the embryo components. It can be compared with the process of the coming together of the aboral organ and the blastopore in the ontogeny of lower Spiralia. Usually we select the rudiment of the aboral organ as the fixed focal point, and we speak of the mobility of the blastopore. In Rotifera, the blastopore moves insignificantly from the point of its original position, while the material of the animal pole traverses a significant path. Obviously, in both cases we are dealing with the same process, which takes place through the expansion of the material of the *D* quadrant of the embryo and reflects the establishment of bilateral symmetry in development.

Adult rotifers have a number of important features that resemble the morphology of turbellarians. Thus, the nervous system in rotifers is represented by a compact ganglion that corresponds to the endonal brain in members of Turbellaria. From the endonal brain in rotifers there extends a pair of basic lateral nerves, which proceed along the entire body to the foot (Martini, 1912; Dehl 1934). There is also an additional nerve that innervates the sensory organs and mastax. In rotifers with a reduced foot (*Asplanchna*), the lateral nerves are less developed. The rotifer nervous system is thus fully comparable with one of the most widespread types of the turbellarian orthogonal nervous system (Figure 96a).

Morphologically, rotifers are more advanced than turbellarians. The rotifer body is clad in cuticle, which represents regenerated surface layers of syncytial, integumental epithelium (Clement 1977). Cilia are preserved only in the corona. All organs of rotifers are constructed of highly specialized polarized tissue. Primitive, nonpolarized parenchyma is reduced, and in its place a widened schizocoel has developed. Musculature of rotifers is represented by separate bundles of circular and longitudinal cells of the obliquely striated type (Clement 1977). There is constancy of cellular composition in relation to the small body size and high degree of tissue specialization in rotifers. The digestive apparatus is differenti-

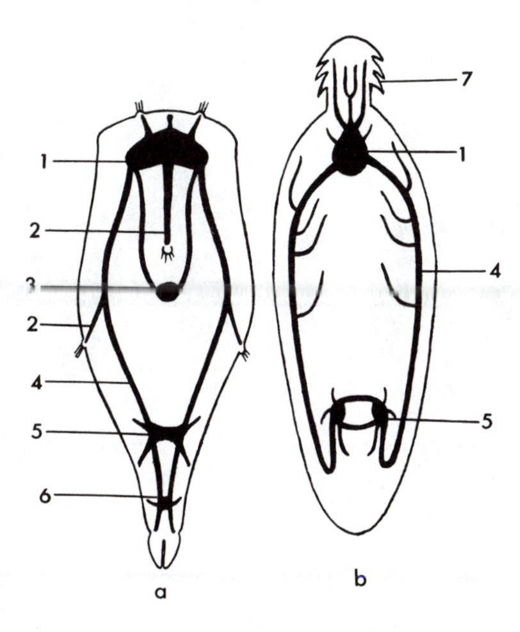

Figure 96. Diagram of the nervous system of a rotifer (a) and an acanthocephalan (b). 1, brain; 2, nerves of the sensory tentacles; 3, ganglion of mastax; 4, lateral nerve trunks; 5, cloacal ganglion; 6, pedal ganglion; 7, proboscis.

ated into a gullet with a special masticatory apparatus (mastax), stomach, and posterior intestine. The posterior intestine in rotifers opens into the cloaca, which also contains the apertures for the reproductive and secretory systems. The cloacal opening does not originate with the blastopore and it represents a new formation. It is possible that in the course of evolution it could have developed at the base of the openings for the reproductive or secretory systems of the turbellarianlike ancestors of rotifers.

Acanthocephala

Acanthocephalan ova have an elongated form. The polar bodies are formed near one of the narrow ends of the ovum, but Meyer (1928), using the peculiarities of division as a basis, believes that the true animal-vegetal axis is situated perpendicularly to the long axis of the envelope. Analysis of acanthocephalan cleavage reveals that it is substantially sim-

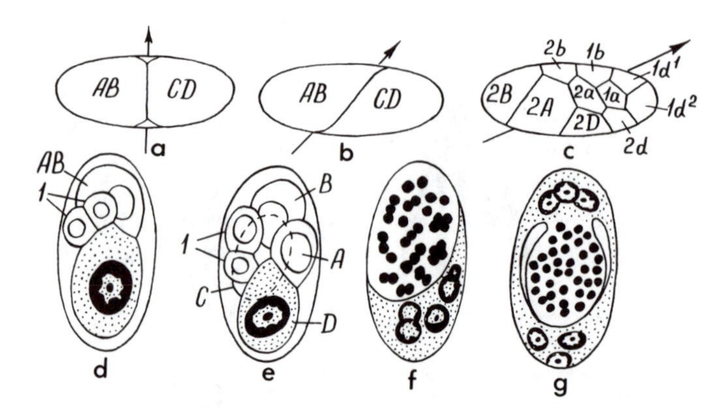

Figure 97. Embryonic development of acanthocephalans (after Ioffe 1979 [a–c]; after Meyer 1931 [d–g]): a–c, rotation of embryo within envelope of ova during division; d–g, division and gastrulation in *Neoechinorhynchus*.

ilar to the cleavage of rotifers and can be inferred from the spiral quartet cleavage (Ioffe 1979). As in rotifers, the acanthocephalan embryo rotates within the envelope of the ovum during the cleavage process, so that, in the later stages, the animal-vegetal axis of the embryo, marking the respective positions of the micro- and macromeres, almost coincides with the long axis of the envelope (Figure 97). The acanthocephalan embryo becomes syncytial early in development. At the vegetal pole of the embryo (situated close to one of the narrow ends of the envelope), the nuclei of the macromeres submerge, providing the beginning of the gonads and ligament, and at the animal pole, the nuclei of the micromeres submerge deeply, forming the rudiment of the brain ganglion and musculature. In the development of acanthocephalans, it is difficult to see the analogy of the merger between the substance of the animal pole and the area of the blastopore that takes place in rotifer development. However, for *Neoechinorhynchus* it is known that a large part of the embryonic integument comes from blastomere *D* of the "basic quartet" (Meyer 1931); logically, this is comparable with the ontogeny of the *D* quartet of Spiralia.

The structures of a number of Acanthocephala organ systems contain important traits similar to those of rotifers and flatworms. The acanthocephalan nervous system is represented by a brain ganglion (homologous with the endon of Turbellaria) and a pair of lateral trunks (Yarzhinskiy 1868; Kaiser 1893; Harada 1931; Bogoyavlenskii and Ivanova 1978). The nerve apparatus of acanthocephalans appears to be fully comparable with that of rotifers (Figure 96b). The presence of a urogenital cloaca is characteristic of acanthocephalans, as it is of rotifers.

Anatomically, acanthocephalans are at approximately the same level of complexity as rotifers. The acanthocephalan body is clad with cuticle, which is a transformed apical zone of syncytial epidermis and is similar in ultrastructure to the cuticle of rotifers (Hammond 1967; Byram and Fischer 1973; Beermann et al. 1974; Bogoyavlenskii and Ivanova 1978). All acanthocephalan organs are constructed of highly specialized, polarized tissue. Parenchyma is fully reduced and there is a widened schizocoel. In a number of acanthocephalan organ systems there is a constancy of cellular composition; this is not only evidence of a high level of tissue differentiation, but also indicates the original small size of the free-living ancestors of the Acanthocephala.

The parasitic form of life makes a deep imprint on the whole organization of acanthocephalans. One of the consequences of the parasitic form of life is the reduction of the intestine, the last remnant of which can be considered to be a band of ligament. It is interesting, in this connection, to note that, in the males of some rotifers, the intestine is also reduced and turns into a ligamentous band that contains the testicles, as in acanthocephalans.

Special features of embryonic development and some anatomical features of adults suggest that rotifers and acanthocephalans are two relatively close groups of pseudocoelomate worms (Meyer 1928, 1932; Haffner 1950; Remane et al. 1976; Ioffe 1979). In the pro-morphological sense, rotifers and acanthocephalans are close to turbellarians, and for both there should be a search for a common ancestor.

Nematoda and Gastrotricha

The significance of similarities and differences in ovum polarity between nematodes and gastrotrichs is unclear. The yolk is distributed evenly in ova of gastrotrichs and nematodes. The polar bodies appear on various parts of the ovum, and often the point of their appearance varies even within the same species. Ova of gastrotrichs and of most nematodes have an elongated form; more rarely (and only in nematodes) the ova are spherical. The long axis of the ovum of nematodes and gastrotrichs always coincides with the anterior-posterior axis. The spindle of the first division in gastrotrich ova and in the long ova of nematodes is always oriented along the anterior-posterior axis; consequently, one of the two blastomeres is anterior and the other posterior. In spherical ova of nematodes the spindle of the first cleavage can coincide with the anterior-posterior axis, but may correspond to the dorsoventral axis. Let us note that the differences in the shape of the ova and orientation of the spindle of the first cleavage is characteristic of closely related species.

Thus, in human ascarids (*Ascaris lumbricoides*), the ova are ellipsoidal, and the spindle of the first division is oriented along the anterior-posterior axis, and, in equine ascarids (*Parascaris equorum*), the ova are spherical, and the spindle of the first division is oriented along the dorsoventral axis. None of the criteria usually used in determining polarity provides a basis for any reliable determination of the position of the animal-vegetal axis. Probably, such animal-vegetal polarity (i.e., the appearance of cellular polarity in ova) has been lost in the case of nematodes and gastrotrichs. The cleavage of gastrotrich and nematode ova clearly proceeds according to a specific pattern, but it is difficult to compare it with the types of ovum cleavage encountered in other groups of the animal kingdom. Thus, numerous attempts have been made to compare the cleavage of nematodes and gastrotrichs with spiral cleavages. At present, all possible views have been expressed. The cleavage of nematodes and gastrotrichs has been interpreted as modified spiral quartet cleavage (Davydov 1914; Dawydoff 1928; Ioffe 1979), as modified duet cleavage (Ivanova-Kazas 1959, 1979), and finally as solo cleavage (Malakhov 1975, 1976).

In the case of interpreting the cleavage of gastrotrichs and nematodes as a derivative of spiral quartet cleavage, the position of the animal-vegetal axis is understood to be perpendicular to the direction of the first cleavage spindle. The first four blastomeres of the embryo are regarded in this case as homologous to the four blastomeres of a basic quartet (Ioffe 1979). However, neither the arrangement of the first four blastomeres nor their prospective potentials has anything in common with spiral quartet cleavage. Also, in the subsequent cleavages of the blastomeres of nematodes and gastrotrichs, there are no indications of spiral quartet cleavage.

There is greater justification for comparing the cleavage of nematodes and gastrotrichs with duet cleavage. The animal-vegetal axis in nematode and gastrotrich ova in this instance also should be understood to be perpendicular to the direction of the first cleavage spindle. At the four-blastomere stage in the cleavage of gastrotrichs and in nematodes of the order Enoplida, tetrahedral figures are formed that are similar to those of duet cleavage. However, prospective potentials of blastomeres here also differ from those of duet cleavage in that the endoderm material is connected not with both blastomeres interpreted as homologous to the first duet of macromeres, but only with one of them. Further, there are also observed large deviations from the duet type both in the geometry of blastomere arrangement and in the prospective potentials.

Considering cleavage in nematodes and gastrotrichs as derived from solo cleavage, the animal-vegetal axis is understood to occur along the direction of the first cleavage spindle; this is contrary to the arrangement of rudiments at the later stages of development. The anterior blastomere here must be interpreted as the first solo of micromeres, and the posterior blastomere as the first solo of macromeres. In some cases (as in nematodes of the subclasses Chromadoria and Rhabditia and gastrotrichs of the superfamily Macrodasyoidea) there is a remarkable correspondence between the prospective potentials of blastomeres with what should be observed in solo cleavage (Malakhov 1975, 1976). However, prospective potentials of the blastomeres in the development of nematodes of the subclass Enoplia and gastrotrichs of the superfamily Chaetonotoidea flagrantly contradict the pattern of solo cleavage: the endoderm that should be related to the posterior blastomere of the embryo and hypothetically homologous to the first solo macromeres comes from the anterior blastomere that is homologous to the first solo micromeres. Thus, none of the attempts to relate cleavage in gastrotrichs and nematodes with any form of spiral cleavage has been successful. This is because cleavage of nematodes and gastrotrichs is not a simple derivative from any definite form of spiral cleavage.

Primitive spiral cleavage is characterized by radial arrangement of blastomeres relative to the animal-vegetal axis. Ectodermal material is localized near the animal pole, ectomesenchymal material is localized more in the vegetal pole, and endoderm and endomesoderm occupy an area of the vegetal pole of the ovum. These characteristics of determinant ova in primitive Spiralia mean that the cleavage grooves that travel the longitudinal planes lead to the formation of qualitatively similar blastomeres of a basic quartet or a duet. The grooves traveling in the latitudinal planes provide qualitatively different blastomeres (macromeres and micromeres). The isolation of rudiments in spiral cleavage also occurs in the process of depositing micromeres. Such a pattern of determinant ova in the lower Spiralia has been shown in experiments on blastomere isolation (Wilson 1903, 1904; Zeleny 1904; Hörstadius 1937; Boyer 1971).

What is the reason for the formation of radial symmetry in the ova of lower Spiralia? Lower Metazoa are radially symmetrical organisms. There is no doubt that radial symmetry preceded bilateral symmetry in the evolution of animals. The gastrulalike stages in the ontogeny of lower Spiralia are fundamentally radially symmetrical, and larvae of lower Spiralia preserve many elements of radial symmetry. The radial symmetry of the ovum of lower Spiralia, it may be supposed, was formed as a consequence of the line of development in the radially symmetrical postembryonic stage. The specific geometry of spiral cleavage (both quartet and duet) was also formed as a consequence of development in the radial symmetry of pelagic larvae.

However, among most present-day Spiralia, the ovum more or less produces a bilateral larva. Therefore, the basic direction in the evolution of Spiralia ontogeny is the transference of definitive bilateral symmetry to all earlier stages of development. This is primarily expressed only in the concentration of endomesodermal material in the dorsal sector of the ovum; that is, in blastomere *D* of the basic quartet.

In most recent evolution, the predominance of the *D* quadrant is becoming still more clear. Among most annelids and mollusks, blastomere *D* significantly exceeds the rest of the blastomeres of the basic quartet, and bilateral symmetry from the beginning is expressed in the geometry of the dividing ovum.

The origin and evolution of nematodes and gastrotrichs are related to a sharp reduction in body size. From the first steps of their evolution the ancestors of gastrotrichs and nematodes were among the organisms of the meiobenthos, and this led, among other organizational transformations, to the loss of pelagic larvae. Small animals of the meiobenthos lay a small number of eggs; this is observed in reality among gastrotrichs and free-living nematodes. This circumstance imposed a kind of restriction on pelagic development relative to the huge mortality rate for plankton larvae. The ancestors of nematodes and gastrotrichs evidently lost radially symmetrical pelagic larvae and made a transition to direct development; this led to the development of specific radially symmetrical forms of laying down blastomeres. The geometry of early cleavage among gastrotrichs and primitive nematodes is characterized by a set of very primitive configurations that are encountered in the cleavage of lower Metazoa. Thus, for example, the tetrahedral, rhombic, and T-shaped figures noted in enoplid development are characteristic of the cleavage of hydroid medusae (Metschnikof 1886; Korsakova 1949). Early transition to direct development and setting the development for the bilaterally symmetrical postembryonic stage were also the basis for very early establishment of bilateral symmetry in the cleavage of most nematodes and gastrotrichs. Therefore, tracing the evolution of ontogeny in the gastrotrich-nematode branch, we do not encounter the gradual appearance of bilateral symmetry in the background of very complete radial geometry of cleavage that took place in a number of living Spiralia.

The absence of clear indications of polarity and the uniqueness of the geometry of cleavage among gastrotrichs and nematodes complicate the analysis of the pro-morphology of these groups. At the early gastrula stage, the morphology of the embryo of nematodes and gastrotrichs is close to that of the Spiralia gastrula. At the later stages, the blastopore of nematodes and gastrotrichs stretches out and takes on a slitlike form. The mouth is formed at the anterior end of the blastopore and the anus at the posterior end. The anterior-posterior axis of the embryo is formed in a direction that is perpendicular to the oral-aboral axis of the gastrula axis. The larger part of the ventral side of nematodes and gastrotrichs is from growth of the blastopore side of the gastrula, and the dorsal side of the body is derived from the aboral surface of the gastrula. There has been no success in establishing any more detailed relationships in the development of nematodes and gastrotrichs.

Gastrotrichs are a more primitive group than are nematodes. The blastopore surface of gastrotrichs serves as a crawling surface. It is very characteristic that in gastrotrichs, along the blastopore side, a ciliary belt—neuratrochoid—is preserved. The gastrotrich nervous system can be understood as a derived circumblastoporal nerve plexus. The most powerful dorsal commissure of the gastrotrich nerve ring conforms to the arch of the circumblastoporal plexus. The ventral trunks of the gastrotrichs border on the blastoporal surface and represent the lateral trunks, constituting the circumblastoporal nerve plexus.

Nematodes lose primary bilateral symmetry, and the blastoporal side of the body serves as the crawling surface. Changes in means of locomotion, namely that nematodes flex the body not in the frontal plane as do most Bilateria, but in the sagittal plane, cause the restructuring of the nervous system. Paired ventrolateral trunks merge into a single ventral trunk. From processes of neurons situated in the ventral trunk there is a development of dorsal nerves that oppose the motor nerves of the ventral trunk. For more even innervation of the muscle layer, additional submedial motor nerves are formed. Despite the significant transformation of the nematode nervous system, it is easy to recognize in it elements of a circumblastoporal nerve plexus, which composes the central nervous system of nematodes. The dorsal commissure of the circumpharyngeal nerve ring is homologous to the anterior arch of the circumblastoporal plexus; the ventral nerve trunk, which begins as a pair of rootlets and bifurcates anterior to the anus, is homologous to the merged pair that compose the circumblastoporal plexus; and the circumrectal commissure is homologous to the posterior arch of the circumblastoporal plexus.

In the nervous systems of nematodes and gastrotrichs there are no traces of the endonal ganglion (primary brain) that develops at the base of the aboral organ and is characteristic of most Spiralia. Once again, this underscores the uniqueness of the gastrotrich and nematode nervous systems.

The integuments of gastrotrichs and nematodes are cuticularized. The musculature of the body wall is represented essentially by longitudinal muscle. The intestine is complete and has a suctorial gullet with a three-sided lumen. The body cavity is developed to varying degrees. All gastrotrichs and nematodes are characterized by internal fertilization.

The traits that bring nematodes and gastrotrichs together have been discussed earlier, and they are evidence of the phylogenetic relationship of these two groups and of the isolated position of these classes among other Bilateria; the ancestors of nematodes and gastrotrichs were isolated early from the general Bilateria branch. Nematodes and gastrotrichs are closer to Spiralia than to other groups of the animal kingdom, but the evolutionary routes of Spiralia, on the one hand, and of nematodes and gastrotrichs on the other, evidently separated very early.

Priapulida, Kinorhyncha, Nematomorpha, and Loricifera

Priapulida make up one of the smallest classes of the animal kingdom, uniting two orders (Priapulomorpha and Seticoronaria) and a total of

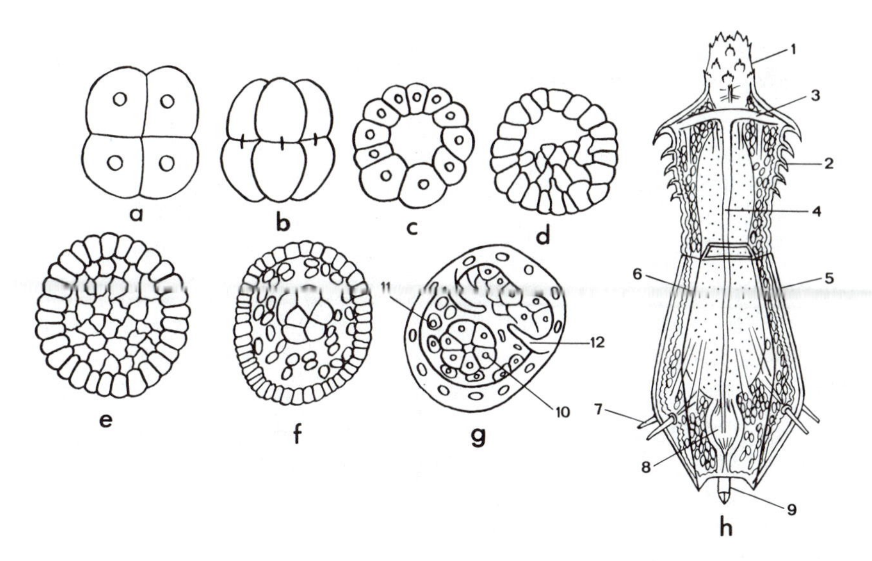

Figure 98. Embryonic and larval development of *Priapulus caudatus* (original [h]; after Zhinkin 1955 [a–g]): a, four-blastomere stage; b, eight-blastomere stage; c, blastula; d, gastrulation; e, parenchymula; f, isolation of gut primordium; g, septum formation; h, larva. 1, oral cone; 2, proboscis; 3, nerve ring; 4, abdominal nerve trunk; 5, test; 6, midgut; 7, sensory setae; 8, posterior gut; 9, foot; 10, intestinal primordium; 11, mesenchyme; 12, septum.

only eleven species. The representatives of this class are marine animals that lead either a burrowing (Priapulomorpha) or a sessile (Seticoronaria) form of life.

The embryonic development of Priapulida has not been fully studied. We refer to the data of Zhinkin and his associates (Zhinkin 1949, 1955; Zhinkin and Korsakova 1953). The ova of priapulids undergo complete and almost even cleavage. The grooves of the first two cleavages occur in the meridional planes. The groove of the third cleavage occurs in the lateral plane (Figure 98). The geometry of the eight-cell and subsequent stages of cleavage is very like that in radial cleavage. Priapulid cleavage is rather multicellular and is probably indeterminate. Gastrulation occurs by the growth of a dense complex of cells within the embryo from the vegetal pole. The blastocoel is completely filled with cells of the phagocytoblastic rudiment, and the embryo takes on the appearance of parenchyma (Figure 98e). Later, the phagocytoblastic rudiment differentiates into an internal cell complex—the rudiment of the endoderm—and a more peripheral layer of mesenchymal cells. Then, out of the ectoderm within the embryo, a partition (septum) grows that separates the rudiment of the proboscis section (preseptum) from the torso (postseptum). The oral opening appears at the anterior end of the embryo without any connection with the blastopore. The anus is formed at the posterior end of the embryo, possibly at the location of the former blastopore.

Priapulid larvae have been described repeatedly (Hammarsten 1915; Lang 1963; Por 1972; Salvini-Plawen 1974; Kirsteuer 1976; Malakhov 1980). The priapulid larval body is contained in a cuticular test. In members of *Priapulus* and *Halicryptus* the test consists of wide dorsal and ventral plates and four pairs of narrow plates on the sides of the body: subdorsal, dorsolateral, ventrolateral, and subventral (Figure 98h). With the extended proboscis, the larval test has an almost square form in transverse section. With the extension of the proboscis, the dorsal and ventral plates come together, the whole test becomes more dense, and the lateral plates assume an accordionlike shape. With the proboscis section completely extended in priapulid larvae, the following parts of the proboscis can be distinguished: the oral cone formed by the extended glottis; a middle part equipped with scalids, directed backward; and a cervical section without scalids (Figure 99). The oral cone is equipped with several corollae of cuticular plates directed forward. In each corolla are five cuticular plates. The plates of the adjacent corollae alternate with one another. On the plates themselves are one to three rows of cuticular teeth. The middle part of the proboscis has thorns—scalids—that are bent back. The anterior corolla contains eight large scalids and the rest of the corollae have twenty-five each. The posterior (cervical) area of the proboscis section is clad in scalidic cuticle and is directly united with the plates of the test (Figure 99).

The oral opening is situated at the apex of the oral cone. The intestine is divided into the glottis, midgut, and posterior intestine. The anus opens terminally at the posterior end of the body. Near the anus is situated a cuticular process that bears the ducts of the caudal glands.

The nervous system of the larvae is represented by a circumglottal nerve ring that lies in the proboscis section at the base of the oral cone, and by the ventral nerve trunk. The body walls are formed by a complex cuticular structure, hypodermal epithelium, and a layer of circular and longitudinal musculature. The larval body cavity contains a large quan-

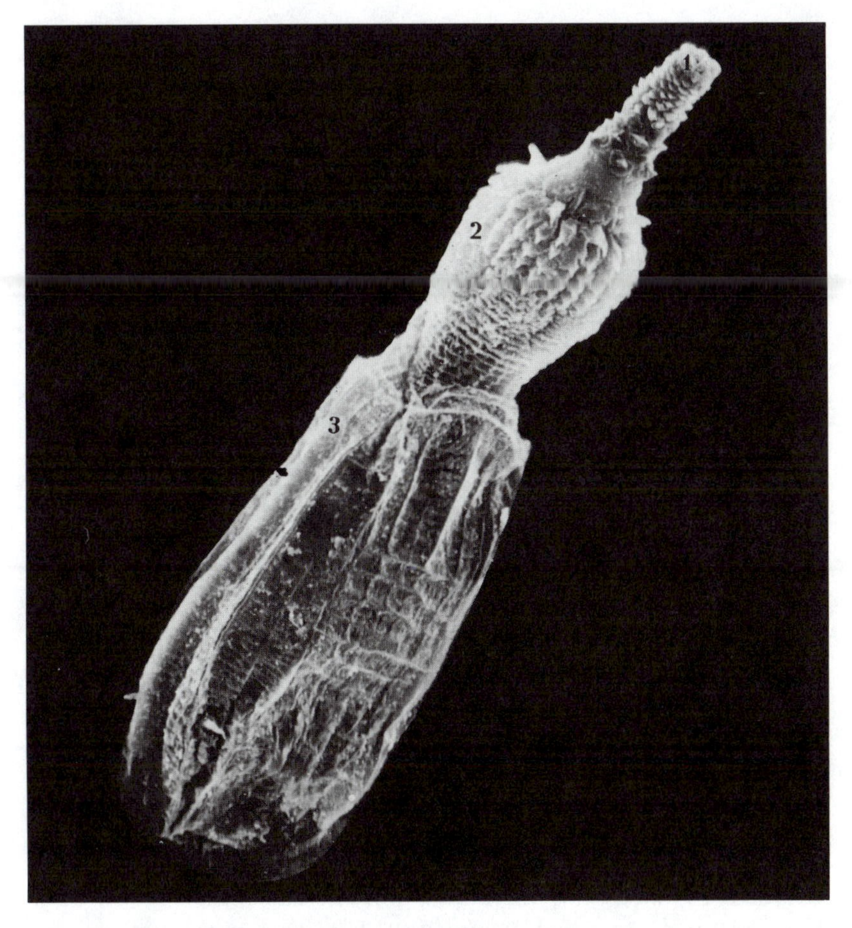

Figure 99. *Priapulus caudatus* larva. Magnification = 170. 1, oral cone; 2, proboscis; 3, torso.

tity of mesenchymal cells that move together with the liquid of the body cavity (Figure 98h).

During metamorphosis the larvae shed the test. The oral cone becomes more dense and turns into a circumoral field equipped with cuticular plates directed forward. Along the edge of the circumoral field is the nerve ring, from which the ventral nerve trunk extends. One of the most discussed questions of priapulid morphology concerns the nature of their body cavity. Some authors regard the priapulid body cavity as a coelom (Hammarsten 1915; Zhinkin 1955), while other researchers deny its coelomic nature (Hyman 1951; Lang 1963; Por and Bromley 1974). Electron microscopy reveals that the body walls of adult priapulids are formed by relatively thin homogenous cuticle, under which lies a layer of hypodermis, and, mesad to it, a layer of circular musculature (Malakhov 1980). The cells of the longitudinal musculature do not form a solid layer but are situated in groups of three to five cells, between which are spaces. In cross sections of the priapulid body wall, as observed with the light microscope, these cells can be seen to be an intermittent coelomic covering. The fluid of the priapulid body cavity directly contacts the longitudinal musculature of the body wall and, in some sections, cells of the circular musculature as well. The intestine is surrounded by a layer of longitudinal muscles that directly adjoins the basal membrane of the intestine. These muscle cells are also in direct contact with the body cavity fluids. Rounded mesenchymal cells are suspended in the priapulid body cavity. Thus, the body cavity structure of larval and adult priapulids has nothing in common with a true coelom.

Kinorhyncha are one of the most poorly studied animal groups. Practically nothing is known about their embryonic development. The body of adult kinorhynchs is divided into two basic sections, the torso and the proboscis. The proboscis section can be divided into three parts: the oral cone, which has cuticular processes directed forward (styli); a middle part equipped with spines that are bent back (scalids); and a smooth cervical area. The trunk is contained in a bilaterally symmetrical cuticular test, divided into separate zonites. The cuticular covering of each zonite is formed by three plates of stiff homogenous cuticle, one dorsal and two subventral (Figure 100). The plates of the cuticular test are united with ligaments of fibrous elastic cuticle. Such ligaments are also present between the plates of adjacent zonites. The cellular hypodermis underlies the cuticle, and the nuclei of the hypodermal cells are gathered into dorsal and ventrolateral rows. The musculature is represented by four longitudinal muscle bands and paired dorsoventral muscles that unite the dorsal and ventral plates of each zonite (Figure 101).

The kinorhynch nervous system is represented by a circumglottal nerve ring that is situated in the proboscis section at the base of the oral cone and by a ventral nerve trunk. Nerve cells are concentrated anterior and posterior to the nerve ring and also lie in metameric groups in the ventral nerve trunk. Ultrastructural research has shown the presence of four additional thin nerves that are connected with and innervate lon-

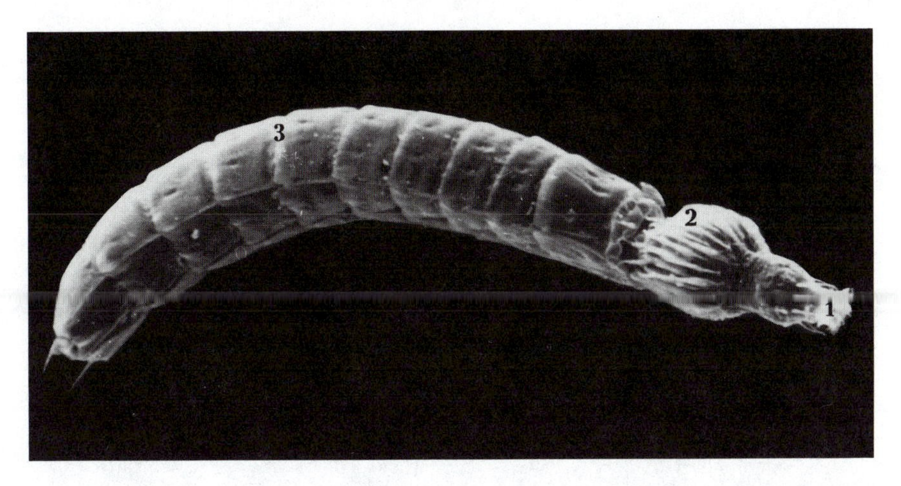

Figure 100. The kinorhynch *Pycnophyes kielensis.* Magnification = 200. 1, oral cone; 2, proboscis; 3, torso.

gitudinal bands of muscles. These nerves contain only nerve fibers, but they receive neuromuscular processes of the muscle cells (Figure 101).

Between the body wall and the internal organs is a small body cavity filled with a dark grainy substance. No cellular elements have been found in the body cavity. Research on kinorhynch anatomy conducted with the light microscope reveals a wide body cavity with transparent contents. Prior to the body cavities, these areas were probably occupied by the sarcoplasmic bellies of muscle cells, which were inflated and poor in organelles (Figure 101).

An outstanding feature of kinorhynchs is the presence of metameric organization of the outer musculature and nervous system. The principal manifestation of this metamerization is the separation of the cuticular body covering into zonites, and this separation was brought about by the necessity for a flexible body wall. Naturally, the muscle bands uniting the plates of adjacent zonites and the nerve elements in the abdominal nerve trunk that innervate the musculature are also metameric.

The cleavage among representatives of Nematomorpha has no evidence of similarity to any definite type of cleavage. The position of the animal-vegetal axis is indefinite. The first cleavage proceeds, as a rule, in the direction of the future anterior-posterior axis of the embryo. After

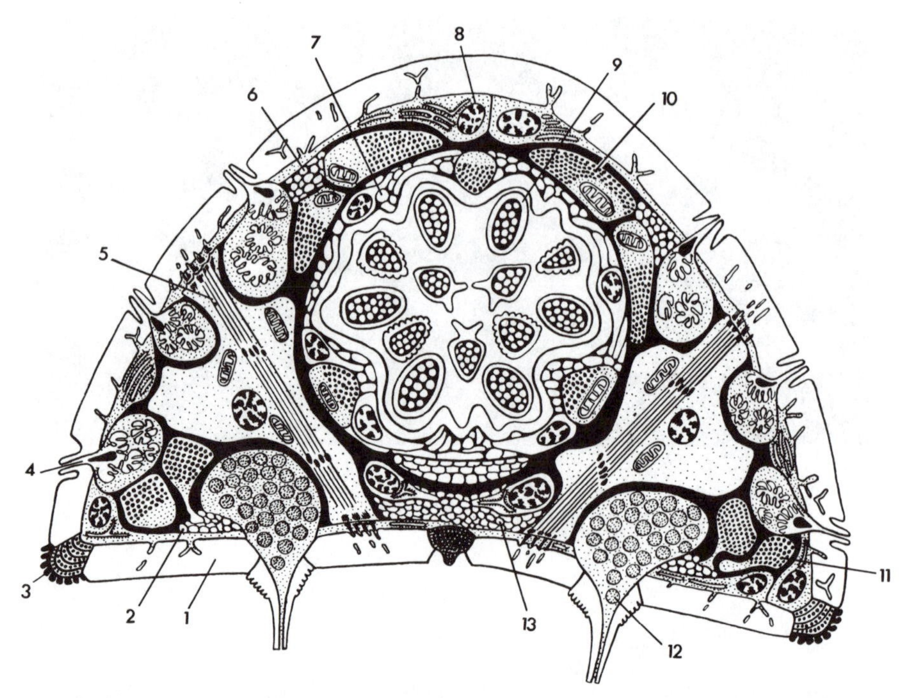

Figure 101. Diagram of a cross section of the kinorhynch *Pycnophyes kielensis*. 1, cuticle; 2, subventral nerves; 3, fibrous connection between separate plates of zonites; 4, glands; 5, dorsoventral muscles; 6, subdorsal nerves; 7, circumglottal nerve ring; 8, hypodermis; 9, scalids of introverted proboscis; 10, subdorsal longitudinal muscles; 11, subventral longitudinal muscles; 12, ventral nerve trunk; 13, attachment seta.

the second cleavage, there are tetrahedral figures and, more rarely, flat rhombic figures. This would seem to make nematomorph cleavage similar to that of the gastrotrichs and primitive nematodes. However, nematomorph cleavage is indeterminate, and the whole further course of embryogeny is completely unlike that of nematodes and gastrotrichs. Cleavage of the nematomorph ovum results in a parenchymula or coeloblastula that soon becomes a parenchymula. An archenteron results from invagination of the endoderm at the posterior end of the embryo. The anus eventually forms at the site of the blastopore, and the mouth forms at the anterior end of the embryo independent of the blastopore. Growth of ectoderm at the anterior end of the embryo forms the pro-

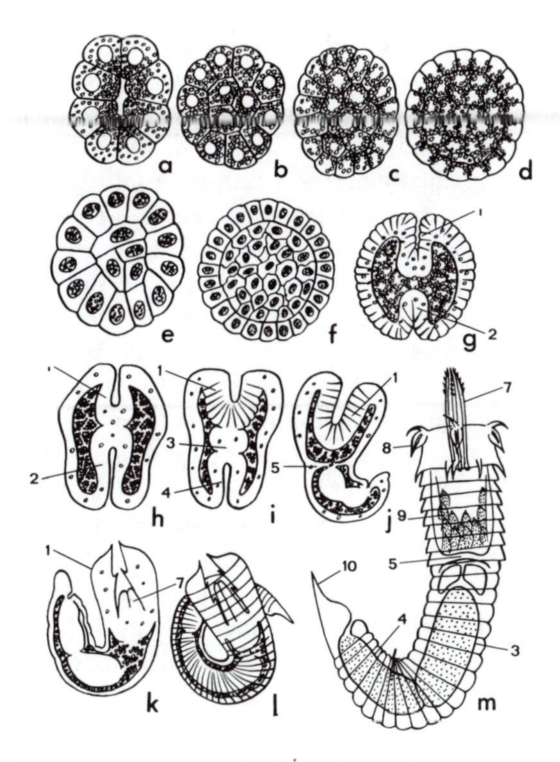

Figure 102. Embryonic development of *Gordius* sp. (Nematomorpha) from *Turkmenia*. (All are optical sections except e and f, which are histological sections.) a, sixteen-cell blastula; b, c, formation of parenchymula; d, yolk reduction in outer cells; e, early parenchymula; f, late parenchymula; g, beginning of archenteron; h, embryo with two patches; i, isolation of larval gland primordium; j, septum formation; k–l, appearance of primordium of oral cone; m, hatched larva. 1, primordial proboscis; 2, archenteron; 3, primordial anterior gut bladder; 4, primordial posterior gut bladder; 5, septum; 6, proboscis; 7, oral cone; 8, scalids of proboscis; 9, nerve cells in proboscis section; 10, caudal spike.

boscis section, which is independent of the origin of the torso rudiment that grows from the ectodermal septum (Figure 102).

The nematomorph larva consists of two basic sections, the proboscis and the torso. The proboscis section is distinguished by the often-mentioned proboscis. The oral cone is constructed of three cuticular plates with an oral opening at the apex surrounded by two or three tiers of mobile cuticular hooks (Figure 102m). The middle part of the proboscis has three tiers of scalids directed backward, the first and second tiers containing six scalids each and the third, seven. The posterior part of the proboscis gradually makes a transition into the torso. The nematomorph larva twists and turns its proboscis section while moving. This is achieved, however, not by pumping liquid from the torso into the proboscis but by shortening the muscles of the proboscis section.

The rudiment of the circumglottal nerve ring of the larva, according to the data of Shepot'yev (1908), is situated in the proboscis section at the base of the oral cone. Along the ventral side of the larva is the rudiment of the abdominal nerve trunk (Mühldorf 1914).

Among adults of Nematomorpha, the proboscis section is not expressed externally. The septum that separates the rudiment of the proboscis section from the torso is preserved only among the primitive marine Nectonematida. The nervous system of adult nematomorphs is represented by the circumglottal nerve ring and the ventral nerve trunk. It is remarkable that the nerve ring among adults of Nectonematida is localized in the preseptum—the homolog of the proboscis section.

Nematomorphs are parasitic organisms. Aquatic arthropods swallow the larvae, which actively burrow through the intestinal wall and become situated in the muscles and other tissues of the host. The development of the larva is completed in one or, more rarely, two hosts, the second host usually being another predatory arthropod. In the end, a large worm is formed, completely unlike the larva.

The body of adult nematomorphans is cylindrical in section and covered by a dense, fibrous cuticle. Under the cuticle lies a layer of hypodermis with a single ventral fold through which passes the ventral nerve trunk. Under the hypodermis is situated a layer of longitudinal musculature. The intestine in adults of Nematomorpha is rudimentary and does not function. The wide space between the intestine and body wall is filled with parenchyma in which there are significant lacunar spaces. The parenchyma is composed of alveolar cells, between which there is a strongly developed fibrous intercellular substance. Adults of

Nematomorpha possess typical undulating locomotion (although the body is flexed in the frontal plane), based on the interaction of cuticle and longitudinal musculature. The taut state of the cuticle is maintained by high turgor of the parenchyma cells.

Representatives of the three classes are very different. Priapulids are burrowing or stationary organisms, kinorhynchs are marine interstitial forms, and nematomorphs are parasites of arthropods. Nevertheless, it should be noted that there is a fundamental similarity in the basic structure of all three groups. The body, at least at the larval stage, is subdivided into two basic sections: the torso and the proboscis. The proboscis is a morphologically isolated anterior part of the body and, in turn, is divided into an oral cone, a middle section equipped with recurved scalids, and a cervical area. In embryogeny (where observed), the proboscis section is isolated from the torso by a septum. Unquestionable similarity is observed in the structure of the central nervous systems of priapulids, kinorhynchs, and nematomorphs. In all three groups it is formed by the circumglottal nerve ring, which lies in the proboscis section at the base of the oral cone, and the ventral nerve trunk (Figure 103).

At the embryological level, we are able to compare only priapulids and nematomorphs. It is significant, in this respect, that representatives of groups so diverse in external morphology and ecology have characteristics of clear similarity. Cleavage in both groups is indeterminate and has nothing in common with the spiral type of cleavage. The embryo undergoes an early transition to a parenchymalike condition. The anus appears at the blastopore, and the mouth breaks through at the anterior end of the embryo without any connection to the blastopore; that is, there is a secondary oral aperture. Rudiments of the proboscis and torso sections are separated by a septum.

All this leads to the thought that priapulids, kinorhynchs, and nematomorphs are relatively close groups of organisms that share common ancestry, the structure of which to some degree is recapitulated in the organization of their larvae. The probable common ancestor of these three groups was a parenchymatose organism whose locomotion was accomplished by eversions of a hook-bearing proboscis, a simple differentiation of the anterior section of the body. The closest to this prototype are the priapulid larvae. Their proboscis section is differentiated to only a slight degree, and the oral cone is simply an everted section of the glottis. With extension of the proboscis, the priapulid oral cone turns inward like the rest of the proboscis section. In larvae of Kinorhyncha

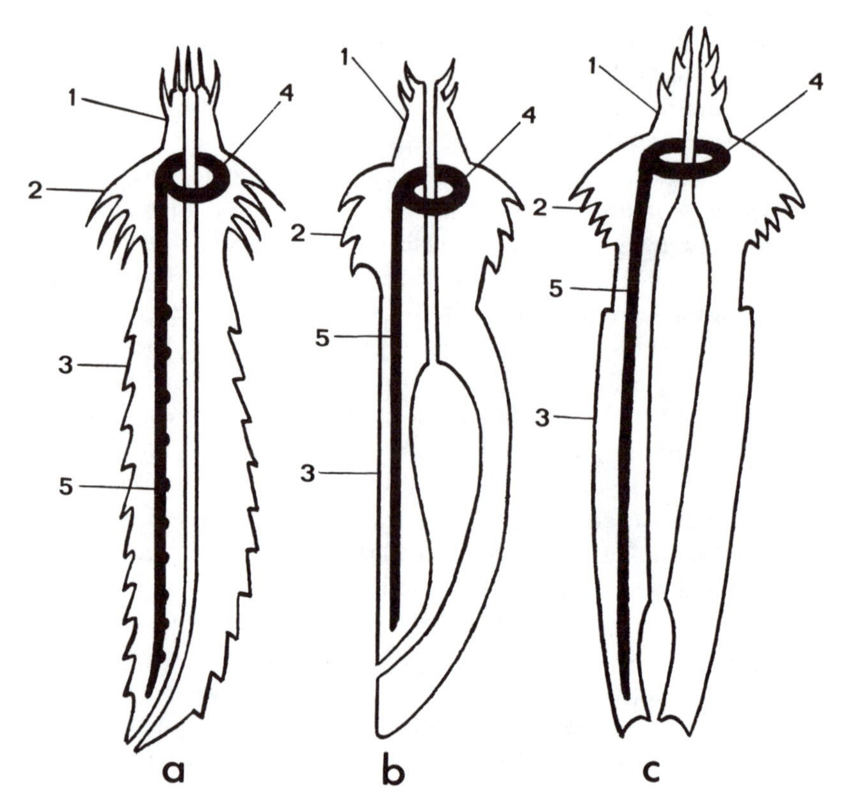

Figure 103. Structure of larvae of Kinorhyncha (a), Nematomorpha (b), and Priapulida (c). 1, oral cone; 2, proboscis; 3, torso; 4, nerve ring; 5, abdominal nerve trunk.

and in Nematomorpha, the oral cone is already more differentiated, and with turning in of the proboscis section, it does not turn in but only stretches out. The difference in the armament of the oral cone and the rest of the proboscis is understood from the different functions of these sections. The middle part of the proboscis section, which is simply a modified anterior portion of the body, fulfills an exclusively locomotor function, and the backward-directed scalids serve to anchor the anterior end of the body in the sediment. The oral cone, which is a modified part of the glottis in priapulid and kinorhynch larvae, serves to seize food particles. In larval nematomorphs, which do not feed, the armament of the oral cone is poorly developed.

The eversion and retraction of the proboscis originally took place wholly by muscular contraction, as takes place among nematomorph larvae and partially among those of kinorhynchs. Nematomorphs preserve the parenchyma with a strongly developed intercellular substance. Among priapulids, in connection with an increase in body size there is a transition to stronger hydraulic locomotion. Parenchyma is diminished, and the mesenchymal cells are simply suspended in the body cavity. While priapulids evolved into larger marine organisms leading primarily a burrowing form of life, and more rarely a sessile form of life, kinorhynchs underwent specialization toward interstitial conditions of existence. A sharp decrease in body size led to the reduction of parenchyma and reduction of the body cavity to slitlike spaces between organs. Nematomorphs underwent a transition to the parasitic form of life, and this, as might be expected, had a dramatic effect on the morphology of adults.

Nonspiral and indeterminate cleavage, absence of curvature of the animal-vegetal axis resulting from growth of the dorsal quadrant of the gastrula, and formation of the oral aperture from secondary invagination of the blastula sharply separate the types of development of the three groups from that of Spiralia and all other groups of pseudocoelomic worms. The morphology of larvae and adults provides no basis for putting members of Priapulida, Kinorhyncha, and Nematomorpha together with other groups of pseudocoelomate worms and Spiralia. Evidently these three classes make up an independent branch of the animal kingdom, originating from ancient ancestors of Bilateria.

Recently, representatives of Loricifera, a new group from the meiobenthos, have been described (Kristensen 1983). These are close to Priapulida, Kinorhyncha, and Nematomorpha according to many morphological characteristics (Figure 104). The only known representative of Loricifera, *Nanaloricus mysticus,* is a microscopic creature 235 m in length and 90 m in width. The sections of the body are the torso and proboscis (Figure 104). The proboscis has an oral cone surrounded by eight oral stylets, a middle section equipped with nine rows of scalids, and a cervical area with two rows of plates bearing cervical scalids. The torso is enclosed in a cuticular test consisting of six plates (one dorsal, two lateral, two intermediate, and one ventral). The plates of the test

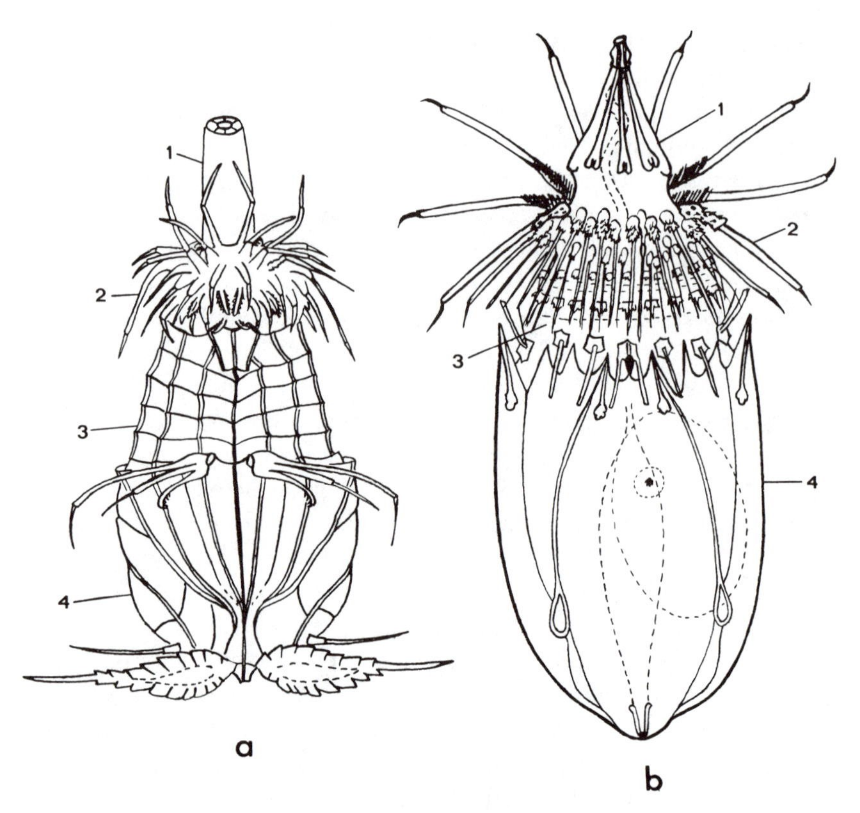

Figure 104. Larval (a) and adult (b) stages of *Nanaloricus mysticus* (after Kristensen 1983). 1, oral cone; 2, proboscis region equipped with scalids; 3, cervical section of proboscis; 4, test covering torso.

have delicately pitted sculpture. Glandular ducts open through pores on the test surface. Besides simple pores, there are bordered pores—flosculi.

The mouth is situated terminally at the apex of the oral cone and leads to a twisting buccal tube connected to the muscular pharyngeal bulb. Beyond the bulb is a short esophagus, a straight midgut, and a short posterior intestine that opens with an anus situated in the anal cone. The reproductive system consists of paired ovaries (females) or paired testicles (males). The nervous system consists of a brain, which occupies the anterior half of the middle part of the proboscis. The brain is situated dorsally, has two lobes, and is connected to a ring of eight ganglia that innervate the first circle of scalids. There is a ventral ganglion in the cervical area, and a suggestion of the presence of several ganglia in the region of the torso, of which only the presence of a ventral anal ganglion has been established precisely. The larvae of *N. mysticus* differ from adults in having an unarmed oral cone, longitudinal plications on the test, and bilaterally arranged locomotor and sensory bristles (Figure 104a).

Morphology of members of Loricifera is very close to that of the larvae of Priapulida and Kinorhyncha. The only difference consists of a dorsal arrangement of a brain ganglion (although it is connected to the circumoral ring and lies just at the base of the oral cone) and stronger ganglionization of the ventral nerve trunk. Evidently, Loricifera is the fourth class in the group that also includes Priapulida, Kinorhyncha, and Nematomorpha.

Pseudocoelomate Classification

Pseudocoelomate worms are not a natural group of organisms. The pseudocoelomate level of organization was independently attained by various representatives of this composite group. The possibility of parallel and independent attainment of the pseudocoelomate organization reflects the general pattern of tissue evolution. As shown by Beklemishev (1925), the basic trend in the evolution of tissues in lower Bilateria is the replacement of primitive nonpolarized tissues by polarized tissues. First to acquire epithelial polarized structure was the kinoblast, which turned into ectodermal epithelium. Originally, only the entoderm was epithelialized in the phagocytoblast, whereas the mesoderm remained in a nonpolarized state of parenchyma. Lower Bilateria are parenchymatose animals. The further evolution of histological structure can lead either to development of epithelium from mesoderm, as in coelomate Bilateria, or to the reduction of nonpolarized parenchyma in place of which arises the schizocoel. The latter process occurs in pseudocoelomate Bilateria, to which belong the group of animals discussed in this chapter. This process may be brought about by various causes. In some instances it is a sharp reduction in body size, which leads to a paucity of cells in the whole organism because there is simply no space left for nonspecialized parenchyma. If subsequently the body size grows again (for example in making a transition to parasitism), an expanded schizocoel arises in place of the reduced parenchyma. In other instances, the schizocoel arises with the dilution of the basic substance that ties together the cells of parenchyma, and then the mesenchymal cells simply float in the body cavity. This route to schizocoel formation is illustrated by Priapulida.

Different groups of pseudocoelomate worms came from various ancestors. Thus, rotifers and acanthocephalans came from some turbellarianlike form, and they belong to the Spiralia branch. Nonetheless, the uniqueness of these groups is so great that each of them should be regarded as a separate phylum.

Undoubtedly, nematodes and gastrotrichs are closely related groups of animals that can be united into one phylum, Nemathelminthes, which came from ancient ancestors of Bilateria close to the common Spiralia branch but independent of the latter in further evolution.

Priapulids, kinorhynchs, and nematomorphs make up a special group. The special features of structure and development sharply separate this group from other pseudocoelomate worms. The fundamental similarity in the structure of all three groups and their isolated position in the system of the animal kingdom permit putting these three classes into a new type of proboscis-head worms, or Cephalorhyncha. The greatest element of doubt for this assemblage concerns the kinorhynchs, the embryonic development of which is still unknown. In the meantime, the structural characteristics of the adults exclude the possibility of assigning them to any phylum other than the Cephalorhyncha. Proboscis-head worms do not have any relationship to Spiralia. Indeterminate cleavage, multicellular rudiments, secondary oral state, and certain other peculiarities of development and structure bring them close to another basic branch of Bilateria that provided origin to Chaetognatha, Tentaculata, and Deuterostomia. However, all these groups unite coelomate animals. The proposition is not without probability that Cephalorhyncha represents a specialized side branch from that trunk of the animal kingdom that in its later development provided the origin of coelomate Chaetognatha, Tentaculata, and Deuterostomia.

Loricifera probably also belongs to the phylum Cephalorhyncha, although its members are distinguished from other representatives of this phylum by greater ganglionization of the nervous system.

It is now appropriate to diagnose the phyla that have been identified.

Diagnosis of phylum Rotifera Cuvier, 1798: Body clad in cuticle.

Cilia present only on corona. Musculature longitudinal and circular. Nervous system with endonic brain, paired lateral nerves, and additional nerves to various organs. Schizocoel present. Gut complete and mastax present. Excretory system consists of protonephridia. Digestive, reproductive, and excretory systems open to exterior by way of cloaca. Sexes separate. Cleavage derived from spiral quartet; oral opening develops from blastopore and anus develops from secondary invagination. One class: Rotatoria Ehrenberg, 1838.

Diagnosis of phylum Acanthocephala Skryabin and Schulz, 1931: Body clad in cuticle and ciliary covering fully reduced. Proboscis present and equipped with cuticular hooks. Longitudinal and circular musculature present. Nervous system consists of endonic brain and paired lateral nerves. Schizocoel present. Gut, mouth, and anus reduced. Protonephridia or special excretory glands present. Sexes separate. Digestive, reproductive, and excretory systems open to exterior by way of cloaca. Cleavage derived from spiral quartet. One class: Acanthocephala Rudolphi, 1808.

Diagnosis of phylum Nemathelminthes Schneider, 1866: Body covered with cuticle. Ciliary covering retained only on ventral side, or is fully reduced. Musculature mainly longitudinal. Nervous system represented by circumglottal nerve ring from which extend a pair of ventrolateral nerves or a merged ventral nerve trunk. In the latter case there are additional extended nerves. Schizocoel developed to varying degrees. Two classes: Gastrotricha Metschnikoff, 1864; Nematoda Rudolphi, 1808.

Diagnosis of phylum Cephalorhyncha Malakhov, 1980: Body divided into proboscis and torso. Proboscis region subdivided into oral cone, midsection equipped with recurved scalids, and cervical area. Body clad in cuticle; ciliary integument fully reduced. Longitudinal and circular musculature present. Nervous system composed of circumglottal nerve ring (sometimes dorsal brain ganglion) situated in proboscis section at base of oral cone, and ganglionated ventral nerve trunk. Schizocoel present, but degree of development variable. Excretory system of protonephridial type, or absent. Buccal orifice developed from secondary invagination of blastula. Embryonic precursor of proboscis and torso divided by septum. Four classes: Priapulida Delage and Herauerd, 1897; Kinorhyncha Reinhard, 1887; Gordiacea von Siebold, 1848; and Loricifera Kristensen, 1986).

In 1908, the Austrian zoologist K. Grobben divided all bilateral animals with three germ layers into two groups: Protostomia (primary mouth) and Deuterostomia (secondary mouth). This division existed for several decades, although facts were gradually accumulated that undermined the very bases of this division. In particular, it turned out that the fate of the blastopore (referred to in the name of each group) did not serve as a reliable criterion for setting the limits of Protostomia and Deuterostomia. These and other contradictions have led many zoologists in the 1970s and 1980s to cease to accept the division of bilaterally symmetrical animals into two groups based primarily upon the origin of the oral opening from primary and secondary invaginations of the blastula. Nevertheless, the evolutionary development of Bilateria can undoubtedly be divided into two chief branches, although they do not coincide with Protostomia and Deuterostomia in the classical sense. One of these branches—Spiralia, or Spiraloblastica—unites phyla that have characteristic spiral cleavage; the origin of endomesoderm (coelomic mesoderm) from two mesoblasts, which, in turn, develop from blastomere 4 *d*; and the formation of a definitive brain with the inclusion of material for an aboral nerve center. This group includes Platyhelminthes, Nemertini (?), Entoprocta, Sipuncula, Echiura, Mollusca, and Articulata. Another branch—Radialoblastica (the name Radialia cannot be used because it has been preempted to designate radially symmetrical multicellular animals)—unites phyla characterized by radial cleavage of the ovum; the origin of mesoderm as a multicellular primordium from the wall of the primary intestine by immigration, delamination, or invagination; and the reduction of the aboral nerve center and development of a secondary definitive brain. This group should include Chaetognatha, Phoronida, Bryozoa, Brachiopoda, Hemichordata, Echinodermata, Pogonophora, and Chordata. The heterogeneity of pseudocoelomate worms clearly stands out relative to these basic branches of Bilateria. Rotifers and acanthocephalans, as shown above, probably belong to Spiraloblastica, although their embryonic development has undergone significant changes. Nemathelminthes does not belong either to Spiraloblastica or Radialoblastica and represents an independent evolutionary branch originating from common ancestors of Bilateria. Proboscis-head worms also cannot be included in either Spiraloblastica or Radialoblastica, but as indicated above, they are closer to the latter.

Having completed the present work, we would like to pay a warm tribute of recognition to the living memory of my teacher, Professor K.

V. Beklemishev, under whose direction this research was accomplished. I also acknowledge with great warmth Yu. M. Frolov, under whose direction I began work in the field of nematology. I am deeply grateful to T. A. Platonova and Ya. I. Starobogatov for help on a number of questions arising in the present work. The conclusions and statements of this work depend significantly on electron microscope data obtained through the collective assistance of the Laboratory for Electron Microscopy of the Biology Faculty of Moscow State University.

Bibliography

Aboul-Eid, H. Z. 1969. Electron Microscope Studies on the Body Wall and Feeding Apparatus of *Longidorus macrosoma*. *Nematologica* 11:451–463.

Adams, J. R., P. E. Holbert, and A. J. Forgash. 1965. Electron Microscopy of the Contact Chemoreceptors of the Stable Fly, *Stomoxys calcitrans* (Diptera: Muscidae). *Annals of the Entomological Society of America* 58:909–917.

Albertson, D. G., and J. N. Thomas. 1976. The Pharynx of *Caenorhabditis elegans*. *Philosophical Transactions of the Royal Society of London* 275: 299–325.

Alekseyev, V. M., and A. D. Naumova. 1977. New Species of Nematodes from Lake Khasan. *Zoologicheskii Zhurnal* [Zoological Journal] 56(2): 292–295.

Andrassy, I. A. 1974. A Nematodak evolucioja es rendszerezese. *Magyar Tudomanyos Akademia Biologiai Tudomanyok Osztalyanak Kozlemenyei* 17: 13–58.

———. 1976. *Evolution as a Basis for the Systematization of Nematodes*. 228 pp. Budapest: Akadémiai Kiadó.

Anisimov, A. P., and N. P. Tokmakova. 1973. Proliferation and Growth of the Intestinal Epithelium of *Ascaris suum* (Nematoda) in Post-natal Ontogeny. I. *Ontogenez* [Ontogeny] 4(2): 373–382.

Anisimov, A. P., and L. N. Usheva. 1973. Proliferation and Growth of the Intestinal Epithelium of *Ascaris suum* (Nematoda) in Post-natal Ontogeny. II. *Ontogenez* [Ontogeny] 4(3): 416–420.

Ansari, J., and M. Basir. 1964. The Histological Anatomy of *Setaria cervi*. *Indian Journal of Helminthology Monograph Supplement* 1:1–85.

Anya, A. O. 1966. The Structural and Chemical Composition of the Nematode Cuticle. Observations on Some Oxyurids and *Ascaris*. *Parasitology* 56:179–198.

Auber-Thomay, M. 1964. Structure et innervation des cellules musculaires de nématodes. *Journal de Microscopie* 3:105–113.

Baccetti, D., R. Dallai, S. de Zio Grimaldi, and A. Marinari. 1983. The Evolution of the Nematode Spermatozoon. *Gamete Research* 8:309–323.

Badolkhodzhayev, I. 1969. Observation of Embryonic Development in *Panagrellus silusioides* Tsalolichin, 1965 (Nematoda). *Vestnik Leningradskogo Universiteta, Seriya Biologii* [Journal of Leningrad University, Biology Series] 9:12–22.

———. 1970. Embryonic Development of *Rhabditis elegans* (Nematoda). *Vestnik Leningradskogo Universiteta, Seriya Biologii* [Journal of Leningrad University, Biology Series] 5:7–17.

Baldwin, J. G., and H. Hirschmann. 1973. Fine Structure of Cephalic Sense Organs in *Meloidogyne incognita* Males. *Journal of Nematology* 5:285–302.

Bastian, H. C. 1865. Monograph on the Anguillulidae, or Free Nematoids, Marine, Land and Fresh-water; with Descriptions of 100 New Species. *Transactions of the Linnean Society of London* 25:73–184.

———. 1866. On the Anatomy and Physiology of the Nematoids, Parasitic and Free, with Observations on Their Zoological Position and Affinities to the Echinoderms. *Philosophical Transactions of the Royal Society, London* 156:545–638.

Baylis, H. A. 1924. On the Systematic Position of the Nematoda. *Annals and Magazine of Natural History* 9:165–173.

———. 1938. Helminth and Evolution. In *Evolution. Essays on Aspects of Evolutionary Biology Presented to Prof. E. S. Goodrich,* ed. G. R. de Beer, 249–270. Oxford: Clarendon Press.

Beams, H. W., and S. S. Sekhon. 1972. Cytodifferentiation during Spermiogenesis in *Rhabditis pellio. Journal of Ultrastructure Research* 38:511–527.

Beauchamp, P. M. 1929. Le développement des Gastrotriches. *Bulletin de la Société Zoologique de France* 54:549–558.

———. 1956. Le développement de *Ploesoma hudsoni* (Imholf) et l'origine des feuillets chez rotifères. *Bulletin de la Société Zoologique de France* 81:374–383.

Beermann, I., H. Arai, and J. Costerson. 1974. The Ultrastructure of the Lemnisci and Body Wall of *Octospinifer macilentus* (Acanthocephala). *Canadian Journal of Zoology* 52:553–555.

Beklemishev, K. V. 1976. Alternative for the Evolutionary Tree of Metazoa. In *Evolyutsionnaya Morfologiya Besposvonochnuykh Zhivotnykh* [Evolutionary Morphology of Invertebrate Animals], 4–5. Leningrad: Zoologicheskii Institut Akademii Nauk SSSR [Leningrad: Zoological Institute of the Academy of Sciences of the USSR].

Beklemishev, V. N. 1925. Morphological Problem of Animal Structures (A Critique of Some Basic Concepts in Histology). *Izvestiya Biologicheskogo Nauchno-Issledovatel'skogo Institata Biologicheskoi Stantsii pri Permskom Gosundarstvennom Universitete* [Bulletin of the Biological Research Institute and of the Biological Station of Perm State University] 3 addendum: 1–74.

———. 1944. *Osnovy Sravnitel'noy Aantomii Bespozvonochnykh* [Principles of

Comparative Anatomy of Invertebrates]. 491 pp. Moskva: Sovetskaya Nauka [Moscow: Soviet Science].
———. 1964a. *Osnovy Sravnitel'noy Aantomii Bespozvonochnykh* [Principles of Comparative Anatomy of Invertebrates], Vol. 1, 432 pp. Moskva: Nauka [Moscow: Science].
———. 1964b. *Osnovy Sravnitel'noy Anatomii Bespozvonchnykh* [Principles of Comparative Anatomy of Invertebrates], Vol. 2, 446 pp. Moskva: Nauka [Moscow: Science].
Bell, J. de, J. Castillo, and V. Sanchez. 1963. Electrophysiology of the Somatic Muscle Cells of *Ascaris lumbricoides. Journal of Cellular and Comparative Physiology* 62:159–177.
Belogurov, O. I., and L. S. Belogurova. 1975a. Organization of the Head End of Nematodes of Oncholaimidae and the Phenomenon of "Cuticle Splitting." *Biologiya Morya* [Marine Biology] 4:24–35.
———. 1975b. Systematics and Evolution of Oncholaiminae (Nematoda: Oncholaiminae). I. Structure, Evolution, and Taxonomic Significance of the De Manian System. *Biologiya Morya* [Marine Biology] 3:36--47.
———. 1977. Systematics and Evolution of the Nematode Subfamily Oncholaiminae. II. *Biologiya Morya* [Marine Biology] 5:33–39.
Belogurov, O. I., and N. P. Listova. 1977. On the Morphology of the Spinneret and a Discussion of Its Origin Among Nematodes of the Order Enoplida. *Zhurnal Obshchei Biologiya* [Journal of General Biology] 38(4): 582–594.
Beneden, É. Van. 1883. Recherches sur la fécondation et maturation. *Archives de Biologie* 4:265–640.
Beneden, É. Van, and A. Neyt. 1887. Nouvelles recherches sur la fécondation et la division mitosique chez l'Ascaride mégalocéphalae. *Bulletin de l'Academie Royal de Belgique,* ser. 3, 14:215–295.
Bertalanffy, L. von. 1938. A Quantitative Theory of Organism Growth (Inquiry on Growth Laws II). *Human Biology* 10:181–213.
Bird, A. F. 1958. Further Observations on the Structure of Nematode Cuticle. *Parasitology* 48:32–37.
———. 1971. *The Structure of Nematodes.* 381 pp. New York and London: Academic Press.
———. 1980. The Nematode Cuticle and Its Surface. In *Nematodes as Biological Models,* ed. B. M. Zuckerman, 2:213–236. New York and London: Academic Press.
Bird, A. F., and G. E. Rogers. 1965. Ultrastructure of the Cuticle and Its Formation in *Meloidogyne javanica. Nematologica* 11:224–230.
Bogolepova, I. N. 1976. *Sravnitel'Naya Funktsional'naya Morfologiya Pishchevaritel'noy Sistemy Nematod* [Comparative and Functional Morphology of the Digestive System in Nematodes]. 371 pp. Dissertatsiya Doktora Biologicheskikh Nauk. Leningrad: Zoologicheskii Institut Akademii Nauk SSSR Dissertation for the degree of Doctor of Biological Sciences. Leningrad: Zoological Institute of the Academy of Sciences of the USSR].
Bogoyavlenskii, Yu. K. 1973. *Struktura i Funktsii Pokrovnykh Tkanei Parazi-*

ticheskikh Nematod [Structure and Function of the Integuments of Parasitic Nematodes]. 206 pp. Moscow: Nauka [Moscow: Science].

Bogoyavlenskii, Yu. K., and G. V. Ivanova. 1978. *Mikrostruktura Tkaney Skrebney* [Microstructure of the Tissue of Acanthocephala]. 271 pp. Moscow: Nauka [Moscow: Science].

Bogoyavlenskii, Yu. K., G. V. Ivanova, and A. A. Spasskii. 1974. *Nervnaya Sistema Paraziticheskikh Nematod* [Nervous System of Parasitic Nematodes]. 190 pp. Kishinev: Shtnintsa.

Boguta, K. K. 1976. Morphology of the Nervous System in *Convoluta convoluta* (Turbellaria, Acoela) and Its Changes Under the Effect of Long-term Starvation. *Zoologicheskii Zhurnal* [Zoological Journal] 55(6): 815–822.

Bömmel, A. von. 1894. Über Cuticular-bildungen bei einigen Nematoden. *Arbeiten aus dem zoologisch-zootomischen Institut in Würzburg* 10:189–212.

Bonfig, R. 1925. Die Determination der Hauptrichtungen des Embryos von *Ascaris megalocephala. Zeitschrift für wissenschaftliche Zoologie* 124:407–456.

Bonner, T. P., and P. P. Weinstein. 1972. Ultrastructure of Cuticle Formation in the Nematodes, *Nippostrongylus brasiliensis* and *Nematospiroides dubius. Journal of Ultrastructure Research* 40:261–271.

Bonner, T. P., M. G. Menefee, and F. J. Etges. 1970. Ultrastructure of the Cuticle Formation in a Parasitic Nematode, *Nematospiroides dubius. Zeitschrift für Zellforschung und Mikroskopische Anatomie* 104:193–204.

Bonnevie, K. 1901. Über Chromatindiminution bei Nematoden. *Jenaische Zeitschrift für Naturwissenschaft* 36:275–288.

Boveri, T. 1887. Die Bildung der Richtungskörper bei *Ascaris megalocephala* und *Ascaris lumbricoides. Zell Studien* 1:1–93.

———. 1888. Die Befruchtung und Teilung des Eies von *Ascaris megalocephala. Zell Studien* 2:1–198.

———. 1899. Die Entwickelung von *Ascaris megalocephala* mit besonderer Rücksicht auf die Kernverhältnisse. In *Festschrift zum siebenzigsten Geburtstag Carl von Kupffer,* 383–-430. Jena: G. Fischer.

———. 1910. Die Potenzen der *Ascaris*-Blastomeren bei abgeänderter Furchung: Zugleich ein Beitrag zur Frage qualitativ ungleicher Chromosomenteilung. In *Festschrift zum sechzigsten Geburtstag für Richard Hertwigs,* 3:131–214. Jena: Fischer.

Boyer, B. C. 1971. Regulative Development in a Spiralian Embryo as Shown by Cell Deletion Experiments on the Acoel, *Childia. Journal of Experimental Zoology* 176:97–106.

Brandes, G. 1892. Über das Nervensystem von *Ascaris megalocephala. Bericht über die Sitzungen der Naturforschenden Gesellschaft zu Halle* 106–107.

———. 1899. Das Nervensystem der als Nemathelminthes zusammengefassten. Wurmtypen. *Abhandlungen der Naturforschenden Gesellschaft zu Halle* 21: 271–299.

Brien, P. 1961. *Eléments de Zoologie et notions d'anatomie comparée. II. Les Coelomates Hyponeriens.* 715 pp. Paris: Librairie Maloine.

Brunold, E. 1954. Zur Morphologie, Biologie und Bakterien freien Zuchtung des Nematodes *Panagrellus zymosiphilis. Zeitschrift für Morphologie und Ökologie der Tiere* 42:374–420.

Bütschli, O. 1874. Beiträge zur Kenntniss des Nervensystems der Nematoden. *Archiv für mikroskopische Anatomie* 10:74–106.

———. 1876. Untersuchungen über freilebenden Nematoden und die Gattung *Chaetonotus. Zeitschrift für wissenschaftliche Zoologie* 26:363–413.

Byerly, L., R. S. Cassada, and R. L. Russel. 1976. The Life Cycle of the Nematode *Caenorhabditis elegans*. I. Wild Type Growth and Reproduction. *Developmental Biology* 51:23–33.

Byers, J. R., and R. V. Anderson. 1972. Ultrastructural Morphology of the Body Wall, Stoma, and Stomatostyle of the Nematode, *Tylenchorhynchus dubuis* (Bütschli, 1873) Filipjev, 1936. *Canadian Journal of Zoology* 50:457–465.

Byram, I., and F. Fischer. 1973. The Absorption Surface in *Moniliformis dubuis* (Acanthocephala). 1. Fine Structure. *Tissue and Cell* 5:553–579.

Calcoen, J., and D. Roggen. 1973. The Form of Moving Nematodes. *Nematologica* 19:408–410.

Cameron, T. W. 1962. Host Specificity and the Evolution of Helminthia Parasites. *Advances in Parasitology* 2:156–230.

Castillo, J., W. Mello, and T. Morales. 1963. The Physiological Role of Acetylcholine in the Neuromuscular System of *Ascaris lumbricoides. Archives Internationales de Physiologie et de Biochimie* 71:741–757.

Chabaud, A. G. 1957. Spécificité parasitaire chez les nématodes parasites des vertebrés. In *Proceedings of the First Symposium on Host Specificity among Parasites of Vertebrates,* 230–243. Neuchatel.

———. 1974. *Keys to the Nematode Parasites of Vertebrates.* 17 pp. Farnham Royal, England: Commonwealth Agricultural Bureaux.

Chitwood, B. G. 1933. A Revised Classification of the Nematoda. *Journal of Parasitology* 20:131.

———. 1937. A Revised Classification of the Nematoda. In *Papers in Helminthology: Published in Commemoration of the Thirtieth Year Jubileum of K. J. Skrjabin, and of the Fifteenth Anniversary of the All-Union Institute of Helminthology,* 69–80. Moscow: Izdatel'stvo Akademii Nauk SSSR [Moscow: Publishing House of the Academy of Sciences of the USSR].

———. 1958. The Designation of Official Names for Higher Taxa of Invertebrates. *Bulletin of Zoological Nomenclature* 15:860–895.

Chitwood, B. G., and M. B. Chitwood. 1933. The Histological Anatomy of *Cephalobellus papilliger,* Cobb, 1920. *Zeitschrift für Zellforschung und mikroskopische Anatomie* 19:309–355.

———. 1950. *Introduction to Nematology.* 213 pp. Baltimore, Md.: Monumental Printing Company. Reprint, 1974, with revisions. 334 pp. Baltimore, London, and Tokyo: University Park Press.

Chitwood, B. G., and E. Wehr. 1934. The Value of Cephalic Structures as Characters in Nematode Classification, with Special Reference to the Superfamily Spiruroidea. *Zeitschrift für Parasitenkunde* 7:273–335.

Christie, J. R. 1936. Life History of *Agamermis decaudata,* a Nematode Parasite of Grasshoppers and Other Insects. *Journal of Agricultural Research* 52: 161–198.

Chuang, S. A. 1962. The Embryonic and Post-embryonic Development of *Rhabditis teres* (A. Schneider). *Nematologica* 7:317–330.

Clark, R. B. 1964. *Dynamics in Metazoan Evolution. The Origin of the Coelom and Segments.* 342 pp. Oxford: Clarendon Press.

Clark, R. B., and J. B. Cowey. 1958. Factors Controlling the Change of Shape of Certain Nemertean and Turbellarian Worms. *Journal of Experimental Biology* 35:731–748.

Clark, S. A., and A. M. Shepherd. 1977. Structure of the Spicules and Caudal Sensory Equipment in the Male of *Aphelenchoides blastophthorus* (Nematoda, Tylenchida, Aphelenchina). *Nematologica* 23:103–111.

Clark, S. A., A. M. Shepherd, and A. Kempton. 1973. Spicule Structure in Some *Heterodera* spp. *Nematologica* 19:242–247.

Clement, P. 1977. Ultrastructural Research on Rotifers. *Archiv für Hydrobiologie* 8:270–297.

Cobb, J. L. S. 1967. The Innervation of the Ampulla of the Tubefoot in the Starfish *Astropecten irregularis. Proceedings of the Royal Society of London, Series B* 168:91–99.

Cobb, N. A. 1930. The Demanian Vessels in Female Nemas of the Genus *Oncholaimus. Journal of Parasitology* 16:159–161.

Colam, J. B. 1971a. Studies on Gut Ultrastructure and Digestive Physiology in *Rhabdias buffonis* and *R. sphaerocephala* (Nematoda: Rhabdita). *Parasitology* 62:247–258.

———. 1971b. Studies on Gut Ultrastructure and Digestive Physiology in *Cyathostoma lari* (Nematoda: Strongylida). *Parasitology* 62:273–283.

Coninck, L. de. 1965. Némathelminthes. (Nématodes). In *Traité de Zoologie. Anatomie, Systématique, Biologie,* ed. P. P. Grassé, 4(2): 1–217. Paris: Masson et Cie.

Crofton, H. D. 1966. *Nematodes.* 160 pp. London: Hutchinson University Library.

Croll, N. A. 1976. *The Organization of Nematodes.* 439 pp. London and New York: Academic Press.

Croll, N. A., and J. M. Smith. 1974. Nematode Setae as Mechanoreceptors. *Nematologica* 20:291–296.

Croll, N. A., and K. A. Wright. 1976. Observations on the Movements and Structure of the Bursa of *Nippostrongylus brasiliensis* and *Nematospiroides dubius. Canadian Journal of Zoology* 54:1466–1480.

Croll, N. A., A. A. F. Evans, and J. M. Smith. 1975. Comparative Nematode Photoreceptors. *Comparative Biochemistry and Physiology, Part A,* 51: 139–149.

Davydov, K. N. 1914. *Kur Embriologii Bespozvonochnykh* [The Course of Invertebrate Embryology]. 512 pp. Kiev.

Dawydoff, C. 1928. *Traité d'embryologie compareé des invertébres.* 930 pp. Paris: Masson et Cie.

Dehl, E. 1934. Morphologie von *Lindia tecusa. Zeitschrift für wissenschaftliche Zoologie* 145:244–298.

Dethier, V. G. 1963. *The Physiology of Insect Senses.* 430 pp. London: Methuen.

Deyneka, D. I. 1912. Nervous System of *Ascaris megalocephala* (Cloq). Histological Research. *Trudy Imperatorskoe Sankt-Peterburgskoe Obshchestvo Estestvoispytatelei* [Transactions of the Saint Petersburg Naturalists' Society], 101–359.

Dick, T. A., and K. A. Wright. 1973. The Ultrastructure of the Cuticle of the Nematode *Syphacia obvelata* (Rudolphi, 1802). II. Modifications of the Cuticle in the Head End. *Canadian Journal of Zoology* 51:197–202.

———. 1974. The Ultrastructure of the Cuticle of the Nematode *Syphacia obvelata* (Rudolphi, 1802). III. Cuticle Associated with Male Reproductive Structures. *Canadian Journal of Zoology* 52:179–182.

Drozdovskiy, E. M. 1967. The Use of Embryonic Development in Nematode Taxonomy. *Trudy Gel'mintologicheskoi Laboratorii, Akademiya Nauk SSSR* [Transactions of the Helminthological Laboratory, of the Academy of Sciences of the USSR] 18:22–29.

———. 1968. Contribution to the Comparative Study of the Initial Stages of Ovum Cleavage in Nematodes. *Doklady Akademii Nauk SSSR* [Reports of the Academy of Sciences of the USSR] 180:750–753.

———. 1969. Contribution to the Analysis of the Embryogeny of Certain Adenophorea (Nematoda). *Doklady Akademii Nauk SSSR* [Reports of the Academy of Sciences of the USSR] 186(12): 720–723.

———. 1975. Ovum Division Among Species of *Eudorylaimus* and *Mesodorylaimus* (Nematoda; Dorylaimida) and the Role of Cleavage in Determining the Composition of Subclasses of Nematodes. *Doklady Akademii Nauk SSSR* [Reports of the Academy of Sciences of the USSR] 222(4): 1102–1108.

———. 1977. The Characteristics of Ovum Division and the Significance of the Preblastula in Nematode Embryogeny. *Arkhiv Anatomi, Gistologii, i Embriologoii* [Archives of Anatomy, Histology, and Embryology] 71(9): 88–94.

———. 1978. The Structure and Formation of the Preblastula as an Indication of the Phylogenetic Relationships and Taxonomic Position of Various Groups of Nematodes, 69–78. In *Fitgel'minthologicheskiye Issiedovaniya* [Phytohelminthological Research]. Moskva: Nauka [Moscow: Science].

———. 1980. On the Division of the Class Nematoda into Subclasses on the Basis of Data from Comparative Anatomy and Embryology, 58–59. In *Tezisnye Doklady Konference Vsesoyuznogo Obshchestva Gel'minthologov* [Thesis Reports of the Conference of the All-Union Society of Helminthologists]. Moskva: Nauka [Moscow: Science].

Dunschen, F. 1929. Inverseutwicklung und Mosaikfrage bei *Ascaris megalocephala* Cloq. *W. Roux' Archiv für Entwicklungsmechanik der Organismen* 115:237–333.

Eckert, J., and R. Schwarz. 1965. Zur Struktur der Cuticula invasionsfähiger Larven einiger Nematoden. *Zeitschrift für Parasitenkunde* 26:116–142.

Ehrenstein, G. von, and E. Schierenberg. 1980. Cell Lineages and Development of *Caenorhabditis elegans* and other Nematodes. In *Nematodes as Biological Models,* ed. B. Zuckerman, 1:1–73. New York: Academic Press.

Endo, B. Y., and W. P. Wergin. 1977. Ultrastructure of Anterior Sensory Organs of the Root-knot Nematode, *Meloidogyne incognita. Journal of Ultrastructure Research* 59:231–249.

Epstein, J., J. Castillo, S. Himmelhoch, and B. M. Zuckerman. 1971. Ultrastructural Studies on *Caenorhabditis briggsae. Journal of Nematology* 3: 69–78.

Ernst, A. 1976. Die Ultrastruktur der Sinneshaare auf den Antennen von *Geophilus longicornis* Leach (Myriapoda, Chilopoda). 1. Die Sensilla Trichoidea. *Zoologische Jahrbücher Abteilung für Anatomie und Ontogenie der Tiere,* 2, 96:586–604.

Fauré-Frémiet, E., and H. Gerault. 1944. Propriétés physiques de l'ascarocollagen. *Bulletin Biologique de la France et de la Belgique* 78:206–214.

Favard, K. 1961. Évolution des ultrastructures cellulaires au cours de la spermatogenèse de l'*Ascaris.* (*Ascaris megalocephala,* Schrank = *Parascaris equorum,* Goerze). *Annales des Sciences Naturelles, Serie Zoologie* 12:53–152.

Filip'jev, I. N. 1912. On the Question of the Structure of the Nervous System in Free-living Nematodes. *Trudy Imperatorskogo Sankt-Peterburgskogo Obshchestva Estestvoipytatelei* [Transactions of the Saint Petersburg Naturalists' Society] 43: 205–215.

———. 1918. Free-living Marine Nematodes in the Vicinity of Sevastopol. *Trudy Osoboi Zoologicheskoi Laboratorii i Sevastopolskoi Biologi Stantsii Rossiysk Akademii Nauk* [Bulletin of the Special Zoological Laboratory and the Sevastopol Biological Station of the Russian Academy of Sciences] 41(4): 1–352.

———. 1921. Free-living Marine Nematodes in the Vicinity of Sevastopol. *Trudy Osoboi Zoologicheskoi Laboratorii i Sevastopolskoi Biologi Stantsii Rossiysk Akademii Nauk* [Bulletin of the Special Zoological Laboratory and the Sevastopol Biological Station of the Russian Academy of Sciences] 41(4): 353–614.

Filip'jev, I. N., and E. Michailova. 1924. Zahl der Entwicklungsstadien bei *Enoplus communis* Bastian. *Zoologischer Anzeiger* 59:212–219.

Flood, P. R. 1966. A Peculiar Mode of Muscular Innervation in *Amphioxus:* Light and Electron Microscopic Studies of the So-called Ventral Roots. *Journal of Comparative Neurology* 125:181–227.

Fogg, L. C. 1930. A Study of Chromatin Diminution in *Ascaris* and *Ephestia. Journal of Morphology* 50:413–444.

Foor, W. E. 1970. Spermatozoan Morphology and Zygote Formation in Nematodes. *Biology of Reproduction, Supplement* 2:177–202.

Frenzen, K. 1954. Untersuchungen zur Morphologie und mikroskopischen Anatomie von *Ascaridia galli* Schrank, 1788. *Zoologische Jahrbücher Abteilung für Anatomie und Ontogenie der Tiere,* 2, 73:395–424.

Fülleborn, F. 1923. Über den "Mundstachel" der Trichotracheliden-Larven und

Bemerkungen über die jüngsten Stadien von *Trichocephalus trichiuris. Archiv für Schiffs-und Tropenhygiene* 27:421–425.

Gerlach, S. A., and F. Riemann. 1973. The Bremenhaven Checklist of Aquatic Nematodes. A Catalogue of Nematoda Adenophorea Excluding the Dorylaimida. *Veröffentlichungen des Instituts für Meeresforschung in Bremerhaven, Supplement* 4:1–404.

———. 1974. The Bremenhaven Checklist of Aquatic Nematodes. A Catalogue of Nematoda Adenophorea Excluding the Dorylaimida. *Veröffentlichungen des Instituts für Meeresforschung in Bremerhaven, Supplement* 4:405–734.

Gilyarov, M. S. 1964. Contemporary Theories on Homology. *Uspekhi Sovremennoi Biologii* [Progress in Contemporary Biology] 57(2): 300–316.

Goette, A. 1882. *Untersuchungen zur Entwicklungsgeschichte der Würmer. II. Entwicklungsgeschichte der Rhabditis nigrovenosa.* 104 pp. Leipzig.

Goldschmidt, R. 1903. Histologische Untersuchungen an Nematoden. I. Die Sinnesorgane von *Ascaris lumbricoides* L. und *A. megalocephala* Cloque. *Zoologische Jahrbücher Abteilung für Anatomie und Ontogenie der Thiere,* 2, 18:1–57.

———. 1904. Histologische Untersuchungen an Nematoden I. *Zoologisches Zentralblatt* 11:517–530.

———. 1905. Über die Cuticula von *Ascaris. Zoologischer Anzeiger* 28:259–266.

———. 1906. Mitteilungen zur Histologie von *Ascaris. Zoologischer Anzeiger* 29:719–737.

———. 1908. Das Nervensystem von *Ascaris lumbricoides* und *megalocephala.* Ein Versuch, in den Aufbau eines einfachen Nervensystems einzudringen. Teil I. *Zeitschrift für wissenschaftliche Zoologie* 90:73–136.

———. 1909. Das Nervensystem von und *Ascaris lumbricoides* und *megalocephala.* II. *Zeitschrift für wissenschaftliche Zoologie* 92:306–357.

———. 1910. Das Nervensystem von *Ascaris lumbricoides* und *megalocephala.* III. In *Festschrift zum sechzigsten Geburtstag für Richard Hertwigs,* 2:253–354. Jena: Fischer.

Golovin, Ye. P. 1901. *Nablyudeniya Nad Nematodami. I. Fagotsitarnyye Organy* [Observations of Nematodes. I. Phagocytic Organs]. 191 pp. Kazan.

Grassé, P. de, R. Poisson, and O. Tuzet. 1961. *Zoologie. 1. Invertébrés.* 919 pp. Paris: Masson et Cie.

Gray, J. 1968. *Animal Locomotion.* 479 pp. London: Weidenfeld and Nicolson.

Gray, J., and H. Lismann. 1964. The Locomotion of Nematodes. *Journal of Experimental Biology* 41:135–154.

Greeff, R. 1869. Untersuchungen über einige merkwürdige Thiergruppen des Arthropoden- und Wurm-Typus. *Archiv für Naturgeschichte* 35(1): 71–121.

Grobben, K. (1908) 1909. Die systematische Einteilung des Tierreiches. *Verhandlungen der k.-k. zoologisch-botanischen Gesellschaft in Wien* 58(10): 491–511.

Grotaert, P., and P. L. Lippens. 1974. Some Ultrastructural Changes in Cuticle and Hypodermis of *Aporcelaimellus* during the First Moult (Nematoda,

Dorylaimoidea). *Zeitschrift für Morphologie und Ökologie der Tiere* 79: 269–282.

Gulyayev, D. V. 1976. Electron Microscopic Observations of Contacts of Nerve and Epthelial-Muscular Cells of *Hydra vulgaris* (Pall.). *Zhurnal Obshchei Biologii* [Journal of General Biology] 37(2): 304–309.

Haffner, K. 1950. Organisation und systematische Stellung der Acanthocephalen. *Zoologischer Anzeiger* 145:243–274.

Hammarsten, O. 1915. Zur Entwicklungsgeschichte von *Halicryptus spinulosus* (von Siebold). *Zeitschrift für wissenschaftliche Zoologie* 112:525–571.

Hammond, R. 1967. The Fine Structure of the Trunk and Praesoma Wall of *Acanthocephalus ranae* (Schrank, 1788) Lühe, 1911. *Parasitology* 57:475–486.

Hanson, J., and J. Lowy. 1961. The Structure of the Muscle Fibres in the Translucent Part of the Adductor of the Oyster *Crassostrea angulata. Proceedings of the Royal Society of London, Series B,* 154:173–196.

Harada, I. 1931. Das Nervensystem von *Bolbosoma turbinella* (Diesing). *Japanese Journal of Zoology* 3:161–199.

Harris, J. E., and H. D. Crofton. 1957. Structure and Function in Nematodes: Internal Pressure and Cuticular Structure in *Ascaris. Journal of Experimental Biology* 34:116–130.

Hattingen, R. 1956. Beiträge zur Entwicklungsgeschichte von *Sphaerularia bombi* Dufour, 1837. *W. Roux' Archiv für Entwicklungsmechanik der Organismen* 148:494–503.

Herla, V. 1894. Étude de la variation de la mitose chez l'*Ascaride megalocephala. Archives de Biologie* 13:424–530.

Hesse, R. 1892. Über das Nervensystem von *Ascaris megalocephala. Zeitschrift für wissenschaftliche Zoologie* 54:548–568.

Heumann, H. G., and E. Zebe. 1967. Über Feinbau und Funktionsweise der Fasern aus dem Hautmuskelschlauch des Regenwurms, *Lumbricus terrestris* L. *Zeitschrift für Zellforschung und mikroskopische Anatomie* 78:131–150.

Hinz, E. 1963. Elektronenmikroskopische Untersuchungen an *Parascaris equorum.* (Integument, Isolationsgewebe, Muskulatur und Nerven). *Protoplasma* 56:202–241.

Hogue, M. 1910. Über die Wirkung der Centrifugalkraft auf die Eier von *Ascaris megalocephala. W. Roux' Archiv für Entwicklungsmechanik der Organismen* 29:109–145.

Hope, W. D. 1969. Fine Structure of the Somatic Muscles of the Free-living Marine Nematode *Deontostoma californicum* Steiner and Albin, 1933 (Leptosomatidae). *Proceedings of the Helminthological Society of Washington* 36:10–29.

———. 1974. *Deontostoma timmerchioi* N. Sp., A New Marine Nematode (Leptosomatidae) from Antarctica, with a Note on the Structure and Possible Function of the Ventromedian Supplement. *Transactions of the American Microscopical Society* 93:314–324.

Hope, W. D., and S. L. Gardiner. 1982. Fine Structure of a Proprioceptor in the Body Wall of the Marine Nematode *Deontostoma californicum* Steiner and Albin, 1933 (Enoplida: Leptosomatidae). *Cell and Tissue Research* 225: 1–10.

Hörstadius, S. 1937. Experiments on Determination in the Early Development of *Cerebratulus lacteus. Biological Bulletin Woods Hole* 73:317–342.

Hoyle, G. 1964. Muscle and Neuromuscular Physiology. In *Physiology of Mollusca,* eds. K. M. Wilbur and C. M. Yonge, 1:313–351. New York and London: Academic Press.

Hsu, Hsi-Fan. 1929. On the Oesophagus of *Ascaris lumbricoides. Zeitschrift für Zellforschung mikroskopische Anatomie* 9:313–326.

Hyman, L. H. 1951. *The Invertebrates. 3: Acanthocephala, Aschelminthes and Entoprocta.* 572 pp. New York, Toronto, and London: McGraw-Hill.

Inglis, W. 1964a. The Structure of the Nematode Cuticle. *Proceedings of the Zoological Society of London* 143:465–502.

———. 1964b. The Marine Enoplida (Nematoda): A Comparative Study of the Head. *Bulletin of the British Museum (Natural History)* 11:265–376.

Inglis, W. G. 1965. Patterns of Evolution in Parasitic Nematodes. In *Evolution of Parasites. Symposium of the British Society of Parasitology,* ed. A. E. R. Taylor, 79–124. Oxford: Blackwell Scientific Publications.

Ioffe, B. I. 1979. Comparative-Embryological Analysis of the Development of Pseudocoelomate Worms. *Trudy Zoologicheskogo Instituta Adakemii Nauk SSSR* [Transactions of the Zoological Institute of the Academy of Sciences of the USSR] 84:39–62.

Ivanov, P. P. 1937. *Obshchaya i Sravnitel'naya Embriologiya* [Fundamental and Comparative Embryology]. 809 pp. Leningrad: Biomedgiz [Leningrad: Biomedical Publishing House].

Ivanov, V. P. 1969. Ultrastructural Organization of Insect Chemoreceptors. *Trudy Vsesoyuznogo Entomologicheskogo Obshchestva* [Transactions of the All-Union Entomological Society] 53:301–333.

Ivanova-Kazas, O. M. 1959. On the Origin and Evolution of Spiral Cleavage. *Vestnik Leningradskogo Universiteta, Seriya Biologii* [Journal of Leningrad University, Biology Series] 9:56–67.

———. 1975. *Sravnitel'naya Embriologiya Bespozvonochnykh Zhivotnykh. Prosteyshiye i Nizshiye Mnogokletochnyye.* [The Comparative Embryology of Invertebrate Animals. Protozoa and Lower Metazoa]. 369 pp. Novosibirsk: Nauka [Novosibirsk: Science].

———. 1979. An Analysis of Cleavage among Nematoda and Gastrotricha. *Zoologicheskii Zhurnal* [Zoological Journal] 58(12): 1765–1777.

Ivashkin, V. M. 1978. The Exchange of Hosts in the Life Cycles of Nematodes in Vertebrates. In *Nauchnyye i Prikladnyye Problemy Gel'mintologii* [Scientific and Applied Problems of Helminthology], 43–48. Moscow: Nauka [Moscow: Science].

Jägerskiöld, L. A. 1901. Weitere Beiträge zur Kenntnis der Nematoden. *Kungliga Svenska Vetenskapsakademiens Handlingar* 35:1–91.

Jamuar, M. P. 1966. Electron Microscope Studies on the Body Wall of the Nematode *Nippostrongylus brasiliensis. Journal of Parasitology* 52:209–232.

Jarman, J. 1959. Electrical Activity in the Muscle Cells of *Ascaris lumbricoides*. *Nature* 184:1244.

Jenning, J. B., and J. B. Colam. 1971. Gut Structure, Digestive Physiology and Food Storage in *Pontonema vulgaris* (Nematoda: Enoplida). *Journal of Zoology* 161:211–221.

Jennings, H. S. 1896. The Early Development of *Asplanchna herrickii* De Guerne. *Bulletin of the Museum of Comparative Zoology, Harvard College* 30:1–117.

Johnson, P. W., S. D. Van Gundy, and W. W. Thomson. 1970. Cuticle Formation in *Hemicycliophora arenaria, Aphelenchus avenae* and *Hirschmanniella gracilis*. *Journal of Nematology* 2:59–79.

Kaestner, A. 1965. *Lehrbuch der speziellen Zoologie. I. Bd. 1. Wirbellose.* 845 pp. Stuttgart: G. Fischer.

Kaiser, H. 1977. Untersuchungen zur Morphologie, Biometrie, Biologie und Systematik von Mermithiden. Ein Beitrag zum Problem der Trennung morphologisch schwer unterscheidbarer Arten. *Zoologische Jahrbücher Abteilung für Systematik, Ökologie und Geographie der Tiere,* 3, 104:20–71.

Kaiser, J. 1893. Die Acanthocephalen und ihre Entwicklung. 1. *Bibliotheca Zoologica* 7:1–137.

Kalamkarova, M. B., and M. Ye. Kryukova. 1966. Ultrastructural Organization and the Peculiarities of Yolk Components of the Contractile Apparatus of the Adductor Anodonts. *Biofizika* [Biophysics] 11(1): 61–68.

Kawaguti, S. 1962. Arrangement of Myofilaments in the Oblique Striated Muscles. In *Proceedings of the 5th International Congress of Electron Microscopy,* 2:11.

Kawaguti, S., and N. Ikemoto. 1959. Electron Microscopic Patterns of Earthworm Muscle in Relaxation and Contraction Induced by Glycerol and Adenosine Triphosphate. *Biological Journal of Okayama University* 5:57–72.

———. 1960. Electron Microscopy on the Adductor Muscle of the File Shell, *Promantellum hirasei. Biological Journal of Okayama University* 6:43–52.

Kirsteuer, E. 1976. Notes on Adult Morphology and Larval Development of *Tubiluchus corallicola* (Priapulida), Based on In Vivo and Scanning Electron Microscopic Examinations of Specimens from Bermuda. *Zoologica Scripta* 5:239–255.

King, R. L., and H. W. Beams. 1938. An Experimental Study of Chromatin Diminution in *Ascaris. Journal of Experimental Zoology* 77:425–444.

Korotkova, G. P., and L. A. Agafonova. 1975. Experimental Morphological Research on Regenerative Capabilities of the Nematode *Pontonema vulgaris* (Bastian, 1865). *Arkhiv Anatomi, Gistologii, i Embriologii* [Archives of Anatomy, Histology, and Embryology] 70(3): 90–98.

Korsakova, G. F. 1949. Early Stages of Division in *Tiaropsis. Doklady Akademii Nauk SSSR* [Reports of the Academy of Sciences of the USSR] 65(3): 413–415.

Kotikova, Ye. A. 1967. The Histochemical Method for Studying the Morphology of the Nervous Systems of Flat Worms. *Parazitologiya* [Parasitology] 1(1): 79–81.

———. 1973. New Data on the Nervous System of Archiannelida. *Zoologicheskii Zhurnal* [Zoological Journal] 52(8): 1611–1615.

Kreis, H. A. 1934. Oncholaiminae Filipjev, 1916: Eine monographische Studie. *Capita Zoologica* 4:1–270.

Kristensen, R. M. 1983. *Loricifera,* A New Phylum with Aschelminthes Characters from the Meiobenthos. *Zeitschrift für Zoologische Systematik und Evolutionsforschung* 21:163–180.

Kryukova, M. Ye. 1972. Comparative Investigation of the Structure and Function of the Adductor Muscles in Bivalve Mollusks. In *Mekhanismy Myshechnogo Sokrashceniya* [Mechanisms of Muscular Contraction], 36–54. Moscow: Nauka [Moscow: Science].

Lang, K. 1953. Die Entwicklung des Eies von *Priapulus caudatus* Lam. und die systematische Stellung der Priapuliden. *Arkiv für Zoologi* 5:321–348.

———. 1963. The Relation between the Kinorhyncha and Priapulida and Their Connection with the Aschelminthes. In *The Lower Metazoa,* ed. E. C. Dougherty, 247–255. Berkeley: University of California Press.

Lechner, M. 1966. Untersuchungen zur Embryonalentwicklung des Radertieres *Asplanchna girodi* de Guerne. *W. Roux' Archiv für Entwicklungsmechanik der Organismen* 157:117–173.

Lee, D. L. 1962. A Histochemical Study of Esterase Enzymes in the Nervous System of *Ascaris lumbricoides. Journal of Parasitology* 48:26.

———. 1965. *The Physiology of Nematodes.* 154 pp. Edinburgh: Oliver and Boyd.

———. 1966a. The Structure and Composition of Helminth Cuticle. *Advances in Parasitology* 4:187–254.

———. 1966b. An Electron Microscope Study of the Body Wall of the Third Stage Larvae of *Nippostrongylus brasiliensis. Parasitology* 56:127–135.

———. 1968. The Ultrastructure of the Alimentary Tract of the Skin-penetrating Larva of *Nippostrongylus brasiliensis* (Nematoda). *Journal of Zoology* 154: 9–18.

———. 1970. The Ultrastructure of the Cuticle of Adult Female *Mermis nigrescens* (Nematoda). *Journal of Zoology* 161:513–518.

———. 1974. Observations on the Ultrastructure of a Cephalic Sense Organ of the Nematode *Mermis nigrescens. Journal of Zoology* 173:247–250.

Lin, J. P. 1954. The Chromosomal Cycle in *Parascaris equorum* (*Ascaris megalocephala*): Oogenesis and Diminution. *Chromosoma* 6:175–198.

Lindner, E., and A. Fischer. 1964. Zur Feinstruktor nereider und heteronereider Muskulatur von *Platynereis dumerlii. Naturwissenschaften* 51:410.

Ling, E. A. 1969. The Structure and Function of the Cephalic Organ of a Nemertine *Lineus ruber. Tissue and Cell* 1:503–524.

Linstow, O. Von. 1900. Die Nematoden. In F. Romer und F. Schaudinn, *Fauna Arctica. Eine Zusammenstellung der arktischen Tierformen, mit besonderer Berücksichtigung des Spitzbergen-Gebietes auf Grund der Ergebnisse der deutschen Expedition in das Nördliche Eismeer in Jahre 1898,* 1:117–132. Jena.

———. 1905. Neue Helminthen. *Archiv für Naturgeschichte* 71(1): 267–276.

———. 1909. II. Parasitische Nematoden. In *Die Süsswasserfauna Deutschlands Herausgegeben von Brauer,* 15:47–83. Jena: G. Fischer.

Lippens, P. L. 1974a. Ultrastructure of a Marine Nematode, *Chromadorina germanica* (Bütschli, 1874). I. Anatomy and Cytology of the Caudal Gland Apparatus. *Zeitschrift für Morphologie und Ökologie der Tiere* 78:181–192.

———. 1974b. Ultrastructure of a Marine Nematode, *Chromadorina germanica* (Bütschli, 1874). II. Cytology of Lateral Epidermal Glands and Associated Neurocytes. *Zeitschrift für Morphologie und Ökologie der Tiere* 79:283–294.

Lippens, P. L., A. Coomans, A. T. de Grisse, and A. Lagasse. 1974. Ultrastructure of the Anterior Body Region in *Aporcelaimellus obtusicaudatus* and *A. obscurus. Nematologica* 20:242–256.

Looss, A. 1905. The Anatomy and Life History of *Anchylostoma duodenale* Dub. *Records of the Egyptian Government School of Medicine,* Cairo 3:1–158.

Lorenzen, S. 1978a. New and Known Gonadal Characters in Free-living Nematodes and the Phylogenetic Implications. *Zeitschrift für Zoologische Systematik und Evolutionforschung* 16:108–115.

———. 1978b. Discovery of Stretch Receptor Organs in Nematodes Structure, Arrangement and Functional Analysis. *Zoologica Scripta* 7:175–178.

———. 1981a. Entwurf eines phylogenetischen Systems der freilebenden Nematoden. *Veröffentlichungen des Instituts für Meeresforschung in Bremerhaven, Supplement* 7:1–472.

———. 1981b. Bau, Anordnung und postembryonale Entwicklung von Metanemen bei Nematoden der Ordnung Enoplia. *Veröffentlichungen des Instituts für Meeresforschung in Bremerhaven* 19:89–114.

Lukasiak, J. 1930. Anatomische und Entwicklungsgeschichtliche Untersuchungen an *Dioctophyme renale* (Goeze, 1782). *Archives de Biologie de la Société des Sciences et des Lettres de Varsovie* 3:1–100.

Maertens, D., and A. Coomans. 1979. The Function of the Demanian System and an Atypical Copulatory Behaviour in *Oncholaimus oxyuris. Annales de la Société Royale Zoolgique de Belgique* 108:83–87.

Maggenti, A. R. 1963. Comparative Morphology in Nemic Phylogeny. In *The Lower Metazoa,* ed. E. C. Dougherty, 273–282. Berkeley: University of California Press.

———. 1964. Morphology of Somatic Setae: *Thoracostoma californicum* (Nemata: Enoplidae). *Proceedings of the Helminthological Society of Washington* 31:159–166.

———. 1979. The Role of Cuticular Strata Nomenclature in the Systematics of Nemata. *Journal of Nematology* 11:94–98.

Malakhov, V. V. 1974a. Postembryonic Development in the Free-living Marine

Nematode *Pontonema vulgare* Bastian. In *Biologiya Belogo Morya* [Biology of the White Sea] 4:148–154.

———. 1974b. Early Stages in the Embryonic Development of the Free-living Marine Nematode *Anoplostoma vivipara.* In *Biologiya Belogo Morya* [Biology of the White Sea] 4:154–161.

———. 1975. On the Cleavage of the Eggs of Nematoda and Gastrotricha as a Derivative of Uni-radial Spiral Cleavage. *Vestnik Moskovskogo Universiteta, Seriya Biologiya* [Journal of Moscow University, Biology Series] 4(2): 14–17.

———. 1976. Distribution and Solo Division (Uni-radial Division) among Invertebrates. *Zhurnal Obshchei Biologiya* [Journal of General Biology] 37(3): 387–402.

———. 1978a. Structure of the Body Wall of the Free-living Marine Nematode *Pontonema vulgare. Zoologicheskii Zhurnal* [Zoological Journal] 57(1): 5–18.

———. 1978b. Structure of the Nervous System in the Head Region of the Free-living Marine Nematode *Pontonema vulgare. Zoologicheskii Zhurnal* [Zoological Journal] 57(5): 645–652.

———. 1980. Cephalorhyncha—A New Phylum of the Animal Kingdom Uniting Priapulida, Kinorhyncha, and Gordiacea, and a Classification for Pseudocoelomate Worms. *Zoologicheskii Zhurnal* [Zoological Journal] 59(4): 485–499.

———. 1981. Embryonic Development of Free-living Marine Nematodes of the Orders Chromadorida and Desmodorida. *Zoologicheskii Zhurnal* [Zoological Journal] 60(4): 485–495.

———. 1982. The Structure of the Nervous System of the Posterior End of the Body of the Free-living Marine Nematode *Pontonema vulgare* and the Principal Plan of the Nervous System in Nematodes. *Zoologicheskii Zhurnal* [Zoological Journal] 61(10): 1481–1491.

Malakhov, V. V., and L. A. Agafonova. 1978. Structure of the Schizocoel in the Free-living Marine Nematode *Pontonema vulgare* (Bastian, 1865) (Enoplida). *Doklady Akademia Nauk SSSR* [Reports of the Academy of Sciences of the USSR] 241:495–496.

Malakhov, V. V., and M. I. Akimushkina. 1976. Embryonic Development of the Free-living Nematode *Enoplus brevis. Zoologicheskii Zhurnal* [Zoological Journal] 55(12): 1788–1799.

Malakhov, V. V., and S. L. Belova. 1977. The Distribution of Cholinesterase Activity in the Nervous System of the Free-living Marine Nematode *Pontonema vulgare* (Bastian, 1865). *Zhurnal Evolyutsii, Biokhimii, Fiziologii* [Journal of Evolution, Biochemistry, and Physiology] 13:638–639.

Malakhov, V. V., and V. G. Cherdantsev. 1975. Embryonic Development of the Free-living Marine Nematode *Pontonema vulgare. Zoologicheskii Zhurnal* [Zoological Journal] 54(2): 165–174.

Malakhov, V. V., and A. V. Ovchinnikov. 1980. A Study of the Sensory Organs in Free-living Marine Nematodes. I. *Sphaerolaimus balticus* (Monhysterida,

Sphaerolaimidae). *Zoologicheskii Zhurnal* [Zoological Journal] 59(6): 805–809.

Malakhov, V. V., and S. E. Spiridonov. 1981. Embryonic Development of *Gastromermis* (Nematoda, Mermithida). *Zoologicheskii Zhurnal* [Zoological Journal] 60(10): 1574–1577.

———. 1983. Embryonic Development of *Eustrongyloides excisus* (Nematoda, Dioctophymida). *Zoologicheskii Zhurnal* [Zoological Journal] 62(1): 113–117.

Malakhov, V. V., and D. A. Voronov. 1981. Contributions to the Knowledge of the Head Structure of the Free-living Marine Nematodes. In *Evolution, Taxonomy, Morphology and Ecology of Free-living Nematodes,* eds. T. A. Platonova and S. Ya. Tsalolikhin, 52–56. Leningrad: Academy of Sciences of the USSR.

Malakhov, V. V., and V. V. Yushin. 1984. A Study of the Sensory Organs of Free-living Marine Nematodes. 3. *Paracanthonchus macrodon* (Chromadorida, Cyatholaimidae). *Zoologicheskii Zhurnal* [Zoological Journal] 63(8): 1137–1143.

Malakhov, V. V., K. M. Ryzhikov, and M. D. Sonin. 1982. The Classification of the Major Taxa of Nematodes: Subclasses, Orders, and Suborders. *Zoologicheskii Zhurnal* [Zoological Journal] 64(8): 1125–1134.

Man, J. G. de. 1886. *Anatomische Untersuchungen über freilebende Nordsee-Nematoden,* 1–82. Leipzig: Verlag von Paul Frohberg.

Mapes, C. J. 1965a. Structure and Function in the Nematode Pharynx. I. The Structure of the Pharynges of *Ascaris lumbricoides, Oxyuris equi, Aplectana brevicaudata* and *Panagrellus silusiae. Parasitology* 55:269–284.

———. 1965b. Structure and Function in the Nematode Pharynx. II. Pumping in *Panagrellus, Aplectana* and *Rhabditis. Parasitology* 55:583–594.

———. 1966. Structure and Function in the Nematode Pharynx. III. The Pharyngeal Pump of *Ascaris lumbricoides. Parasitology* 56:137–139.

Martini, E. 1903. Über Furchung und Gastrulation bei *Cucullanus elegans. Zeitschrift für wissenschaftliche Zoologie* 74:501–556.

———. 1906. Über Subcuticula und Seitenfelder einiger Nematoden. I. *Zeitschrift für wissenschaftliche Zoologie* 81:699–766.

———. 1907. Über Subcuticula und Seitenfelder einiger Nematoden. II. *Zeitschrift für wissenschaftliche Zoologie* 86:1–54.

———. 1908. Über Subcuticula und Seitenfelder einiger Nematoden. III. *Zeitschrift für wissenschaftliche Zoologie* 91:191–235.

———. 1909. Über Subcuticula und Seitenfelder einiger Nematoden. IV. *Zeitschrift für wissenschaftliche Zoologie* 93:535–622.

———. 1912. Studien über die Konstanz histologischer Elemente. III. *Hydatina senta. Zeitschrift für wissenschaftliche Zoologie* 102:425–645.

———. 1913. Über die Stellung der Nematoden im System. *Deutsche Zoologischen Gesellschaft zu Bremen* 23:233–248.

———. 1916. Die Anatomie der *Oxyuris curvula. Zeitschrift für wissenschaftliche Zoologie* 116:137–534.

Maupas, E. 1900. Modes et formes de reproduction des Nématodes. *Archives de Zoologie Expérimentale et Générale,* Serie 3, 8:463–624.

McLaren, D. J. 1972. Ultrastructural and Cytochemical Studies on the Sensory Organelles and Nervous System of *Dipetalonema viteae* (Nematoda: Filarioidea). *Parasitology* 65:507–524.

———. 1973. The Structure and Development of Spermatozoon of *Dipetalonema viteae* (Nematoda, Filarioidea). *Parasitology* 66:447–463.

———. 1976. Nematode Sense Organs. *Advances in Parasitology* 14:195–265.

Metalnikoff, S. I. 1923. Les quatre phagocytes d'*Ascaris megalocephala* et leure rôle dans l'immunité. *Annales de l'Institut Pasteur, Paris* 37:680–685.

Metschnikoff, E. 1886. *Embryologische Studien an Medusen. Ein Beitrag zu Genealogie der Primitiv Organe.* 159 pp. Wien: A. Hölder.

Meyer, A. 1928. Die Furchung nebst Eibildung, Reifung und Befructung des *Gigantorhynchus gigas.* Ein Beitrag zur Morphologie der Acanthocephalen. *Zoologische Jahrbücher Abteilung für Anatomie und Ontogenie der Tiere,* 2, 50:117–218.

———. 1931. Urhautzelle, Hautbahn und plasmodiale Entwicklung der Larve von *Neoechinorhynchus rutili* (Acanthocephala). (Ein Beitrag zur Entwicklungsmechanik gebahnter Organbildung.) *Zoologische Jahrbücher Abteilung für Anatomie und Ontogenie der Tiere,* 2, 53:103–126.

———. 1932. Acanthocephala. E. Physiognomik I. Voraussetzungen und besondere Schwierigkeiten der Physiognomik bei Parasiten. In *Dr. H. G. Bronn's Klassen und Ordnungen des Tierreiches. Vermes.* 2(2): 1–332.

Meyer, O. 1895. Celluläre Untersuchungen an Nematoden-Eiern. *Zeitschrift für Naturwissenschaften* 29:391–410.

Minier, L. N., and T. P. Bonner. 1975. Cuticle Formation in Parasitic Nematodes: Ultrastructure of Molting and the Effects of Actinomycin-D. *Journal of Ultrastructure Research* 53:77–86.

Moritz, K. B. 1967. Die Blastomerendifferenzierung für Soma und Keimbohn bei *Parascaris equirum* 2. Untersuchungen mittels uv-Bestrahlung und Zentrifugierung. *W. Roux' Archiv für Entwicklungsmechanik der Organismen* 159:203–266.

Mühldorf, A. 1914. Beiträge zur Entwicklungsgeschichte und zu den phylogenetischen Beziehungen der *Gordius* larvae. *Zeitschrift für wissenschaftliche Zoologie* 111:1–75.

Müller, H. 1903. Beiträge zur Embryonalentwicklung der *Ascaris megalocephala. Bibliotheca Zoologica* 7:1–30.

Mulvey, R. H. 1959. Investigations on the Clover Cyst Nematode, *Heterodera trifolli* (Nematoda: Heterodoridae). *Nematologica* 4:147–156.

Nachtwey, R. 1925. Untersuchungen über die Keimbahn, Organogenese und Anatomie von *Asplanchna priodonta. Zeitschrift für wissenschaftliche Zoologie* 126:239–492.

Nagacura, K. 1930. Über die Bau und die Lebensgeschichte der *Heterodera radicola. Japanese Journal of Zoology* 3:95–160.

Narang, H. K. 1970. The Excretory System of Nematodes: Structure and Ultrastructure of the Excretory System of *Enoplus brevis*. *Nematologica* 16: 517–522.

Nassonov, N. 1900. Zur Kenntnis der phagocytären Organe bei den parastischen Nematodenm. *Archiv für mikroskopische Anatomie* 55:488–513.

Nigon, V., P. Guerrier, and H. Monin. 1960. L'architecture polaire de l'oeuf les mouvements des constituants cellulaires au cours des premières étapes du développement chez quelques nématodes. *Bulletin Biologique de la France et de la Belgique* 94:132–202.

Nurseitov, S. K. 1972. On the Question of the Innervation of the Middle Intestine of *Ganguleterakis squamosa*. *Trudy Vsesoyuznogo Gel'mintologicheskogo Obshchestva* [Transactions of the All-Union Helminthological Society], 309–321.

Osche, G. 1952. Systematik und Phylogenie der Gattung *Rhabditis* (Nematoda). *Zoologische Jahrbücher Abteilung für Systematik, Ökologie und Geographie der Tiere,* 3, 81:190–280.

Ospovat, M. F. 1978. Development of the Nervous System in the Ontogeny of *Phyllodoce* (Polychaeta, Phyllodocidae). Trochophora-Metatrochophora. *Zoologicheskii Zhurnal* [Zoological Journal] 57(8): 987–997.

Ott, J. 1976. Brood Protection in Marine Freeliving Nematode with the Description of *Desmodora* (*Croconema*) *ovigera* N. Sp. *Zoologischer Anzeiger* 196: 175–181.

Pagenstecher, H. A. 1875. *Allgemeine Zoologie oder Grundgesetze des thierischen baus und lebens*. 275 pp. Berlin: Wiegandt, Hempel, and Parey.

Pai, S. 1928. Die Phasen des Lebenscyclus der *Anguillula aceti* Ehrbg. und ihre experimentell-morphologische Beeinflussung. *Zeitschrift für wissenschaftliche Zoologie* 131:293–344.

Pal'chikova-Ostroumova, M. V. 1926. Free-living Nematodes as a Subject for the Observation of the Processes of Fertilization and Division. *Uchenye Zapiski Kazanskii Gosudarstvennogo Universitet* [Scientific Papers, Kazan State University] 86(1): 24–39.

Paramonov, A. A. 1937. The Position of Nematodes in the Classification. *Uchenye Zapiski Bol'shevskoi Biologicheskoi Stantsii* [Scientific Papers from the Bolshevo Biological Station] 10:97–117.

———. 1962. *Osnovy Fitogel'mintologii* [Fundamentals of Phytohelminthology]. 480 pp. Moscow: Izdatel'stvo Akademii Nauk SSSR [Moscow: Publishing House of the Academy of Sciences of the USSR].

Pearse, A. S. 1942. *An Introduction to Parasitology*. 375 pp. Baltimore, Md.: C. C. Thomas.

Petrochenko, V. I. 1956. *Akantotsefaly Domashnikh i Dikikh Zhivotnykh* [Acanthocephala in Domestic and Wild Animals]. 435 pp. Leningrad: Izdatel'stvo Akademii Nauk SSSR [Leningrad. Publishing House of the Academy of Sciences of the USSR].

Poinar, G. O., and R. Hess. 1974. Structure of the Pre-parasitic Juveniles of *Filipjevimermis leipsandra* and Some Other Mermithidae (Nematodea). *Nematologica* 20:163–173.

Poinar, G. O., and R. Leutenegger. 1968. Anatomy of the Infective and Normal

Third-stage Juveniles of *Neoplectana carpocapsae* Wieser (Steinernematidae: Nematoda). *Journal of Parasitology* 54:340–350.

Poinar, G. O., R. Leutenegger, and G. M. Thomas. 1970. On the Occurrence of Protein Platelets in the Pseudocoelom of a Nematode (*Hydromermis* sp., Mermithidae). *Nematologica* 16:348–352.

Popova, T. I., and M. A. Valovaya. 1978. Ontogenesis of *Cucullanus cirratus* Müller, 1977 (Nematoda, Cucullanata), 16–17. In *IV International Congress of Parasitology,* Warszawa.

Por, F. D. 1972. Priapulida from Deep Bottoms Near Cyprus. *Israel Journal of Zoology* 21:525–528.

Por, F. D., and H. J. Bromley. 1974. Morphology and Anatomy of *Maccabeus tentaculatus* (Priapulida: Seticoronaria). *Journal of Zoology* 173:173–197.

Pray, F. A. 1965. Studies of the Early Development of the Rotifer *Monostyla cornuta* Müller. *Transactions of the American Microscopical Society* 84: 210–216.

Pybus, M. J., L. S. Uhazy, and R. C. Anderson. 1978. Life Cycle of *Truttaedacnitis stelmioides* (Vessichelli, 1910) (Nematoda: Cucullanidae) in American Brook Lamprey (*Lampetra lamottenii*). *Canadian Journal of Zoology* 56: 1420–1429.

Rachor, E. 1969. Das de Mansche Organ der Oncholaimidae, eine genito-intestinale Verbindung bei Nematoden. *Zeitschrift für Morphologie und Ökologie der Tiere* 66:87–166.

Raski, D. J., N. O. Jones, and D. R. Roggen. 1969. On the Morphology and Ultrastructure of the Esophageal Region of *Trichodorus allius* Jensen. *Proceedings of the Helminthological Society of Washington* 36:106–118.

Rauther, M. 1906. Beiträge zur Kenntnis von *Mermis albicans* V. Sieb. mit besonderer Berücksichtigung des Haut-Nerven-Muskelsystems. *Zoologische Jahrbücher Abteilung Anatomie und Ontogenie der Tiere,* 2, 23:1–76.

———. 1907. Über den Bau des Oesophagus und die Lokalisation der Nierenfunktion bei freilebenden Nematoden. *Zoologische Jahrbücher Abteilung Anatomie und Ontogenie der Tiere,* 2, 23:703–740.

———. 1909. Morphologie und Verwandtschaftsbeziehungen der Nematoden und einiger ihnen nahe gestellter Vermalien. *Ergebnisse und Fortschritte der Zoologie* 1:491–596.

Remane, A. 1935. Gastrotricha und Kinorhyncha (s. Echinodera). In *Dr. H. G. Bronn's Klassen und Ordnungen des Tierreiches. Vermes.* 4(2): 1–160.

———. 1936. Gastrotricha und Kinorhyncha (s. Echinodera). In *Dr. H. G. Bronn's Klassen und Ordnungen des Tierreiches. Vermes.* 4(2): 161–385.

———. 1956. *Die Grundlagen des natürlichen Systems, der vergleichenden Anatomie und der Phylogenetik.* 364 pp. Leipzig: Geest und Portig.

Remane, A., V. Storch, and V. Welsch. 1976. *Systematische Zoologie: Stämme des Tierreiches.* 678 pp. Stuttgart: Fischer.

Richter, S. 1971. Zum Feinbau von Mermithiden (Nematoda). I. Der Bohrapparat der vorparasitischen Larven von *Hydromermis contorta* (Linstow, 1889) Hagmeier, 1912. *Zeitschrift für Parasitenkunde* 36:32–50.

Rieger, G. E., and R. M. Rieger. 1977. Comparative Fine Structure of the Gastrotrich Cuticle and Aspects of Cuticle Evolution within the Aschelminthes. *Zeitschrift für Zoologische Systematik und Evolutionsforschung* 15:81–124.

Rieger, R. M. 1984. Evolution of the Cuticle in Lower Eumetazoa. In *Biology of the Integument,* eds. J. Breiter-Hahn, A. G. Matolsky, and K. S. Richards, 1:389–399. Berlin: Springer.

Riemann, F. 1972. Corpus gelatum und ciliäre Strukturen als lichtmikroskopisch sichtbare Bauelemente des Seitenorgans freilebender Nematoden. *Zeitschrift für Morphologie und Ökologie der Tiere* 72:46–76.

———. 1976. Meeresnematoden (Chromadorida) mit lateralen Flossensäumen (Alae) und dorsoventraler Abplattung. *Zoologische Jahrbücher Abteilung für Systematik Ökologie und Geographie der Tiere,* 3, 103:290–308.

———. 1977a. On the Excretory Organ of *Sabatieria* (Nematoda, Chromadorida). *Veröffentlichungen des Instituts für Meeresforschung in Bremerhaven* 16:263–267.

———. 1977b. Oesophagusdrüsen bei Linhomoeidae (Monhysterida, Siphonolaimoidea). Beitrag zur Systematik freilebender Nematoden. *Veröffentlichungen des Instituts für Meeresforschung in Bremerhaven* 16:105–116.

Riemann, F., and M. Schrage. 1978. The Mucus-trap Hypothesis on Feeding of Aquatic Nematodes and Implications for Biodegradation and Sediment Texture. *Oecologia* 34:75–88.

Riemann, F., E. Rachor, and I. Freudenhammer. 1970. Das Seitenorgan von *Halalaimus.* Zur Morphologie eines vermutlich sensorischen Organs von freilebenden Nematoden. *Veröffentlichungen des Instituts für Meeresforschung in Bremerhaven* 12:429–441.

Riemann, F., W. von Thun, and S. Lorenzen. 1971. Über den phylogenetischen Zusammenhang zwischen Desmoscolecidae und Leptolaimidae (freilebende Nematoden). *Veröffentlichungen des Instituts für Meeresforschung in Bremerhaven* 13:147–152.

Riley, J. 1973. Histochemical and Ultrastructural Observations on Digestion in *Tetrameres fissispina* Diesing 1861 (Nematoda, Spiruridea) with Special Reference to Intracellular Digestion. *International Journal of Parasitology* 3:157–164.

Roggen, D. R. 1970. Functional Aspects of the Lower Size-limit of Nematodes. *Nematologica* 16:532–536.

———. 1973. Functional Morphology of the Nematode Pharynx. I. Theory of the Soft-walled Cylindrical Pharynx. *Nematologica* 19:349–365.

Roggen, D. R., D. J. Raski, and N. O. Jones. 1967. Further Electron Microscopic Observations of *Xiphinema index. Nematologica* 13:1–16.

Rohde, E. 1892. Muskeln und Nerven bei Nematoden. *Sitzungsberichte der Königlich Preussischen Akademi der Wissenschaften zu Berlin* 28:515–526.

Rosenbluth, J. 1963. Fine Structure of Body Muscle Cells and Neuromuscular Junctions in *Ascaris lumbricoides. Journal of Cell Biology* 19:82–97.

———. 1965a. Ultrastructural Organization of Obliquely Striated Muscle Fibers in *Ascaris lumbricoides. Journal of Cell Biology* 25:495–515.

———. 1965b. Ultrastructure of Somatic Muscle Cells in *Ascaris lumbricoides.*

II. Intermuscular Junctions, Neuromuscular Junctions and Glycogen Stores. *Journal of Cell Biology* 26:579–591.
———. 1967. Obliquely Striated Muscle. III. Contraction Mechanism of *Ascaris* Body Muscle. *Journal of Cell Biology* 34:15–33.
———. 1968. Obliquely Striated Muscle. IV. Sarcoplasmic Reticulum, Contractile Apparatus and Endomysium of the Body Muscle of a Polychaete, *Glycera,* in Relation to Its Speed. *Journal of Cell Biology* 36:245–259.
Rubtsov, I. A. 1978. *Mermitidy. Klassifikatsiya, Znacheniye, Ispol'zovaniye* [Mermithidae: Classification, Importance, and Application]. 207 pp. Leningrad: Nauka [Leningrad: Science].
Sacks, M. 1955. Observations on the Embryology of a Gastrotrich *Lepidodermella squamata. Journal of Morphology* 96:473–495.
Salensky, W. 1872. Beiträge zur Entwicklungsgeschichte der *Brachionus urceolaris. Zeitschrift für wissenschaftliche Zoologie* 22:455–466.
Salvini-Plawen, L. von. 1974. Zur Morphologie und Systematik der Priapulida: *Chaetostephanus praeposteriens,* der Vertreter einer neuen Ordnung Seticoronaria. *Zeitschrift für Zoologische Systematik un Evolutionsforschung* 12:31–54.
Samoiloff, M. R. 1973. Nematode Morphogenesis: Pattern of Transfer of Protein to the Cuticle of Adult *Panagrellus silusiae* (Cephalobidae). *Nematologica* 19:15–18.
Samoiloff, M. R., and J. Pasternak. 1969. Nematode Morphogenesis: Fine Structure of the Molting Cycles in *Panagrellus silusiae* (de Man, 1913) Goodey, 1945. *Canadian Journal of Zoology* 47:639–643.
Schepotieff, A. 1908a. Die Desmoscoleciden. *Zeitschrift für wissenschaftliche Zoologie* 90:181–204.
———. 1908b. *Trichoderma oxycaudatum* Greeff. (Untersuchungen über einige wenig bekannte freilebenden Nematoden. I.). *Zoologische Jahrbücher Abteilung für Systematik, Geographie und Biologie der Tiere,* 3, 26:385–392.
Schmidt, K., and W. Gnatzy. 1972. Die Feinstruktur der Sinneshaare auf den *Cerci* von *Gryllus bimaculatus* Deg. (Saltatoria, Glyllidae). III. Die kurzen Borstenhaare. *Zeitschrift für Zellforschung und mikroskopische Anatomie* 126:206–222.
Schneider, A. 1866. *Monographie der Nematoden.* 357 pp. Berlin: G. Reimer.
———. 1869. Noch ein wort über die Muskeln der Nematoden. *Zeitschrift für wissenschaftliche Zoologie* 19:284–286.
Schönberg, M. 1944. Histologische Studien zu den Problemen der Zellkonstanz an *Rhabditis longicauda fertilior. Biologia Generalis* 17:338–366.
Schuurmans-Stekhoven, J. H. 1959. Nematodes. In *Dr. H. G. Bronn's Klassen und Ordnungen des Tierreiches. Vermes.* 4(2): 661–879.
Seck, P. 1937. Zur Entwicklungsmechanik des Essigälchens. *W. Roux' Archiv für Entwicklungsmechanik der Organismen* 137:57–85.
Seeliger, G. 1977. Der Antennendzapfen der tunesischen Wüstenassel *Hemilpistus reaumuri,* ein komplexes Sinnesorgan (Crustacea, Isopoda). *Journal of Comparative Physiology* 113:95–103.
Seurat, L. G. 1920. *Histoire naturelle des nématodes de la Berbérie. 1. Mor-*

phologie, developpement, ethologie et affinitiés des nematodes. 221 pp. Alger: Imprimerie S. Stamel.

Severtsev, A. N. 1967. *Glavnye Napravleniya Evolyutsionnogo Protsessa. Morfobiologicheskaya Teoriya Evolyutsii* [Principal Trends in the Evolutionary Process: Morphobiological Theory of Evolution]. 101 pp. Moskva: Izdatel'stvo Moskovskogo Universiteta [Moscow: Publishing House of Moscow University].

Sheffield, H. 1962. Electron Microscopy of the Bacillary Cells of *Trichuris muris* (Schrank, 1788). *Journal of Parasitology* 48:42–48.

Shepherd, A. M., S. A. Clark, and P. J. Dart. 1972. Cuticle Structure in the Genus *Heterodera. Nematologica* 18:1–17.

Shepot'yev, A. S. 1908. *O Nematodakh i Blizkikh k Nim Gruppakh* [On Nematodes and Related Groups]. 91 pp. Sankt-Peterburgskogo [Saint Petersburg].

Shimkevich, V. M. 1898. On Certain Applications of Dyes of Methylene Blue. *Trudy Imperatorskoe Sankt-Peterburgskoe Obshchestvo Estestvoispytatelei* [Tranasctions of the Saint Petersburg Naturalists' Society] 29:320–323.

Shishov, B. A. 1971. On Cholinesterase Activity in the Nematodes *Eustrongylus excisus* Jägersiöld, 1909 (Diocotophymidae). *Parazitologiya* [Parasitology] 5(5): 341–343.

Shishov, B. A., G. Koyshibayeva, and T. N. Timmofeyeva. 1973. Cholinesterase Activity in the Nematode *Stephanofilaria stilesi* and the Nematode *Setaria labiatopapillosa. Trudy Gel'mintologicheskoi Laboratorii, Akademii Nauk SSSR* [Transactions of the Helminthology Laboratory of the Academy of Sciences of the USSR] 21:123–130.

Siddiqui, I. A., and D. R. Viglierchio. 1970. Fine Structure of Photoreceptors in *Deontostoma californicum. Journal of Nematology* 2:274–276.

———. 1977. Ultrastructure of the Anterior Body Region of Marine Nematode *Deontostoma californicum. Journal of Nematology* 9:56–82.

Skryabin, K. I. 1941. The Phylogenetic Interrelationship between Nematodes of the Subclass Phasmidia. *Zoologicheskii Zhurnal* [Zoological Journal] 20(3): 327–340.

———. 1946. *Stroitel'stvo Sovetskoy Gel'mintologii* [The Development of Soviet Helminthology]. 211 pp. Moskva: Izdatel'stvo Akademii Nauk SSSR [Moscow: Publishing House of the Academy of Sciences of the USSR].

Skryabin, K. I., and R. S. Schulz. 1931. Gel'minotzy Cheloveka [Helminthosis of Man]. In *Osnovy Meditsinskoy Gel'mintologii* [Fundamentals of Medical Helminthology]. 378 pp. Moscow: Gosmedizdat [Moscow: Medical Publishing House].

Skryabin, K. I., N. P. Shikhobalova, and A. A. Sobolev. 1949. *Opredelitel' Paraziticheskikh Nematod* [A Key to the Parasitic Nematodes], 1:1–519. Moscow-Leningrad: Izdatel'stvo Akademii Nauk SSSR [Moscow-Leningrad: Publishing House of the Academy of Sciences of the USSR].

Slifer, E. H. 1970. The Structure of Arthropod Chemoreceptors. *Annual Review of Entomology* 15:121–137.

Spemann, H. 1895. Zur Entwicklung des *Strongylus paradoxus. Zoologische Jahrbücher Abteilung für Anatomie und Ontogenie der Tiere,* 2, 8:301–317.

Sprent, J. F. 1962. The Evolution of Ascaridoidea. *Journal of Parasitology* 48: 818–824.

Stauffer, H. 1924. Die Lokomotion der Nematoden. Beiträge zur Kausalmorphologie der Fadenwürmer. *Zoologische Jahrbücher Abteilung für Systematik, Geographie und Biologie der Tiere,* 3, 49:1–119.

Steinböck, O. 1958. Schlusswort zur Diskussion Remane-Steinböck. *Verhandlungen der Deutschen Zoologischen Gesellschaft* 21:196–218.

Steiner, G. 1916. Neue und wenig bekannte Nematoden von der Westküste Afrikas I. *Zoologischer Anzeiger* 47:322–351.

———. 1921. Untersuchungen über den allgemeinen Bauplan des NematodenKörpers. Ein Beitrag zur Aufhellung der Stammesgeschichte und der Verwandschafts-verhältnisse der Nematoden. *Zoologische Jahrbücher Abteilung für Anatomie und Ontogenie der Tiere* 43:1–96.

Stevens, N. M. 1909. The Effect of Ultra-violet Light upon the Developing Eggs of *Ascaris megalocephala. W. Roux' Archiv für Entwicklungsmechanik der Organismen* 27:622–639.

Stewart, F. H. 1906. The Anatomy of *Oncholaimus vulgaris* Bastian with Notes on Two Parasitic Nematodes. *Quarterly Journal of Microscopical Science* 50:101–150.

Storch, V., and F. Riemann. 1973. Zur Ultrastruktur der Seitenorgane (Amphiden) des limnischen Nematoden *Tobrilus aberrans* (W. Schneider, 1925) (Nematoda, Enoplida). *Zeitschrift für Morphologie und Ökologie der Tiere* 74:163–170.

Strassen, O. von. 1959. Neue Beiträge zur Entwicklungsmechanick der Nematoden. *Bibliotheca Zoologica* 38:1–142.

Strassen, O. zur. 1892. *Bradynema rigidum* v. Sieb. *Zeitschrift für wissenschaftliche Zoologie* 54:655–747.

———. 1896. Embryonalentwicklung der *Ascaris megalocephala. W. Roux' Archiv für Entwicklungsmechanik der Organismen* 3:27–105, 132–190.

———. 1898. Über die Riesenbildung bei *Ascaris*-Eiern. *W. Roux' Archiv für Entwicklungsmechanik der Organismen* 7:642–676.

———. 1904. *Anthraconema,* eine neue Gattung freilebender Nematoden. *Zoologische Jahrbücher, Supplement* 7:301–346.

Strubell, A. 1888. Untersuchungen über den Bau und die Entwicklung des Rüben-Nematode *Heterodera schachtii. Bibliotheca Zoologica* 2:1–52.

Sulston, J., M. Dew, and S. Brenner. 1975. Dopaminergic Neurons in the Nematode *Caenorhabditis elegans. Journal of Comparative Neurology* 163:215–226.

Sulston, J. E. 1976. Post-embryonic Development in the Ventral Cord of *Caenorhabditis elegans. Philosophical Transactions of the Royal Society of London, Series B* 275:287–297.

Sulston, J. E., and H. R. Horvitz. 1977. Post-embryonic Cell Lineages of the Nematode, *Caenorhabditis elegans. Developmental Biology* 56:110–156.

Sulston, J. E., E. Shierenberg, J. G. White, and J. N. Thomson. 1983. The Embryonic Cell Lineage of the Nematode, *Caenorhabditis elegans. Developmental Biology* 100:64–119.

Tannreuther, G. W. 1920. The Development of *Asplanchna ebbesbornii* (Rotifer). *Journal of Morphology* 33:389–437.

Taylor, C. E., P. R. Thomas, W. M. Robertson, and I. M. Roberts. 1970. An Electron Microscope Study of the Oesophageal Region of *Longidorus elongatus* (de Man). *Nematologica* 16:6–12.

Tessin, G. 1886. Über Eibildung und Entwicklung der Rotatorien. *Zeitschrift für wissenschaftliche Zoologie* 44:273–302.

Teuchert, G. 1968. Zur Fortpflanzung und Entwicklung der Macrodasyoidea (Gastrotricha). *Zeitschrift für Morphologie und Ökologie der Tiere* 63:343–418.

———. 1972. Die Feinstruktur des Protonephridialsystems von *Turbanella cornuta* Remane, einem marinen Gastrotrich der Ordnung Macrodasyoidea. *Zeitschrift für Zellforschung und mikroskopische Anatomie* 136:277–289.

———. 1974. Aufbau und Feinstruktur der Muskelsysteme von *Turbanella cornuta* Remane (Gastrotricha, Macrodasyoidea). *Mikrofauna Meeresbodens* 39:1–26.

———. 1975. Differenzierung von Spermien bei dem marinen Gastrotrich *Turbanella cornuta* Remane (Ordnung Macrodasyoidea). *Verhandlungen der Anatomischen Gesellschaft* 69:743–748.

———. 1976a. Elektronenmikroskopische Untersuchung über die Spermatogenese und Spermatohistogenese von *Turbanella cornuta* Remane (Gastrotricha). *Journal of Ultrastructure Research* 56:1–14.

———. 1976b. Sinneseinrichtungen bei *Turbanella cornuta* Remane (Gastrotricha). *Zoomorphologie* 83:193–207.

———. 1977a. The Ultrastructure of the Marine Gastrotrich *Turbanella cornuta* Remane (Macrodasyoidea) and Its Functional and Phylogenetical Importance. *Zoomorphologie* 88:189–246.

———. 1977b. Leibeshöhlenverhaltnisse von dem marinen Gastrotrich *Turbanella cornuta* Remane (Ordnung Macrodasyoidea) und eine phylogenetische Bewertung. *Zoologische Jahrbücher Abteilung für Anatomie und Ontogenie der Tiere,* 2, 97:586–596.

Thust, R. 1968. Submikroskopische Untersuchungen über die Morphogenese des Integumentes von *Ascaris lumbricoides* Linnaeus, 1758. *Zeitschrift für wissenschaftliche Zoologie* 178:1–39.

Tietjen, J. H., and J. J. Lee. 1973. Life-history and Feeding Habits of the Marine Nematode *Chromadorina macrolaimoides* Steiner. *Oecologia* 12: 303–314.

Timm, R. W. 1953. Observations on the Morphology and Histological Anatomy of a Marine Nematode, *Leptosomatum acephalatum* Chitwood, 1936, New Combination (Enoplidae, Leptosomatinae). *American Midland Naturalist* 49:229–248.

Toldt, K. 1899. Über den feineren Bau der Cuticula von *Ascaris megalocephala* Cloquet nebst Bemerkungen über die Subcuticula desselben Tieres. *Arbeiten aus den Zoologischen Instituten der Universität Wien* 11:1–38.

———. 1904. Die Saftbahnen in der Cuticula von *Ascaris megalocephala* Cloqu. *Zoologischer Anzeiger* 27:728–730.

———. 1905. Über die Differenzierungen in der Cuticula von *Ascaris megalocephala* Cloqu. *Zoologischer Anzeiger* 28:539–542.

Türk, F. 1903. Über einige im Golfe von Neapel freilebende Nematoden. *Mitteilungen aus der Zoologischen Station zu Neapel* 16:281–348.

Twarog, B. M. 1967. The Regulation of Catch in Molluscan Muscle. *Journal of General Physiology* 50:157–169.

Valovaya, M. A. 1978. Embryonic and Post-embryonic Development in *Cucullanus cirratus* Müller, 1877 (Nematoda, Cucullanidae). *Parazitologiya* [Parasitology] 12(5): 426–433.

Valovaya, M. A., and V. V. Malakhov. 1978. Certain Patterns of Post-embryonic Nematode Development. In *Nauchnyye i Prikladnyye Problemy Gel'mintologii* [Scientific and Applied Problems of Helminthology], 24–29. Moscow: Nauka [Moscow: Science].

Vinnikov, Ya. A. 1971. *Tsitologicheskiye i Molekulyarnye Osnovy Retseptsii* [The Cytological and Molecular Bases of Reception]. 372 pp. Leningrad: Nauka [Leningrad: Science].

Voltzenlogel, E. 1902. Untersuchungen über den anatomischen und histologischen Bau des Hinterendes von *Ascaris megalocephala* und *Ascaris lumbricoides. Zoologische Jahrbücher Abtheilung Anatomie und Ontogenie der Thiere,* 2, 16:481–510.

Wagner, G., and K.-A. Seitz. 1981. Funktionsmorphologische Untersuchungen am männlichen Kopulationsapparat von *Pelodera strongyloides* (Schneider, 1866) (Nematoda, Rhabditidae). *Zoomorphology* 97:121–132.

Walton, A. C. 1918. The Oogenesis and Early Embryology of *Ascaris incurva* Werner. *Journal of Morphology* 30:527–603.

Ward, S. 1973. Chemotaxis by the Nematode *Caenorhabditis elegans:* Identification of Attractants and Analysis of the Response by Use of Mutants. *Proceedings of the National Academy of Science, United States of America* 70:817–821.

Ward, S., N. Thomson, J. G. White, and S. Brenner. 1975. Electron Microscopical Reconstruction of the Anterior Sensory Anatomy of the Nematode *Caenorhabditis elegans. Journal of Comparative Neurology* 163:215–226.

Ware, R. W., D. Clark, K. Crossland, and R. L. Russel. 1975. The Nerve Ring of the Nematode *Caenorhabditis elegans:* Sensory Input and Motor Output. *Journal of Comparative Neurology* 162:71–110.

Watson, B. 1965a. The Fine Structure of the Body-wall and the Growth of the Cuticle in the Adult Nematode *Ascaris lumbricoides. Quarterly Journal of Microscopical Science* 106:83–91.

———. 1965b. The Fine Structure of the Body-wall in a Free-living Nematode *Euchromadora vulgaris. Quarterly Journal of Microscopical Science* 106: 75–81.

Wen, G. Y., and T. A. Chen. 1976. Ultrastructure of the Spicules of *Pratylenchus penetrans. Journal of Nematology* 8:69–74.

Wessing, A. 1953. Histologische Studien zu den Problemen der Zellkonstanz:

Untersuchungen an *Rhabditis anomala*. P. Hertwig. *Zoologische Jahrbücher Abteilung für Anatomie und Ontogenie der Tiere,* 2, 73:69–102.

Wharton, D. A. 1979. Oogenesis and Egg-shell Formation in *Aspiculuris tetraptera* Schulz (Nematoda: Oxyuroidea). *Parasitology* 78:131–143.

White, J. G., E. Southgate, J. N. Thomson, and S. Brenner. 1976. The Structure of the Ventral Nerve Cord of *Caenorhabditis elegans. Philosophical Transactions of the Royal Society of London, Series B* 275:327–348.

Wieser, W., and J. Kanwisher. 1960. Growth and Metabolism in a Marine Nematode, *Enoplus communis* Bastian. *Zeitschrift für vergleichende Physiologie* 43:29–36.

Wilson, E. B. 1903. Experiments on Cleavage and Localization in the Nemertine Egg. *W. Roux' Archiv für Entwicklungsmechanik der Organismen* 16:411–460.

———. 1904. Experimental Studies on Germinal Localization. I. The Germ Regions in the Egg of *Dentalium. Journal of Experimental Zoology* 1:1–72.

Wisse, E., and W. T. Daems. 1968. Electron Microscopic Observations on Second-stage Larvae of the Potato Root Eelworm *Heterodera rostochiensis. Journal of Ultrastructure Research* 24:210–231.

Wright, K. A. 1964. The Fine Structure of the Somatic Muscle Cells of Nematode *Capillaria hepatica* (Bancroft, 1893). *Canadian Journal of Zoology* 42:483–490.

———. 1965. The Histology of the Oesophageal Region of *Xiphinema index* Thorne and Allen, 1950, as Seen with the Electron Microscope. *Canadian Journal of Zoology* 43:689–700.

———. 1968a. Structure of the Bacillary Band of *Trichuris myocastoris. Journal of Parasitology* 54:1106–1110.

———. 1968b. The Fine Structure of the Cuticle and Interchordal Hypodermis of the Parasitic Nematodes, *Capillaria hepatica* and *Trichuris myocastoris. Canadian Journal of Zoology* 46:173–179.

———. 1972. The Fine Structure of the Oesophagus of Some Trichuroid Nematodes. I. The Stichosome of *Capillaria hepatica, Trichuris myocastoris* and *Trichuris vulpis. Canadian Journal of Zoology* 50:319–324.

———. 1974a. The Fine Structure of the Esophagus of Some Trichuroid Nematodes. II. The Buccal Capsule and Anterior Esophagus of *Capillaria hepatica* (Bancroft, 1893). *Canadian Journal of Zoology* 52:47–58.

———. 1974b. Cephalic Sense Organs of the Parasitic Nematode *Capillaria hepatica* (Bancroft, 1893). *Canadian Journal of Zoology* 52:1207–1213.

———. 1975. Cephalic Sense Organs of the Rat Hookworm, *Nippostrongylus brasiliensis*—Form and Function. *Canadian Journal of Zoology* 53:1131–1146.

———. 1978. Structure and Function of the Male Copulatory Apparatus of the Nematodes *Capillaria hepatica* and *Trichuris muris. Canadian Journal of Zoology* 56:651–662.

———. 1980. Nematode Sense Organs. In *Nematodes As Biological Model,* ed. B. M. Zuckerman, 2:237–295. New York: Academic Press.

Wright, K. A., and R. Carter. 1980. Cephalic Sense Organs and Body Pores of *Xiphinema americanum* (Nematoda: Dorylaimoidea). *Canadian Journal of Zoology* 58:1439–1451.

Wright, K. A., and W. D. Hope. 1968. Elaborations of the Cuticle of *Acanthonchus duplicatus* Wieser, 1959 (Nematoda: Cyatholaimidae) as Revealed by Light and Electron Microscopy. *Canadian Journal of Zoology* 46:1005–1011.

Wright, K. A., T. A. Dick, and G. S. Hamada. 1972. The Identity of Pseudocoelomic Membranes and Connective Tissue in Some Nematodes. *Zeitschrift für Parasitenkunde* 39:1–16.

Yamaguti, S. 1961. *Systema Helminthum. 3. The Nematodes of Vertebrates,* Pt. 1. 679 pp. New York and London: Interscience Publications.

———. 1963. *Systema Helminthum. 3. The Nematodes of Vertebrates,* Pt. 2. 1,261 pp. New York and London: Interscience Publications.

Yarzhinskiy, T. 1868. Investigation of the Structure of the Nervous System in Echinorhynchen. *Trudy Pervogo s'yezda Russkogo Estestvoispytatelei Otdel Zoologii* [Transactions of the First Congress of Russian Naturalists, Department of Zoology], 298–310.

Yuen, P.-H. 1967. Electron Microscopical Studies on *Ditylenchus dipsaci* (Kühn). I. Stomatal Region. *Canadian Journal of Zoology* 45:1019–1033.

Yuksel, H. S. 1960. Observations on the Life-cycle of *Ditylenchus dipsaci* on Onion Seedling. *Nematologica* 5:289–296.

Yushin, V. V. 1988. *Comparative Morphological Analysis of the Fine Structure of the Cuticle of Free-living Marine Nematodes.* 121 pp. Ph.D. Dissertation. Vladivostock.

Yushin, V. V., and V. V. Malakhov. 1989. Cuticle Formation in Embryogenesis of the Free-living Marine Nematode *Enoplus demani. Doklady Akademii Nauk SSSR* [Reports of the Academy of Sciences of the USSR] 308:497–499.

Zacharias, O. 1913. Über den feineren Bau der Eiröhren von *Ascaris megalocephala,* insbesondere über zwei ausgedehnte Nervengeflechte in denselben. *Anatomischer Anzeiger* 43:193–211.

Zavarzin, A. A. 1976. *Osnovy Chastnoy Tsitologii i Sravnitel'noy Gistologii Mnogokletochnykh Zhivotnykh* [The Fundamentals of Particular Cytology and Comparative Histology of Multicellular Animals]. 410 pp. Leningrad: Nauka [Leningrad: Science].

Zeleny, C. 1904. Experiments on the Localization of Developmental Factors in the Nemertine Egg. *Journal of Experimental Zoology* 1:293–329.

Zhinkin, L. N. 1949. Early Stages in the Development of *Priapulus caudatus. Doklady Akademii Nauk SSSR* [Reports of the Academy of Sciences of the USSR] 65(4): 409–412.

———. 1955. Developmental Characteristics and the Systematic Position of *Priapulida. Uchenye Zapiski Leningradskogo Pedagogicheskii Institut im. Gertsena* [Scientific Papers of the Herzen Pedagogical Institute] 110:129–139.

Zhinkin, L. N., and G. F. Korsakova. 1953. Early Stages in the Development of *Halicryptus spinulosus*. *Doklady Akademii Nauk SSSR* [Reports of the Academy of Sciences of the USSR] 88(2): 571–573.

Zhuchkova, N. I., and B. A. Shishov. 1979. Constancy and Conservatism in the Cellular Composition of the Nematode Nervous System: Catecholaminergic Neurons. *Zhurnal Obshchei Biologii* [Journal of General Biology] 15(2): 259–270.

Ziegler, H. E. 1895. Untersuchungen über die ersten Entwicklungsvorgänge der Nematoden. Zugleich ein Beitrag zur Zellenlehre. *Zeitschrift für wissenschaftliche Zoologie* 60:351–410.

Zmoray, I., and A. Guttekovà. 1972. The Ecological Aspect of the Study of the Intestinal Ultrastructure of Nematodes. *Zeitschrift für Parasitenkunde* 39: 127–135.

Zoja, R. 1896. Untersuchungen über die Entwicklung der *Ascaris megalocephala*. *Archiv für mikroskopische Anatomie* 47:218–260.

Zuckerman, B. M., S. Himmelhoch, and M. Kisiel. 1973. Fine Structure Changes in the Cuticle of Adult *Caenorhabditis briggsae* with Age. *Nematologica* 19: 109–112.

Index

Abos, 180
Acanthocephala, 24, 225, 229, 230, 231, 251
Acanthonchus, 5
Acronema, 180
Adenophorea, 176, 177
Adenophori, 177
Agnatha, 199
Ananus, 180
Anaplectus granulosus, 107
Ancylostomatoidea, 186
Anisakis sp., 107
Anisakoidea, 109, 198
Anisakoides, 108
Annelida, 44, 202
Anomonema, 194
Anonchus, 195; *mirabilis,* 107
Anoplostoma, 29; *rectospiculum,* 38
Anoplostomatidae, 96, 100
Anticoma, 29, 131, 132
Anticomidae, 124
Anticyathus, 100
Anticyclus, 105, 122, 123, 193
Aphanolaimus, 120
Aphasmidia, 176, 177
Aphelenchina, 186
Aphelenchoides, 81, 85, 89; *blastophorus,* 89, 90
Aponchiidae, 130
Araeolaimida, 3, 100, 101, 120, 123, 124, 126, 130, 159, 160, 161, 167, 168, 178, 183, 192, 194
Araeolaimidae, 130, 183
Araeolaimoides, 124
Araeolaimus elegans, 88
Arhynchohelminthes, 225
Arthropoda, 65, 170, 221
Articulata, 252
Ascaridae, 164
Ascaridata, 127
Ascaridia, 86, 173; *galli,* 58, 67, 68
Ascaridida, 31, 121, 124, 161, 163, 168, 178, 186, 198
Ascaridina, 127, 187
Ascaris, 1, 9, 14, 18, 44, 54, 57, 96, 102, 112, 119, 173; *lumbricoides,* 142, 232; sp., 107
Aschelminthes, 224, 225
Aspiculuris, 135
Asplanchna, 227, 228
Axonolaimidae, 126, 128, 183
Axonolaimus, 18, 29, 87, 96, 159; *paraspinosus,* 160

Bathyeurystomina, 18
Bathylaimus, 29
Bathyodontoidea, 180
Bathyonchus embryonophorus, 91
Bilateria, 24, 105, 110, 115, 129, 169, 219, 220, 221, 236, 247, 249, 250, 252

Brachiopoda, 252
Bryozoa, 252

Caenorhabditis, 10, 58, 67, 81, 85, 86, 88; *briggsae,* 14; *elegans,* 75, 79, 80, 119, 164, 165
Camacolaimus, 26, 124
Camallanidae, 121, 198
Camallanina, 187
Camallanus, 164; *lacustrus,* 107
Capillaria, 66; *hepatica,* 73, 74, 85
Cephalobina, 185
Cephaloboidea, 186
Cephalorhyncha, 250, 251
Ceramonematina, 183
Chaetognatha, 202, 250, 252
Chaetonotoidea, 209, 210, 211, 213, 214, 216, 217, 218, 219, 233
Chordata, 207, 252
Chordodasys, 213
Chromadora, 29, 108, 122; *macrolaimoides,* 170; *quadrilinea,* 107
Chromadoria, 13, 37, 42, 43, 65, 74, 86, 88, 93, 106, 109, 123, 154, 177, 178, 181, 182, 183, 184, 187, 192, 193, 194, 195, 196, 201, 233
Chromadorida, 3, 5, 12, 24, 29, 67, 100, 101, 120, 126, 128, 130, 132, 133, 154, 159, 161, 167, 168, 177, 178, 182, 192, 194, 201, 218, 219
Chromadorina, 182
Chromadorina germanica, 66, 67; sp., 88
Chromadoropsis, 29
Chromaspirina, 100
Cnidaria, 42
Cobbia, 99
Coelenterata, 44
Comesomatidae, 130, 183
Criconematidae, 21, 117
Cucullanidae, 121, 198
Cucullanus, 106, 122; *chabouidi,* 198; *cirratus,* 91, 198
Cyatholaimidae, 192
Cyatholaimina, 182
Cyclostomata, 118, 199, 200
Cystoopsoidea, 181

Dalyelloidea, 117, 118
Daptonema, 18, 26; setosum, 77, 159
Deontostoma, 12, 29, 112; *californicum,* 67, 88, 92
Desmodora, 96; *serpentulus,* 157
Desmodorida, 3, 5, 24, 29, 99, 100, 101, 120, 126, 128, 130, 157, 159, 160, 161, 163, 167, 168, 178, 183, 192, 194
Desmodorina, 183
Desmolaimus, 29
Desmoscolecida, 21, 100, 101, 104, 117, 130, 178, 184, 192, 194, 216
Desmoscolecoidea, 184
Deuterostomia, 250, 252
Diaphanocephaloidea, 186
Dioctophymida, 101, 123, 126, 147, 148, 150, 166, 178, 181, 191, 201
Dioctophymidae, 181
Dipetalonema, 89, 135; *vitae,* 89, 139
Diphtherophorina, 180
Diplogaster, 142
Diplogasteridae, 130
Diplopeltidae, 183
Diplopeltis, 124
Diplopeltula, 108
Diptera, 202
Ditylenchus, 80
Dolicholaimus, 26
Dorylaimida, 3, 5, 24, 67, 72, 120, 123, 124, 126, 127, 130, 152, 154, 166, 178, 180, 189, 190, 191, 201
Dorylaimina, 180
Draconematidae, 104, 105, 216
Draconematina, 183
Drilonematina, 185

Echinodermata, 44, 180, 189, 202, 252
Echiura, 252
Ectoprocta, 117, 118
Eleutherolaimus, 122, 123, 193
Enchelidiidae, 29, 100
Enoplia, 13, 37, 43, 65, 68, 73, 86, 88, 93, 101, 106, 109, 142, 177, 178, 179, 180, 181, 187, 189, 191, 192, 193, 195, 196, 201, 233
Enoplida, 3, 5, 12, 24, 26, 29, 36, 37, 43, 44, 46, 53, 55, 57, 68, 72, 92, 93, 95, 96, 99, 100, 101, 105, 106, 110, 112, 114, 119, 120, 121, 123, 124, 125,

128, 130, 132, 142, 144, 150, 165, 166, 167, 168, 177, 178, 179, 187, 189, 191, 193, 201, 217, 218, 219, 233
Enoplidae, 95, 96, 100, 121
Enoplina, 179
Enoploides, 99
Enoplolaimus, 99
Enoplus, 3, 4, 5, 6, 9, 12, 18, 29, 39, 40, 96, 97, 137; *communis,* 88; *demani,* 15, 19, 20, 136, 148
Entoprocta, 252
Euchromadora, 5
Eurystomina, 122
Eustrongylides excisus, 148, 149

Filariidae, 121
Filariina, 42, 127, 187

Gastromermis, 73, 152, 164; *hibernalis,* 153
Gastrotricha, 24, 34, 93, 105, 117, 118, 129, 169, 193, 202, 204, 205, 207, 208, 209, 213, 225, 231, 251
Gnathostoma, 137
Gnathostomata, 199
Gordiacea, 251
Gordius sp., 243
Greeffielloidea, 184

Halichoanolaimus, 18, 29
Halicryptus, 238
Haliplectoidea, 184
Hammerschmidtiella diesingi, 141; sp., 107
Hemichordata, 252
Hemicycliophora, 18
Heterakina, 187
Heterakis, 86; *gallinarum,* 66, 67, 68, 81, 83, 84
Heterodera, 80, 90, 142
Hypodontolaimus, 29, 156; *inaequalis,* 155, 156
Hystriognathoidea, 186

Kinorhyncha, xiii, 24, 117, 118, 129, 169, 225, 236, 240, 245, 246, 247, 249, 251

Lepidodermella squamata, 210
Leptolaimidae, 194
Leptolaimoidea, 184
Leptosomatidae, 93, 95, 99, 103, 123, 124, 189
Linhomoeidae, 120, 126
Linhomoeina, 183, 193
Linhomoeus, 29
Lophophorates, 117
Loricifera, 236, 247, 249, 250, 251

Macrodasyoidea, 93, 118, 203, 204, 205, 206, 207, 208, 209, 210, 211, 213, 215, 216, 218, 219, 222, 233
Macrodasys, 213
Macroposthonia, 80
Manunema, 194
Marimermis, 180
Marimermithida, 154, 178, 179, 189, 191, 201
Megadesmolaimus, 18
Meloidogyne, 16, 80
Mermis, 12, 73, 96, 122
Mermithida, 3, 5, 73, 86, 101, 106, 115, 123, 126, 152, 154, 167, 178, 180, 190, 191, 201
Mermithoidea, 181
Mesacanthion, 29, 137
Metalinhomoeus, 26
Metaparoncholaimus longispiculum, 33
Metastrongyloidea, 186
Metazoa, 21, 133, 169, 219, 234, 235
Meylioidea, 184
Microlaimidae, 130
Mollusca, 252
Monhystera, 124, 126
Monhysterida, 3, 5, 12, 24, 26, 29, 56, 77, 99, 100, 101, 103, 105, 106, 120, 123, 124, 126, 128, 130, 133, 158, 159, 161, 167, 168, 178, 182, 192, 193, 194, 201
Monhysteridae, 120
Monhysterina, 183
Monocelidae, 132
Mononchida, 106, 120, 123, 124, 126, 130, 151, 152, 154, 166, 178 180, 189, 191, 201
Mononchoidea, 180
Mononchus, 122

Monoposthia, 9, 12; *costata,* 7, 10, 41
Monoposthiidae, 183
Muspecioidea, 181

Nanaloricus mysticus, 247, 248, 249
Nectonematida, 244
Nemathelminthes, 224, 225, 250, 251, 252
Nematoda, 43, 53, 95, 109, 114, 116, 173, 175, 176, 177, 178, 181, 184, 187, 202, 221, 224, 225, 231, 251
Nematomorpha, xiii, 24, 224, 225, 236, 241, 243, 244, 245, 246, 247, 249
Nematorhyncha, 224
Nematospiroides dubius, 18
Nemertinea, 21, 22
Nemertini, 252
Neoaplectana, 91
Neochromadora, 25, 87
Neoechinorhynchus, 230
Nippostrongylus, 89, 135, 137; *brasiliensis,* 81, 82, 84, 90
Nudora, 99

Octracoderma, 199, 200
Oesophagostomum dentatum, 107
Oncholaimidae, 29, 95, 100, 121, 123, 125, 132
Oncholaimina, 179
Oncholaimium, 131
Oncholaimus, 96, 98, 108, 122; *oxyuris,* 132; *vesicarius,* 72, 73, 86
Oswaldocruzia, 142
Otoplanidae, 117, 118
Oxystomina, 26
Oxystominidae, 96, 100, 123, 124
Oxyurida, 31, 37, 88, 121, 124, 127, 140, 178, 186, 196, 197, 198
Oxyuris, 9, 14, 56, 58; *equi,* 58
Oxyuroidea, 186

Panagrollus, 14
Paracanthonchus, 5, 9, 29, 65, 67, 75, 87; *macrodon,* 11, 41, 66; sp., 78
Paralinhomoeus, 123, 193
Parascaris, 55, 59, 142; *equorum,* 163, 232
Pelodera, 89, 90
Penetrantia, 177
Phanodermopsis, 120; *longisetae,* 107
Phasmidia, 176, 177
Phoronida, 252
Placoderma, 200
Platyhelminthes, 252
Plectida, 3, 26, 90, 106, 120, 121, 124, 126, 128, 130, 159, 161, 163, 167, 168, 178, 184, 192, 194, 195, 196, 201
Plectidae, 126
Plectoidea, 184
Plectus, 160, 161
Pleuromyarii, 177
Pogonophora, 252
Polychaeta, 189
Polycladida, 169
Polygastrophora, 120
Pontonema, 12, 18, 40, 68, 84, 85, 97, 98, 105, 110; *vulgare,* 9, 27, 30, 33, 36, 47, 48, 49, 50, 51, 52, 60, 62, 68, 69, 70, 71, 72, 99, 111, 113, 143, 145, 146, 147, 170
Pratylenchus penetrans, 90
Priapulida, xiii, 24, 34, 180, 189, 225, 236, 237, 246, 247, 249, 250, 251
Priapulomorpha, 236
Priapulus, 238; *caudatus,* 237, 239
Prionchulus, 106; sp., 152
Prismatolaimus, 125
Prochromadorella, 29
Proseriata, 132
Protostomia, 252
Pseudocella, 29
Pteraspidomorpha, 199
Pycnophyes kielensis, 241, 242

Radialia, 252
Radialoblastica, 252
Radopholus, 80
Resorbentes, 177
Rhabdias, 142; *buffonis,* 196
Rhabditia anomala, 170
Rhabditia, 1, 3, 9, 13, 14, 30, 31, 37, 42, 43, 53, 54, 55, 58, 67, 68, 75, 81, 84, 85, 86, 88, 89, 90, 93, 101, 102, 103, 106, 109, 112, 119, 121, 123, 127, 129, 130, 132, 133, 135, 138, 161, 162, 163, 164, 167, 168, 170, 177, 178, 184, 185, 186, 187, 194, 195, 196, 200, 233

Rhabditida, 30, 31, 37, 43, 88, 127, 128, 164, 165, 177, 178, 185, 196, 197, 198, 201
Rhabditidae, 126
Rhabditina, 185
Rhabditis, 96, 108, 122, 220; *dolichura,* 107; *pelio,* 142; *strongyloides,* 107
Rhabditostrongylida, 200
Rhabdodemania, 125
Rhaptothyreus, 180
Rhigonematoidea, 186
Rhynchohelminthes, 225
Rotatoria, 225, 251
Rotifera, 24, 129, 169, 225, 226, 227, 228, 250
Rotylenchus, 80, 131

Sabatieria, 29, 105; *celtica,* 107
Secernentea, 102, 133, 168, 176, 177, 195, 218, 219
Secernentes, 177
Secernentia, 177
Seticoronaria, 236
Siphonolaimidae, 120
Siphonolaimina, 183
Siphonolaimus, 56
Sipuncula, 252
Soboliphymidae, 181
Sphaerolaimidae, 16, 120, 126, 193
Sphaerolaimus, 5, 10, 12, 18, 29, 85, 86, 87, 96, 101; *balticus,* 8, 17, 74, 75, 76, 77
Spiralia, 228, 230, 233, 234, 235, 236, 247, 250, 252
Spiraloblastica, 252
Spirina, 29
Spirinoidea, 183
Spiruria, 177
Spirurida, 124, 161, 163, 168, 177, 178, 187, 198, 199
Spiruridae, 121
Spirurina, 187
Steineria, 100
Strongylida, 37, 121, 124, 126, 128, 161, 164, 170, 178, 186, 196, 197
Strongyloidea, 186
Strongyloides stercoralis, 196
Strongylus, 142
Syngamus, 173

Tardigrada, 117, 118
Tentaculata, 250
Tetradonematoidea, 181
Thelastomatoidea, 186
Thelastomidae, 127
Theristus, 5, 99, 158, 193
Tobrilina, 179, 189
Tobrilus, 131; *aberrans,* 72
Torquentia, 177
Trematoda, 21
Trichocephalida, 67, 68, 73, 85, 90, 101, 123, 126, 150, 151, 178, 181, 190, 191, 201
Trichocephaloidea, 181
Trichocephalus trichurus, 151
Trichostrongyloidea, 186
Tripyla, 122
Tripylidae, 124
Tripylina, 123, 124
Tripyloidina, 86, 179, 189, 192
Trophomera, 180
Truttodacnites stelmoides, 198
Turbanella, 213; *cornuta,* 212
Turbatrix aceti, 128
Turbellaria, 21, 22, 42, 45, 117, 118, 129, 132, 169, 228
Turkmenia, 243
Tylenchida, 14, 30, 37, 43, 80, 81, 85, 88, 126, 128, 162, 164, 168, 178, 185, 196
Tylenchida sp., 107
Tylenchina, 186
Tylenchulus semipenetrans, 90
Tylenchus, 80

Vermes, 224

Xiphinema, 72, 137; *americanum,* 66, 67, 88
Xyalidae, 192